CHEMICAL MODULATORS OF PROTEIN MISFOLDING AND NEURODEGENERATIVE DISEASE

CHEMICAL MODULATORS OF PROTEIN MISFOLDING AND NEURODEGENERATIVE DISEASE

PIERFAUSTO SENECI
Università degli Studi di Milano,
Dipartimento di Chimica,
Milan, Italy

ELSEVIER

AMSTERDAM • BOSTON • HEIDELBERG
LONDON • NEW YORK • OXFORD • PARIS
SAN DIEGO • SAN FRANCISCO • SINGAPORE
SYDNEY • TOKYO
Academic Press is an Imprint of Elsevier

British Library Cataloguing-in-Publication Data
A catalogue record for this book is available from the British Library

Library of Congress Cataloging-in-Publication Data
A catalog record for this book is available from the Library of Congress

ISBN: 978-0-12-801944-3

For Information on all Academic Press publications
visit our website at http://store.elsevier.com/

Typeset by Thomson Digital

Printed and bound in United States of America

Working together
to grow libraries in
developing countries

www.elsevier.com • www.bookaid.org

Dedication

To Hanno, the king of contrarians: I owe you.

Dedication

Contents

5. Targeting Selective Autophagy of Insoluble Protein Aggregates

6. Targeting Assembly and Disassembly of Protein Aggregates

Abbreviations

15-DSG	15-deoxyspergualin
17-DMAG	N,N-dimethylaminoethyl-17-demethoxy geldanamycin
5-HT	serotonin/5-hydroxytryptamine
6-OHDA	6-OH dopamine
AA	aminoacid
AAG	17-allylamino-17-demethoxy geldanamycin
Aβ	amyloid β
ABAD	amyloid-binding alcohol dehydrogenase
Abl	Abelson kinase
ACD	α-crystallin domain
AChE	acetylcholinesterase
AD	Alzheimer's disease
ADAM10	α-secretase disintegrin and metalloproteinase domain 10
ADMET	adsorption, distribution, metabolism, excretion, toxicity
ADNP	activity-dependent neuroprotective protein
ADP	adenosine diphosphate
AE	adverse event
AF	atrial fibrillation
AFM	atomic force microscopy
AICD	APP intracellular domain
Akt	protein kinase B
ALS	amyotrophic lateral sclerosis
AMBRA1	autophagy/beclin 1 regulator 1
AmF	amyloid fiber
AMP	adenosine monophosphate
AMPK	AMP-mediated protein kinase
APs	autophagosomes
APAF-1	apoptosis protease activating factor 1

APIase	aminopeptidyl isomerase
ApoE4	apolipoprotein E4
APP	amyloid precursor protein
ARE	antioxidant response element
Atg	autophagy-related gene
ATM	ataxia telangiectasia mutated
ATP	adenosine triphosphate
ATPZ	aminothienopyridazine
ATR	ataxia telangiectasia mutated and RAD3-related
BACE-1	β-APP cleaving enzyme, β-secretase 1
BAG	Bcl-2-associated athanogene
BARD	BRCA1–associated RING domain protein
BBB	blood–brain barrier
BCR	breakpoint cluster region
BRCA1	breast cancer type 1
BTA	benzothiazole aniline
BuChE	butyl cholinesterase
CAMKKβ	calmodulin-activated kinase kinase β
CamKII	Ca^{2+}/calmodulin-dependent protein kinase II
cAMP	cyclic AMP
cdc37	cell division cycle 37 homolog
cdc4	cell division control protein 4
Cdk	cyclin-dependent kinase
CF	cystic fibrosis
CFTR	cystic fibrosis transmembrane regulator
CHIP	C-terminus of Hsc70 interacting protein
CIA	collagen induced arthritis
cIAP	cellular inhibitor of apoptosis protein
CK2	casein kinase 2
CMA	chaperone-mediated autophagy
CML	chronic myeloid leukemia
CMT	Charcot–Marie–Tooth disease
CNS	central nervous system
COPD	chronic obstructive pulmonary disease
COS-7	CV-1 in origin SV40-carrying
CP	core particle

CRBN	cereblon
CSF	cerebrospinal fluid
CTF	C-terminal fragment
Cul	cullin
Δ^9-THC	Δ^9-tetrahydrocannabinol
D1R	dopamine receptor 1
DDBP1	damaged DNA binding protein 1
DHA	docosahexaenoic acid
DHP	dihydropyrimidine
DLB	dementia with Lewy bodies
DMB	dynein motor binding
DPBP	diphenylbutylpiperidine
DPmT	dual PI3K/mTOR
DUB	deubiquitinating enzyme
DYRK1A	dual specificity tyrosine-phosphorylation-regulated kinase 1A
E	enzymes
E1	UBQ-activating enzyme
E2	UBQ-conjugating enzyme
E3	UBQ ligase
E4	UBQ chain assembly factor
E6AP	E6-associated protein
ECG	epicatechin-3-gallate
EGFR	epidermal growth factor receptor
EGCG	epigallocatechin-3-gallate
EIF4E-BP1	eukaryotic translation initiation factor 4 epsilon binding protein 1
Epac	exchange protein directly activated by cAMP
EpoD	epothilone D
EPR	electron paramagnetic resonance
ER	endoplasmic reticulum
ERK	extracellular-regulated signal kinase
ES	endosome
EXP1	exported protein 1
FCA	Freund's complete adjuvant
FKBP	FK-binding protein
FL	full length

FLICE	FADD-like interleukin-1 beta-converting enzyme
FLIP	FLICE inhibitory protein
Foxp3$^+$	forkhead box P3
FRB	FKBP–rapamycin binding
FTD	frontotemporal dementia
FTLD	frontotemporal lobar degeneration
FUS	fused in sarcoma
GA	gambogic acid
GAP-43	growth-associated protein 43
GFP	green fluorescent protein
GlcNAc	N-acetyl-D-glucosamine
GOF	gain-of-function
GR	glucocorticoid receptor
GSI	γ-secretase inhibitor
GSK-3	glycogen synthase kinase 3
GSM	γ-secretase modulator
GSPE	grape seed polyphenolic extract
GSR	glutathione disulfide reductase
GTP	guanosine triphosphate
HAA	hardwickiic acid
HCC	hepatocellular carcinoma
HCS	high content screen
HCT116	human colon carcinoma
HD	Huntington disease
HDAC	histone deacetylase
HDM2	human double minute 2 homolog
HECT	homologous to the E6AP carboxyl terminus
HEK	human embryonic kidney
hERG	human ether-a-go-go related gene
HGPS	Hutchinson–Gilford progeria syndrome
HIF-1α	hypoxia factor 1α
HIV	human immunodeficiency virus
Hop	Hsp70-Hsp90 organizing protein
HP	hyperphosphorylation, hyperphosphorylated
H2R	histamine receptor 2
Hsc	heat-shock constitutive

HSF1	heat shock factor 1
HSP	heat shock protein
HspB1	Hsp70-binding protein 1
HT	high throughput
Htt	huntingtin
HTS	high-throughput screening
HuR	human antigen R
HX-MS	hydrogen exchange mass spectrometry
I1R	imidazoline 1 receptor
IAP	inhibitor of apoptosis protein
IC	ion channel
i.c.v.	intracerebroventricular
IC_{50}	inhibitory concentration, 50% inhibition
IDE	insulin-degrading enzyme
IFN-β	interferon β
IGF-1	insulin growth factor 1
IKK	I kappa B kinase
IL	interleukin
IND	investigational new drug
iNOS	inducible nitric oxide synthase
i.p.	intraperitoneal
IP3	inositol triphosphate
IQA	indoloquinazoline acetic acid
IRE1	inositol-requiring protein 1
IRS1	insulin receptor signaling 1
ISF	interstitial fluid
ITC	isothermal calorimetry
i.v.	intravenous
JAK	Janus-activated kinase
JNK	c-jun N-terminal kinase
K_D	dissociation constant
kDa	kilodalton
Keap-1	kelch-like ECH-associated protein 1
K_i	inhibition constant
KIR	keap1-interacting region
KO	knockout
LC	liquid chromatography

LC3	light chain 3
LD	lipid droplet
LIR	LC3-interacting region
LMT	leucomethylene blue
LOF	loss-of-function
LPS	lipopolysaccharide
LS	lysosome
MA	macroautophagy
MALDI	matrix assisted laser desorption ionization
MAP	microtubule-associated protein
MAPK	mitogen-activated protein kinase
mAtg	mammalian autophagy-related gene
MB	methylene blue
MCI	mild cognitive impairment
MCL-1	myeloid cell leukemia sequence 1
MEFs	murine embryonic fibroblasts
MEK	MAP2K, MAPKK, mitogen-activated protein kinase kinase
MEKK3	mitogen-activated kinase kinase kinase 3
MIC	minimal inhibitory concentration
MK2	mitogen-activated protein kinase-activated protein kinase 2
μM	micromolar
MMP	matrix metalloproteinase
MOK	MAPK/MAK/MRK-overlapping kinase
MPP^+	1-methyl-4-phenylpyridinium
MPTP	1-methyl-4-phenyl-1,2,3,6-tetrahydropyridine
MS	mass spectrometry
MSA	multiple system atrophy
MT	microtubule
MTBR	MT-binding repeat
mTORC	mammalian target of rapamycin complex
NAD	nicotinamide adenine dinucleotide
NADP	nicotinamide adenine dinucleotide phosphate
NBD	nucleotide-binding domain
NCI	National Cancer Institute
NDD	neurodegenerative disease

NEF	nuclear exchange factor
NEP	neprilysin
NET	norepinephrine transporter
NIH	National Institute of Health
NF-κB	nuclear factor kappa-light-chain-enhancer of activated B cells
NFTs	neurofibrillary tangles
NIH	National Institute of Health
nM	nanomolar
NMDA	N-methyl-D-aspartate
NMR	nuclear magnetic resonance
NO	nitric oxide
Nrf2	NF-E2-related factor 2
NSAID	non-steroidal anti-inflammatory drug
NSCLC	non-small cell lung cancer
OGA	O-GlcNAc hydrolase
OGT	O-GlcNAc transferase
OKA	okadaic acid
OPCA	OPR/PC/AID
OPN	optineurin
PAS	phagophore assembly site
PB1	phox and bem 1p
PcTS	phthalocyanine tetrasulfonate
PD	Parkinson's disease
PDB	Paget's disease of bone
PDE	phosphodiesterase
PDK1	3-phosphoinositide-dependent kinase 1
PDS-95	post-synaptic density protein 95
PDy	pharmacodynamic
PE	phosphatidylethanolamine
PEG	polyethyleneglycol
PES	pifithrin-μ
PET	positron emission tomography
PG	prostaglandin
Pgp	P-glycoprotein
PHF	paired helical filament
PI	phosphatidylinositol

PI3K	phosphatidylinositol-3-kinase
PI3P	phosphatidylinositol-3-phosphate
PiB	Pittsburgh B
PK	pharmacokinetic
PKC	protein kinase C
PKR	protein kinase R
PLA	polylactic acid
PLC	phospholipase C
PLGA	poly-(lactic-co-glycolic acid)
PLpro	papain-like protease
p.o.	*per os*, oral
PolyQ	polyglutamine
PP2A	protein phosphatase 2A
PPARγ	peroxisome proliferator-activated receptor γ
PPI	protein–protein interaction
PQC	protein quality control
PrP	prion protein
Prx	peroxiredoxin
PSEN	presenilin
PSP	progressive supranuclear palsy
PTH	phenylthiazolyl hydrazide
PTM	post-translational modification
QSAR	quantitative structure–activity relationship
R	receptor
Rab	ras-related in brain
Raf	rapid accelerated fibrosarcoma
Rag	ras-related small GTP binding protein
Rag GTPases	ras-related GTPases
RANTES	regulated on activation, normal T cell expressed and secreted
Rap2B	ras-related protein 2B
RBR	RING-in-between-RING
RDS	rhabdomyosarcoma
REDD1	regulation of DNA damage response 1 protein
RET	rearranged during transfection
RFP	red fluorescent protein
RING	really interesting new gene

ROCK	rho-associated coiled-coil kinase
ROS	reactive oxygen species
RP	regulatory particle
S6K1	S6 kinase 1
SAHA	suberoyl anhydride hydroxamic acid
SAR	structure–activity relationship
SARS	severe acute respiratory syndrome
SBD	substrate-binding domain
s.c.	subcutaneous
SCA	spinocerebellar ataxia
SCF	Skp-cullin-F-box-containing
scNB	scrapie-infected neuroblastoma
sHsp	small heat shock protein
siRNA	small interference RNA
SIRT	sirtuin
SMA	spinal muscular atrophy
Smac/DIABLO	second mitochondria-derived activator of caspase/direct inhibitor of apoptosis-binding protein with low pI
SMER	small molecule enhancers of rapamycin
SMIR	small molecule inhibitors of rapamycin
SMN2	survival motor neuron 2
SOD1	superoxide dismutase 1
SPA	scintillation proximity assay
Spautin-1	specific and potent autophagy inhibitor-1
SPECT	single photon emission computer tomography
SPR	surface plasmon resonance
STAT	signal transducer and activator of transcription
TAI	tau aggregation inhibitor
TAR	transactive response
TBB	tetrabromobenzimidazole
TBBt	tetrabromobenzotriazole
TBI	traumatic brain injury
TBK1	TANK-binding kinase 1
TBS	TRAF6 binding site
TEM	transmission electron microscopy

TDP-43	transactive response/TAR DNA-binding protein 43
TDZD	thiadiazolidinone
TG	transgenic
TG2	transglutaminase 2
Th	thioflavin
THP	tetrahydrohyperforin
TIBI	tetraiodobenzimidazole
TMD	transmembrane domain
TNF	tumor necrosis factor
TOF	time of flight
TPPP1	tubulin polymerization-promoting protein 1
TPR	tetratricopeptide
TRAF6	TNF receptor associated factor 6
TRB3	tribbles 3
Tregs	T-regulatory cells
TSA	trichostatin A
TSC	tuberous sclerosis complex
UAF1	USP1-USP1 associated factor 1
UBA	ubiquitin-associated
UBC	ubiquitin-conjugating
UBL	ubiquitin-like
UBQ	ubiquitin
UCH	ubiquitin C-terminal hydrolase
UIS4	up-regulated in sporozoites 4
ULK	UNC-51-like kinase
UPR	unfolded protein response
UPS	ubiquitin–proteasome system
USP	ubiquitin-specific protease
Vps34	vacuolar protein sorting protein 34
WT	wild type
ZZ	zinc finger motif

Chemical Modulators of Protein Misfolding, Neurodegeneration and Tau
What is not Covered Next

Is this book comprehensive? I guess not. This is how the biology-oriented companion book [1] ends, and this is even truer at the end of this book. Here my task apparently was easier, as I planned to cover five pathways (chaperones (HSPs)/Chapter 2, ubiquitin proteasome system (UPS)/Chapter 3, autophagy/Chapter 4, aggrephagy/Chapter 5, anti-aggregation and disassembly of amyloidogenic proteins/Chapter 6) through the modulators of selected targets for each pathway (heat shock protein 27/Hsp27, Hsp70, Hsp90/Chapter 2; C-terminus of Hsc70 interacting protein/CHIP, ubiquitin-specific protease 14/USP14/Chapter 3; mammalian target of rapamycin complex 1/mTORC1/Chapter 4; p62, histone deacetylase 6/HDAC6/Chapter 5; tau antiaggregation and disassembly agents/Chapter 6). More than 250 compounds act on a mix of chemically validated (e.g., Hsp90, mTORC1, and tau aggregation inhibitors) and scarcely exploited targets (e.g., Hsp27, p62, and tau disassembly modulators). While writing the last words of Chapter 6, though, I realized that some well-characterized, putative disease-modifying compounds endowed with validated mechanisms of action against neurodegenerative disease (NDD) would have been missing.

Is it right to cut them out just because their mechanism of action does not strictly comply with my selection rules? I believe that the right answer is no. Thus, I inserted an introductory "miscellaneous" chapter here, which makes this book a more complete chemistry-oriented description of misfolding and NDD. Twelve compounds acting on tau and Aβ (respectively **1.1–1.6**, Figure 1.1 and **1.7–1.11b**, Figure 1.2) are thoroughly described in Chapter 1. While the focus of this book is unchanged, the most prospective putative disease-modifiers that do not act on targets covered in Chapters 2 to 6 are described here.

Chemical Modulators of Protein Misfolding and Neurodegenerative Disease. http://dx.doi.org/10.1016/B978-0-12-801944-3.00001-1

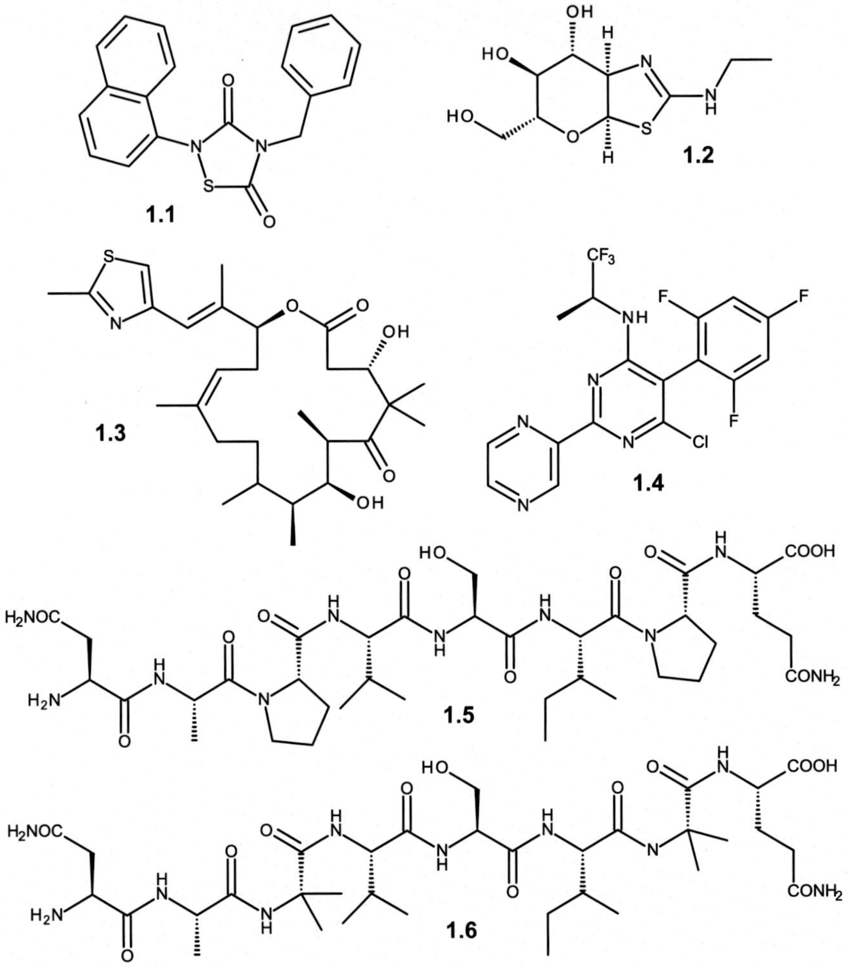

FIGURE 1.1   Small molecule modulators of tau: chemical structures, **1.1–1.6**.

1.1 TAU-TARGETED COMPOUNDS

1.1.1 Tau Kinase Inhibitors

Decreasing tau hyperphosphorylation (HP) is a validated approach against tauopathies [2]. The serine-threonine *glycogen synthase kinase 3 beta (GSK-3β* [3]) is among the most exploited kinase targets against Alzheimer's disease (AD) and tauopathies, and GSK-3β inhibitors have been clinically evaluated. A recent review [4] covers the use of small molecule GSK-3β inhibitors in the central nervous system (CNS).

Thiadiazolidinones (TDZDs) are the most promising synthetic adenosine triphosphate (ATP)-non-competitive GSK-3β inhibitors [5]. 1-Naphthyl TDZD (tideglusib, **1.1**, Figure 1.1) has been clinically developed against AD and tauopathies [6]. It interacts strongly and irreversibly with GSK-3β, without covalently binding to the kinase [7]. It activates the peroxisome proliferator-activated receptor γ (PPARγ) nuclear receptor [8]. Tideglusib (200 mg/kg/daily, orally/p.o., 3 months) in a double amyloid precursor protein (APP) tau transgenic (TG) mouse model provides biochemical (reduction of tau phosphorylation at GSK-3β epitopes, reduction of amyloid load) and functional efficacy (block of neuronal loss in AD-affected brain areas) [9]. Tideglusib is active in a double APP-presenilin (PSEN) TG model [10]. It increases the levels of neurotrophic peptide insulin growth factor 1 (IGF-1), and promotes endogenous hippocampal neurogenesis *in vitro* and *in vivo* through GSK-3β inhibition [11]. Clinical trials with tideglusib show good tolerability (Phase I study, healthy volunteers, oral daily dosages between 50 and 1200 mg, up to 14 days) [12]. A Phase IIa study on 30 AD patients (400 to 1000 mg daily, p.o., 15 weeks) confirms the good tolerability of tideglusib and provides signs of efficacy in mild to moderate AD patients [11]. Two Phase IIb studies on AD (ARGO [13], 306 patients) and progressive supranuclear palsy (PSP, TAUROS [14], 146 patients) do not reach primary and secondary endpoints [15], although signs of reduced brain atrophy are observed in TAUROS patients [16]. Consequently, the development of tideglusib is stopped. Tau kinase inhibitors are not currently in clinical trials against AD or other tauopathies, although tau HP remains a validated and prospective therapeutic target.

1.1.2 Tau O-GlcNAcylation Enhancers

The modification of peptide hydroxy groups with N-acetylglucosamine (OGlcNAc, O-GlcNAcylation) is controlled by two enzymes. The O-GlcNAc transferase OGT [17] introduces a GlcNAc moiety on more than 500 protein substrates, including up to 11 Ser and Thr residues on tau. The O-GlcNAc hydrolase OGA [18] removes GlcNAc from the same residues/substrates in a dynamic equilibrium. O-GlcNAc dynamic cycling influences cell cycle control [19], development [20], signaling [21], and trafficking [22]. O-GlcNAcylation is a regulation mechanism in the brain, and deregulated O-GlcNAcylation is observed in neuronal disorders [23]. A hypo-O-GlcNAcylation/HP pattern is observed on tau in AD brain tissues [24]. Brain-targeted deletion of OGT in mice leads to tau HP and to neuronal death [20]. An increase in O-GlcNAcylation, through small molecule-mediated OGA inhibition in mice, results in a residue-specific reduced level of phosphorylation on tau [25] that does not perturb the microtubule (MT)–tau interaction [26]. Thus, an *increase of O-GlcNAcylation* on tau is a sound therapeutic goal.

Either an increase in OGT activity or a decrease in OGA activity should increase tau O-GlcNAcylation. As enzyme inhibitors are more easily found and rationally designed than enzyme activators, OGA inhibitors seem to be a more achievable goal [27]. Most of them are OGlcNAc analogues, mimicking the substrate-assisted enzymatic mechanism of OGA [28]. The X-ray structure of complexes between human OGA homologues and OGA inhibitors [29] facilitate the rational design of selective OGA-targeted inhibitors [30].

The 2-aminothiazoline thiamet G (**1.2**, Figure 1.1) [31] is a rationally designed OGlcNAc mimic. Its 2-amino function increases its pK_a, strengthens the interaction between thiamet G and OGA, improves its selectivity *vs.* structurally similar hydrolases, and increases its aqueous solubility and stability [31]. Thiamet G reduces tau phosphorylation in PC-12 cells at pathology-related Ser 396 and Thr231 residues. It crosses the blood–brain barrier (BBB) when administered p.o. to healthy rats at 200 mg/kg, and causes similar phosphoepitope reduction patterns [31]. Chronic treatment with thiamet G (500 mg/kg daily, p.o., 36 weeks) shows efficacy on P301L tau-expressing JNPL3 TG mice [32]. Its biochemical (increased motor neuron count, decreased neurofibrillary tangle (NFT) count, decreased neurogenic atrophy of skeletal muscle) and behavioral effects (increased body weight, improved rotarod and cage-hang performance) indicate neuroprotection devoid of toxic effects [32]. Brain samples from TG mice show a thiamet G-dependent increase of OGlcNAcylation, and a phosphorylation state-independent reduction of tau aggregation [32]. The recent agreement between a small biotech and a major pharmaceutical company [33] could lead to the clinical development of thiamet G-related OGA inhibitors as treatments against tauopathies.

1.1.3 Microtubule (MT)-binding Compounds

Tau toxicity stems from *loss-of-function* (LOF) and *gain-of-function* (GOF) of pathological tau species [34]. The latter element refers to pathological interactions between neurotoxic, degradation-resistant aggregated tau species and neuronal components. Lower levels of soluble, functional tau in an aggregation/LOF scenario, conversely, impair its functional interactions. In particular, MTs rely on tau to dynamically adapt to neuronal environments. Tau aggregation-dependent LOF/MT neurotoxicity would be expected [35]. Surprisingly, an MT stabilization, compensatory role for microtubule-associated proteins (MAPs) other than tau is observed in neurons (tau knockout (KO) does not affect MT integrity and functionality [36], an excess of tau— not its reduction—impairs axonal transport [37]) and *in vivo* (three tau KO mice are viable [38] and do not show behavioral deficits [39,40]).

Treatment with the marketed, MT-stabilizing drug taxol/paclitaxel (12 weeks, intraperitoneally/i.p.) ameliorates motor impairment and restores axonal transport in a human tau TG mouse model (9-month-old prion protein/PrP T44 mice) [41]. An increased content of stable MTs in motor neurons is observed. Taxol does not reach the CNS, due to its inability to cross the BBB [41]. Taxol binds to MTs on a site that partially overlaps with the MT–tau binding surface [42]. The BBB-permeable macrocyclic ketone epothilone D (epoD, BMS-241027, **1.3**, Figure 1.1) is CNS-active in a preventive (3 months dosing, 3-month-old mice, weekly i.p. injections) [43] and a therapeutic intervention model (3 months dosing, 9-month-old mice, weekly i.p. injections) on human P301S mutated tau-expressing PS19 TG mice [44]. EpoD ameliorates the behavioral profile of treated TG mice, delays and reduces tau pathology, and improves MT stability and axonal transport. EpoD is neuroprotective at much lower dosages than in cancer treatment (1/10 to 1/100 ratios), so that toxicity should not be a major issue [43,44]. Similar efficacy is reported in a P301L tau-overexpressing Tg4510 mouse model [45]. Surprisingly, low epoD dosages reduce the hyperdynamicity of MTs and cognitive deficits, while higher dosages are ineffective [45]. EpoD (1 mg/kg/daily, i.p., 4 days 100 ratios after toxin treatment) rescues MT defects, restores axonal transport, and attenuates nigrostriatal degeneration in a Parkinson's disease (PD) model induced by 1-methyl-4-phenyl-1,2,3,6-tetrahydropyridine (MPTP) [46]. Unexpectedly, Phase Ib clinical evaluation of epoD (0.003–0.3 mg/kg/weekly, intravenously/i.v., 9 weeks) shows sub-optimal pharmacodynamics (PDy) levels [47].

The poor PDy profile of epoD, its lack of oral bioavailability, and the susceptibility of epothilones to P-glycoprotein (Pgp)-driven cellular efflux [48] encourage the search for orally bioavailable, Pgp-insensitive MT modulators. The phenylpyrimidine **1.4** results from the structural optimization of MT stabilization-targeted, synthetic heterocycles [49]. It is a BBB-permeable, Pgp-insensitive potent MT stabilizer in cellular assays, endowed with good PDy profile after oral administration (10 mg/kg) in wild-type (WT) mice [49]. Its profiling *in vivo* in tauopathy models could confirm its therapeutic usefulness.

Davunetide (NAP, **1.5**) is a neuroprotective octapeptide whose NAPV-SIPQ sequence is found in the activity-dependent neuroprotective protein (ADNP) [50]. ADNP KO is lethal in mice, while partial ADNP deficits lead to tau HP, NFT formation, and aging-associated neurodegeneration [51]. NAP treatment ameliorates the neurodegenerative phenotype of ADNP deficit-driven cellular and *in vivo* models [51]. NAP binds to MTs, stabilizes them without interfering with the MT–tau interaction, stimulates axonal transport, and promotes neurite outgrowth [52]. NAP and taxol show different (sometimes opposed) effects in neuropathological cellular environments [52]. NAP is neuroprotective *in vivo* on triple TG AD mice (APP, double Swedish, K670M/N671L; PSEN1, M146V; tau, P301L), either at early (9-month-old, 0.5 μg/day, intranasal, 3 months' treatment) [53] or at late disease stage (12 to 18-month-old, 2 μg/day, intranasal, 6 months' treatment) [54]. Both schedules reduce HP tau, increase soluble tau, and decrease tau aggregates. Cognitive benefits are observed in the longer treatment schedule on aged triple TG mice [54]. An early reduction of Aβ levels [53] is not observed in aged triple TG mice [54], suggesting a time-dependent activity profile in AD models. NAP is neuroprotective in a pure TG tauopathy model (double P301S/K275T mutant tau, 2-month-old, 0.5 μg/day, intranasal, 10 months' treatment) [55]. It increases soluble tau, reduces HP tau and NFTs, and improves the cognitive behavior of TG mice [55]. NAP protects superoxide dismutase (SOD1)-G93A TG mice in a severe model of amyotrophic lateral sclerosis (ALS) [56]. Chronic treatment (3-month-old mice, 4 or 10 μg daily, i.p., 2–3 months) increases their lifespan, restores their body weight, protects against tau-driven neurotoxicity, and restores basal axonal transport [56].

Intranasal and i.v. administration of NAP in rats and dogs leads to suitable pharmacokinetic (PK)/PDy profiles [57]. NAP is stable, is highly bioavailable in the brain, and shows a good toxicity profile [58]. Clinical evaluation of intranasal (AL-108, single dose, up to 15 mg) and i.v. (AL-208, single dose, up to 300 mg). NAP in Phase I studies shows good PK/PDy and tolerability for both formulations [58]. Two Phase II studies on patients suffering from pre-AD stage mild cognitive impairment (MCI) (AL-108/NCT00422981—144 patients, AL-208/NCT00404014—234 patients) [59] support further clinical development of NAP [60]. Intranasal NAP (15 mg twice daily, 12 weeks' treatment) is safe and well tolerated, and the treatment results in behavioral benefits in two out of five cognitive tests [61].

The Phase II study results determine the selection of PSP as a target tauopathy for further clinical evaluation [61]. A Phase II/III study on PSP entails treatment with intranasal NAP (NCT01110720, 30 mg twice daily, 52 weeks' treatment) on advanced stage PSP patients [59]. Safety and tolerability are confirmed on ≈300 treated patients, but the study results do not show efficacy in each of the primary and secondary clinical outcomes [62]. Several hypotheses could justify the failure of intranasal NAP and support further clinical testing [62], but the clinical development of NAP is discontinued. Further preclinical and clinical testing in suitable NDDs of either NAP itself or of modified analogues (i.e., similarly active isoNAP, **1.6**, Figure 1.1) [63] is desirable. More information on the molecular interactions between NAP, MTs, and tau may even allow the rational design of non-peptidic NAP analogues as innovative, drug-like MT modulators/tau anti-aggregation agents.

1.2 Aβ-TARGETED COMPOUNDS

1.2.1 γ-Secretase Inhibitors (GSIs) and Modulators (GSMs)

APP is a transmembrane protein, processed to the APP C-terminal fragment (CTF) by α- and β-secretases, and to small Aβ oligomeric species by γ-secretases [64]. The γ-secretase multi-subunit proteolytic complexes,

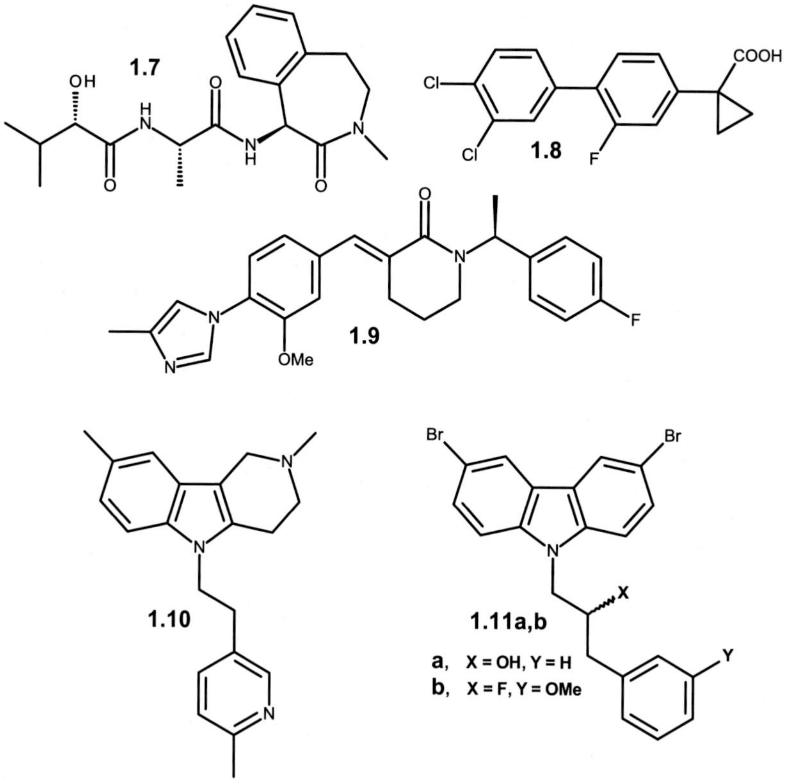

FIGURE 1.2 Small molecule modulators of Aβ: chemical structures, **1.7–1.11b**.

relying on PSEN1 or PSEN2 as catalytic cores, cleave the transmembrane domain (TMD) of ≈100 transmembrane proteins, including APP CTF [65]. When β-secretase cleaves APP, the resulting 99-residue βCTF is cleaved first by γ-secretase to yield Aβ$_{48}$ or Aβ$_{49}$ species (γ-cleavage) [66]. γ-Secretase sequentially removes VIT and TVI tripeptides from Aβ$_{48}$ (ε-cleavage) to yield neurotoxic, aggregation-prone Aβ$_{42}$. Three ε-cleavage steps on Aβ$_{49}$ sequentially cut ITL, VIV, and IAT tripeptides and yield non-neurotoxic, aggregation-resistant Aβ$_{40}$ [66]. Further ε-cleavage of Aβ$_{42}$ and Aβ$_{40}$ species leads to non-neurotoxic, smaller Aβ peptides. An increase of the basal ≈1:10 Aβ$_{42}$:Aβ$_{40}$ ratio, due to mutations in the *PSEN1/2* and *APP* genes or to pathological neuronal environments, increases the risk of AD and accelerates the aggregation of Aβ$_{42}$ [67].

Orally bioavailable, brain-penetrant, extremely potent (sub-nM IC$_{50}$) γ-secretase inhibitors (GSIs) have been known for 20 years [68]. The constrained peptidomimetic semagacestat (LY-450139, **1.7**, Figure 1.2) is an unselective GSI that blocks the processing of most γ-secretase substrates [69]. It is effective in cellular assays [70] and it reduces (3 and 30 mg/kg, oral dosage) the levels of Aβ in brain, cerebrospinal fluid (CSF), and plasma of TG PDAPP mice expressing a V717F mutation [71,72]. Safety and tolerability for semagacestat is confirmed in healthy volunteers (5 to 50 mg/kg daily, p.o., 14 days) [73] and in mild to moderate AD patients (30 mg/kg daily, 1 week, then 40 mg/kg daily, 5 weeks, p.o.) [74]. Both studies show lower Aβ levels in plasma (≈40% reduction), while the Aβ content in CSF does not vary. Similar biochemical results are obtained in a small Phase II study on mild to moderate AD patients (up to 140 mg/kg daily, p.o., 14 weeks) [75]. Namely, a ≈60% reduction of Aβ levels in plasma, no effects on CSF Aβ levels, limited side effects, and absence of cognitive effects (to be expected with such a short treatment schedule) are reported [75]. A large Phase III trial on >1500 mild to moderate AD patients (three groups/placebo, 100 or 140 mg/kg p.o. daily, up to 76 weeks' treatment) reveals a worsening of cognitive deficits, loss of weight, and an increase of adverse events (AEs, especially skin cancers) in drug-treated patients [76]. Lower Aβ levels in plasma (≈40% reduction) and no effects on CSF Aβ levels suggest limited brain bioavailability [76]. The off-target activity of non-selective semagacestat on γ-secretase substrates (especially Notch) could explain its poor AE profile [77]. More selective GSIs, though, show similar effects (cognitive worsening, significant AEs) in mild to moderate AD patients [78].

GSIs decrease the production of small Aβ oligomers, and increase the concentration of APP CTF. It appears, though, that most GSIs decrease the levels of smaller, non-neurotoxic species ($\leq A\beta_{40}$) while increasing the levels of larger oligomers (including neurotoxic $A\beta_{42}$) [79]. A better activity profile for orally bioavailable, brain-penetrant γ-secretase modulators (GSMs) entails the block of γ-secretase cleavage leading to $A\beta_{42}$, without modifying—or even increasing—the levels of smaller Aβ oligomers.

Non-steroidal anti-inflammatory drugs (NSAIDs) are moderately active acidic GSMs that increase selective processing of $A\beta_{42}$ to $A\beta_{38}$ [80]. Poorly BBB-permeable, acidic (R)-flurbiprofen/tarenflurbil is used to a Phase III trial on mild AD patients [81] that shows no cognitive benefits, due to low brain concentrations (lower than the high μM concentration of tarenflurbil needed to lower $A\beta_{42}$ levels in earlier studies) [81]. The second generation, bisaryl carboxylate NSAID CHF5074 (**1.8**) [82,83] shows moderate μM potency in Aβ-overexpressing neuroblastoma cells, with an ≈five-fold selectivity for $A\beta_{42}$ vs. $A\beta_{40}$ and for Aβ vs. Notch. It is more BBB permeable than tarenflurbil (≈0.05 brain/plasma ratio for CHF5074 vs. ≈0.01 for tarenflurbil), and has a reasonable adsorption/distribution/metabolism/excretion/toxicity (ADMET)/PK profile [82,83].

In vivo data on CHF5074 are abundant, especially on Tg2576 APPSwe (K670M/N671L double mutation) TG mice. Acute CHF5074 treatment on plaque-free, neurologically impaired 5-month-old TG mice (subcutaneous injections (s.c.), 10–30–100 mg/kg, 3 or 24 hr before behavioral testing) leads to dose-dependent memory improvement without altering the low

Aβ levels in TG mice [84]. Subchronic treatment ($\approx$60 mg/kg daily, p.o., 4 weeks) shows full memory recovery and unchanged Aβ levels [84]. Similar effects are observed with acute and subchronic treatment of 7-month-old TG mice (30 mg/kg, s.c., 3 or 24 hr before behavioral testing, and for 8 days) [85]. CHF5074 restores pre-synaptic acetylcholinesterase (AChE) release from cortical nerve terminals and improves memory up to WT mice levels, while semagacestat is fully ineffective [85]. Chronic CHF5074 treatment of 6-month-old TG mice ($\approx$60 mg/kg daily, p.o., 9 months) does not show major toxicity and restores basal behavior in animals [86]. CHF5074 treatment produces an increase of neuroblasts in dentate gyrus, higher synaptophysin levels in the cortex, and a reduced number of Aβ plaques in aged animals when compared with vehicle-treated TG mice. CHF5074 treatment slightly reduces $A\beta_{40}$ levels and does not affect $A\beta_{42}$ levels [86]. Longer treatments ($\approx$20 and $\approx$60 mg/kg daily, p.o., 13 months) confirm safety and tolerability of CHF5074, its cognitive benefits (only at the higher dosage), and a lower number of Aβ plaques [87]. The levels of soluble, oligomeric Aβ species are unchanged. The reduction of activated microglia levels (unaffected by GSIs) indicates anti-neuroinflammatory/neuroprotective, γ-secretase-unrelated properties of CHF5074 [87]. A single study reporting shorter CHF5074 treatment of aged TG mice (9.5- to 10.5-month-old, $\approx$60 mg/kg daily, p.o., 17 weeks) shows reduction of soluble $A\beta_{40}$ and $A\beta_{42}$ levels, in addition to a reduction in Aβ plaque burden [88].

CHF5074 is neuroprotective in other AD TG models. Treatment of hAPP TG mice (Swedish K670M/N671L and London V717I mutations) with CHF5074 (6-month-old, $\approx$60 mg/kg daily, p.o., 6 months) improves their behavior, is well tolerated, reduces Aβ plaques and the areas occupied by them, and lowers the levels of activated microglia [89]. The levels of soluble Aβ oligomers and of their APP CTF precursor are unchanged [89]. Treatment with CHF5084 reduces both total and HP tau levels, possibly by increasing the levels of inactive tau kinase GSK-3β [90]. Finally, CHF5074 extends the survival time and reduces deposits of i.p.-injected pathogenic prion protein (PrP^{Sc}) in a mouse model of prion infection [91].

CHF5074 does not bind to the γ-secretase complex, but to its APP CTF substrate [92]. Its binding site is located in the APP intracellular domain (AICD), which contributes to the development of AD pathology [93]. The *in vitro* and *in vivo* profile of CHF5074 combines $A\beta_{42}$-lowering/$A\beta_{38}$-increasing effects with the modulation of AICD transport (reduced nuclear translocation and AICD occupancy) and with transcriptional activation (e.g., reduced activation of the proapoptotic tetraspannin-encoding *KAI1* gene) [92]. Interestingly, AICD-overexpressing TG mice show AD-like tau HP, GSK-3β activation, memory deficits, and neuroinflammation [94,95]— *in vivo* neuroprotection by CHF5074 may at least partially depend on AICD modulation.

A short-term Phase I study of CHF5074 in healthy volunteers (p.o., 200/400/600 mg/day, 2 weeks' treatment) shows absence of relevant AEs, a linear PK profile, limited brain exposure ($\approx 0.6\%$ of plasma concentration), no effects on plasma and brain Aβ levels, and preliminary evidence of reduction of microglia activation [96]. A small Phase II study on MCI patients (p.o., 200/400/600 mg/day, 12 weeks' treatment) shows good tolerability and an acceptable AE profile [95]. CHF5074 causes dose-dependent reduction of biomarkers of neuroinflammation in the CSF of patients, while Aβ and tau levels are unchanged [96]. An open label extension of the study shows a non-significant trend of cognitive improvements, especially in patients carrying the apolipoprotein E4 (ApoE4) gene [97]. A Phase III study on ApoE4-bearing AD patients using CSP-1103 (the new denomination of CHF5074) is planned for the near future [97].

Second generation, more potent, neutral non-NSAID GSMs cause the simultaneous increase of Aβ_{37} (ε-cleavage of Aβ_{40}) and Aβ_{38} (ε-cleavage of Aβ_{42}) [68]. The alkylidenpiperidinone E-2012 (**1.9**, Figure 1.2) does not bind to Aβ_{40} monomers, but binds to Aβ_{40} fibrils on their growing ends and modulates their fibrillation [98]. It acts as a GSM by binding to the PSEN subunits of the γ-secretase complex, as other non-acidic GSMs do, with a preference for PSEN2 [99]. E-2012 reduces both Aβ_{40} and Aβ_{42} levels in neurons (respectively $IC_{50} = 330$ and 92 nM), and reduces Aβ_{42} levels in rat plasma, brain, and CSF (respectively $>95\%$, $>42\%$, and $>47\%$ reduction at 30 mg/kg) [100]. The observed induction of cataracts by E-2012 in a rat preclinical study [101] is not detected in a 13-month study in monkeys [102]. In a Phase I study in healthy volunteers, E-2012 (up to 400 mg, single p.o. dose) reduces Aβ_{40} and Aβ_{42} levels in plasma (respectively $\approx 30\%$ and $\approx 50\%$ reduction), in brain, and CSF (undisclosed percentages) in a dose-dependent manner [103]. The development of E-2012 is halted to progress another non-NSAID GSM, E-2212 (undisclosed structure) [104]. E-2212 is more potent than E-2012 *in vitro* and *in vivo*/preclinical studies, with Aβ_{42} level-lowering effects at doses between 3 and 10 mg/kg [104]. A single dose Phase I study using E-2212 (up to 250 mg) shows better safety and tolerability, a PK profile similar to E-2012, and a $>50\%$ reduction of Aβ_{42} levels at the highest dosage [104].

1.2.2 Multi-targeted Neuroprotective and Proneurogenic Compounds

Histamine receptor-targeted (especially histamine receptor H3 antagonists/reverse agonists [105]) and serotonin/5-hydroxytryptamine receptor-targeted drugs (especially 5-HT6 and 5-HT7 [106,107]) are being repurposed against NDDs, and AD in particular. A non-selective histamine receptor antagonist, the tetrahydropyrido[4,3-b]indole dimebon

TABLE 1.1   Antagonism (Receptors, R) and Inhibition (Enzymes, E, and Ion Channels, IC) of Molecular Targets by Dimebon [109,110]

Molecular targets	Affinity range
Histamine H_1 R, serotonin 5-HT_7 R, adrenergic α_{2B} R	1 nM < aff. ≤ 10 nM
α_{1A} R, α_{1B} R, 5-HT_6 R, α_{2C} R	10 nM < aff. ≤ 50 nM
α_{1D} R, 5-HT_{5A} R, 5-HT_{2A} R, 5-HT_{2C} R, α_{2A} R	50 nM < aff. ≤ 100 nM
Imidazoline I_2 R, H_2 R, dopamine D2 R	100 nM < aff. ≤ 500 nM
D3 R, D1 R, 5-HT_{2B} R	500 nM < aff. ≤ 1 μM
L-type Ca^{2+} channels IC, N-Me-D-aspartate (NMDA) R	1 μM < aff. ≤ 10 μM
5-HT_{1A} R, 5-HT_{1B} R, sigma σ1 R, σ2 R, norepinephrine transporter (NET) R, site 2 Na^+ channel IC, AChE E, butylcholinesterase E	10 μM < aff. ≤ 100 μM

(latrepirdine, dimebolin, **1.10**, Figure 1.2), was developed and marketed in Russia in the 1980s as a treatment for skin allergy and allergic rhinitis [108]. Multi-targeted dimebon modulates a set of molecular targets (receptors, ion channels, and enzymes, see Table 1.1).

The developers of dimebon claim a novel mitochondrial mechanism of action. Beneficial effects of dimebon at nM concentration on Aβ-dependent mitochondrial toxicity in human APP-overexpressing human embryonic kidney (HEK) cells support this claim [111]. A mechanistic hypothesis links the block of NMDA receptors and the stabilization of Ca^{2+}-induced increase of mitochondrial permeability by dimebon at medium–high μM with its effects on mitochondria [112]. The sub-μM potency of dimebon in cells and *in vivo* questions the validity of this hypothesis [113]. Rather, the recently reported sub-μM activation of the energy sensor adenosine monophosphate (AMP)-mediated protein kinase (AMPK) by dimebon could explain its mitochondrial effects [114]. The action of dimebon on molecular targets listed in Table 1.1 (and possibly others), and particularly its potent effects on histamine and serotonin receptors, influences its overall activity [110]. Dimebon shows multiple putative disease-modifying effects in AD and other NDDs, related to Aβ and other amyloidogenic proteins.

Dimebon stimulates autophagy in a yeast model in an autophagy-related gene 8 (Atg8)-dependent manner (see also Chapter 4 of the biology-oriented companion book [1]), promotes the sequestration of green fluorescent protein (GFP)-labeled $A\beta_{42}$ in degrading autophagic vacuoles, and prevents its intracellular accumulation [115]. Dimebon protects yeast cells at low μM concentrations from $A\beta_{42}$ toxicity, being ≈10 times less potent than the autophagy inducer rapamycin (see also section 4.2, Figure 4.1) [115]. Dimebon elicits autophagy-related protective effects (increased protein aggregate clearance, increased cell survival) in an α-synuclein toxicity model in yeast [116]. Conversely, dimebon is ineffective against TAR DNA binding protein 43 (TDP-43)-, fused in sarcoma (FUS)-, and polyGln (polyQ)-induced toxicity in yeast [116], suggesting a protein aggregate-selective induction of autophagy by dimebon. Dimebon alone and in combination with the anti-aggregation agent methylene blue (MB, see also section 6.2, Figure 6.1) prevents the aggregation of aggregation-prone TDP-43 fragments overexpressed in neuroblastoma SH-SY5Y cells [117]. The combination of dimebon and MB shows synergistic effects, and an autophagy-inducing effect of the former compound is compatible with the observed effects of the combination on TDP-43 [117].

A direct effect of dimebon on the aggregation of amyloidogenic proteins, and of Aβ peptides in particular, is postulated [118]. Protein aggregation is a multistep, protein-specific process modulated by small molecules and biologicals. Anti-aggregating agents show neuroprotective or neurotoxic effects, depending on their affinity for monomers, oligomers or aggregates, and on their inhibition or stimulation of neurotoxic aggregation pathways (see Chapter 6). Either an increase [119] or a decrease of Aβ species [120], or even no effects on Aβ levels [121], are reported by research groups following in vitro or in vivo experiments with dimebon. Differences in models, experimental conditions, and detection techniques may justify the conflicting observations. Molecular interactions between any Aβ species (or any other amyloidogenic protein) and dimebon are not yet indisputably proven, and it may be that indirect effects on protein aggregation (e.g., autophagy, ubiquitin-proteasome system (UPS), dimebon target(s)-mediated changes of the cellular environment) modulate the levels of Aβ species.

Neurotoxin-treated rats show behavioral improvements when exposed to dimebon (1 mg/kg daily, i.p., 10 days) [122]. Biochemical evidence justifying the neuroprotective effects of dimebon is not provided. WT rats show memory improvements when submitted to a single dose treatment with dimebon (0.05 to 5 mg/kg, p.o.) [123]. Rats show preferential accumulation in the brain (brain:plasma ratio ≈10). Significant efficacy is observed even at low brain concentrations (1.7 nM for 0.05 mg/kg) [123].

WT mice treated with dimebon (12 mg/kg/day, p.o., 4 months) show a good PK profile and brain bioavailability [121]. Two-month-old double TG TgCRND8 mice bearing the Swedish K670M/N671L and London V717I mutations, when treated with dimebon (12 mg/kg/day, p.o., 4 months), show marginal, non-significant improvements in behavior. Dimebon does not alter their total and soluble Aβ levels with respect to placebo-treated TG mice [121]. Three-month-old TgCRND8 mice treated with dimebon (3.5 mg/kg/day, i.p., 1 month) show restoration of WT cognitive functions [124]. Aβ pathology is only marginally impacted (statistically non-significant reduction of soluble oligomeric Aβ in dimebon-treated TG mice). Autophagy is induced/restored to WT mice levels in dimebon-treated TG mice [124]. Acute treatment of Tg2576 APPSwe mice with dimebon (3.5 mg/kg, i.p.) increases the levels of soluble $A\beta_{40}$ in brain interstitial fluid (ISF) for up to 10 hours post-injection [119]. It is suggested that synaptic activity-dependent dynamic fluctuations in $A\beta_{40}$ levels, and/or a block of insoluble aggregate formation, could explain these unexpected results [119]. Five-week-old 5xFAD TG mice (Swe, London and Florida/I716V hAPP mutations; M146L and L286V PSEN1 mutations) show only marginal behavioral improvements, and no biochemical evidence of either autophagy induction or Aβ pathology amelioration after chronic dimebon treatment ($\approx$11 mg/kg daily, p.o., 8 months) [125]. Aβ pathology is affected in a brain compartment-specific manner by dimebon treatment (1 mg kg daily, i.p., 1.5 months) of 6.5-month-old triple APP (APPSwe)—PSEN1 (M146V)—tau (P301L) TG mice [120]. Namely, dimebon reduces Aβ levels only in the hippocampus of TG mice, possibly accelerating the formation of Aβ plaques (statistically non-significant increase of Aβ plaques). Interestingly, a non-significant reduction of pathological HP tau levels in the hippocampus is also reported [120]. Chronic dimebon treatment ($\approx$11 mg/kg daily, p.o., 6 months) slows the decline of motor functions in tau P301S TG mice, and causes a reduction of pathological HP tau levels in their spinal cords [126]. Dimebon neither increases neuronal counts, nor reduces activation of astroglia, i.e., neuroinflammation. Nor does it reduce HP tau levels *via* autophagy induction, but it likely protects/rescues some neurons from tau pathology [126].

Dimebon ($\approx$1.5 mg/kg daily, p.o., 6 and 9 months' treatment respectively on 6- and 3-month-old TG mice) is partially neuroprotective in a Thy1mγ SN TG mouse model showing pathological α-synuclein inclusions in motor neurons (ALS-like phenotype) [127]. It reduces motor impairments after 6 months of treatment of younger TG mice, and shows a limited lifespan increase. The number of α-synuclein inclusions in the spinal cords is reduced, as is the number of activated astroglia (reduced neuroinflammation) [127]. Conversely, dimebon ($\approx$1.5 mg/kg daily, p.o., 11 months' treatment on 3-month-old mice) does not protect αSyn (1–120) TG mice overexpressing C-terminally truncated α-synuclein in a model of early PD, and does not modulate any of their biochemical abnormalities [128].

Clinical testing of dimebon against NDDs started in Russia, with an open label Phase I/II study on a small number of mild to moderate AD patients [122]. Dimebon treatment (20 mg/kg three times a day, p.o., 8 weeks) significantly improves their cognitive functions, and reduces their neuropsychiatric symptoms. Dimebon is well tolerated, with minor AEs and without pathological changes in hematological and biochemical parameters [122]. A larger double-blind Phase II study in Russia entails a similar dimebon schedule (10 mg/kg three times a day, p.o., 1 weeks' treatment, then 20 mg/kg three times a day, p.o., 25 weeks' treatment, extended for most patients to 52 weeks) [129]. Significant improvements are observed in the psychometric assessments of cognition, function, and behavior for dimebon-treated *vs.* placebo-treated patients at 26 weeks. Additional improvements are observed at 52 weeks. AEs are limited, although previously observed trends towards reduction of neuropsychiatric symptoms are not confirmed [129]. A Phase II study in the UK and the USA using dimebon (10 mg/kg, first day, then 10 mg/kg three times a day, 6 days, then 20 mg/kg three times a day, p.o., in total 12 weeks' treatment) on mild to moderate Huntington disease (HD) patients shows good safety and tolerability [130]. Statistically non-significant cognitive trends are observed, while improvements in motor impairments are not detected (unsurprisingly, due to the short duration of the study) [130].

The results of Phase II trials in AD and HD justified the design and execution of large Phase III trials in these indications. As to HD, the HORIZON study involves >400 mild to moderate HD patients treated with dimebon (20 mg/kg three times a day, p.o., 26 weeks) [131]. While safety and tolerability are confirmed, no signs of efficacy are observed [131]. As to AD, both the CONNECTION (≈600 mild to moderate AD patients, 5 mg/kg or 20 mg/kg three times a day, p.o., 6 months' treatment) [132] and the CONCERT trial (≈1000 mild to moderate AD patients, 5 mg/kg or 20 mg/kg three times a day, p.o., 12 months' treatment) [133] fail to show any sign of clinical efficacy. The interpretation of the two AD failures is complicated by the absence of data regarding the amyloid load in AD patients, by differences between Russian and international patients (history of treatment, concomitant drugs, etc.), and by the surprising lack of deterioration of the placebo groups in the CONNECTION and CONCERT studies [109]. The clinical development of dimebon is stopped due to these results.

The absence of a validated mechanism of action for dimedon prevents the rational design of second generation leads, although structurally similar analogues with NDD-targeted activity are reported in the literature [134]. Namely, a collection of ≈1000 drug-like compounds is pooled into 100 mixtures, each made of 10 compounds, that are screened *in vivo* in adult mice for their capacity to enhance neuron formation in the mice hippocampus [135]. Each mixture is injected intracerebroventricularly (i.c.v., each compound at 10 μM concentration, 7-day infusion with osmotic minipumps) to avoid PK and BBB permeation issues. Out of 10 positive/neuron formation-enhancing pools, and of eight deconvoluted positive compounds, the carbazole P7C3 (**1.11a**) is an i.p.–p.o. bioavailable, BBB-permeable, potent compound [135]. It enhances the formation of viable neurons in the mature hippocampus, and it ameliorates age-dependent cognitive decline in mice at 10 mg/kg. It acts by protecting mitochondrial membrane integrity, as does dimebon at higher dosage in a comparative *in vivo* study. It selectively antagonizes neuronal apoptosis, apparently without apoptosis dysregulation-related AEs [135]. P7C3 shows a drug-like ADMET profile, a good PK profile after i.p. or p.o. administration, and is more selective than dimebon (i.e., its histamine receptor blocking activity is marginal) [136].

A potent P7C3 analogue (P7C3A20, **1.11b**, Figure 1.2) is characterized in mice models of PD [137] and ALS [138]. It prevents 1-methyl-4-phenyl-pyridinium (MPP$^+$)-mediated death of dopaminergic neurons in *C. elegans* and preserves worm mobility (≈80% protection and mobility preservation at 10 μM), with P7C3 being less active (≈50% protection and ≈60% mobility preservation at 10 μM) [137]. The same activity trend is observed in MPP$^+$-treated mice. P7C3A20 causes ≈85% dopaminergic neuron preservation at 20 mg/kg/day i.p. for 21 days, and shows ≈30% preservation at 1 mg/kg/day. P7C3 shows ≈60% preservation at 20 mg/kg/day i.p., and is inactive at 1 mg/kg/day i.p. Dimebon is completely inactive in this model at any dosage [137]. ALS-recapitulating (SOD1)-G93A TG mice treated at disease onset (80 days of age) with P7C3A20 (20 mg/kg, i.p., up to 40 days) show a delayed death of motor neurons/delayed disease progression, while P7C3 is much less potent and dimebon-treated mice do not show any neuroprotective effect [138]. P7C3A20 similarly delays motor deficits in (SOD1)-G93A mice, while P7C3 and dimebon do not. Surprisingly, the spinal cord concentration of ineffective P7C3 and dimebon are similar at 20 mg/kg daily i.p. dosages (> 1000 ng/mL), while effective P7C3A20 at the same dosage reaches only ≈5% of their spinal cord concentration [138]. P7C3A20 shows efficacy *in vivo* in models of traumatic brain injury (TBI, 10 mg/kg daily, i.p., 7 days) [139] and depression (10 mg/kg daily, i.p., 18 days) [140], in both cases due to its proneurogenic activity in the hippocampus of treated mice.

References

[1] Seneci, P. Molecular targets in protein misfolding and neurodegenerative disease. *Academic Press* **2014**, 278 pages.

[2] Martin, L.; Latypova, X.; Wilson, C. M.; Magnaudeix, A.; Perrin, M. -L.; Yardin, C.; Terro, F. Tau protein kinases: involvement in Alzheimer's disease. *Ageing Res. Rev.* **2013**, *12*, 289–309.

[3] Mondragon-Rodriguez, S.; Perry, G.; Zhu, X.; Moreira, P. I.; Williams, S. Glycogen synthase kinase 3: a point of integration in Alzheimer's disease and a therapeutic target? *Int. J. Alzheimer's Dis.* **2012**, 276803.

[4] Eldar-Finkelman, H.; Martinez, A. GSK-3 inhibitors: preclinical and clinical focus on CNS. *Front. Molec. Neurosci.* **2011**, *4*, 32.

[5] Martinez, A.; Alonso, M.; Castro, A.; Perez, C.; Moreno, F. J. First non-ATP competitive glycogen synthase kinase 3 beta (GSK-3 beta) inhibitors: thiadiazolidinones (TDZD) as potential drugs for the treatment of Alzheimer's disease. *J. Med. Chem.* **2002**, *45*, 1292–1299.

[6] Martinez, A.; Gil, C.; Perez, D. I. Glycogen synthase kinase 3 inhibitors in the next horizon for Alzheimer's disease treatment. *Int. J. Alzheimer's Dis.* **2011**, 280502.

[7] Domínguez, J. M.; Fuertes, A.; Orozco, L.; del Monte-Milla, M.; Delgado, E.; Medina, M. Evidence for irreversible inhibition of glycogen synthase kinase-3β by tideglusib. *J. Biol. Chem.* **2012**, *287*, 893–904.

[8] Luna-Medina, R.; Cortes-Canteli, M.; Sanchez-Galiano, S.; Morales-Garcia, J. A.; Martinez, A.; Santos, A.; Perez-Castillo, A. NP031112, a thiadiazolidinone compound, prevents inflammation and neurodegeneration under excitotoxic conditions: potential therapeutic role in brain disorders. *J. Neurosci.* **2007**, *27*, 5766–5776.

[9] Sereno, L.; Coma, M.; Rodriguez, M.; Sanchez-Ferrer, P.; Sanchez, M. B.; Gich, I., et al. A novel GSK-beta inhibitor reduces Alzheimer's pathology and rescues neuronal loss in vivo. *Neurobiol. Dis.* **2009**, *35*, 359–367.

[10] Bolos, M.; Fernandez, S.; Torres-Aleman, I. Oral administration of a GSK3 inhibitor increases brain insulin-like growth factor I levels. *J. Biol. Chem.* **2010**, *285*, 17693–17700.

[11] Morales-Garcia, J. A.; Luna-Medina, R.; Alonso-Gil, S.; Sanz-SanCristobal, M.; Palomo, V.; Gil, C., et al. Glycogen synthase kinase 3 inhibition promotes adult hippocampal neurogenesis in vitro and in vivo. *ACS Chem. Neurosci.* **2012**, *3*, 963–971.

[12] del Ser, T.; Steinwachs, K. C.; Gertz, H. J.; Andres, M. V.; Gomez-Carrillo, B.; Medina, M., et al. Treatment of Alzheimer's disease with the GSK-3 inhibitor tideglusib: a pilot study. *J. Alzheim. Dis.* **2013**, *33*, 205–215.

[13] http://www.clinicaltrials.gov/ct2/show/NCT01350362?term=tideglusib&rank=1&submit_fld_opt=.

[14] http://www.clinicaltrials.gov/ct2/show/NCT01049399?term=tideglusib&rank=2&submit_fld_opt=.

[15] Tolosa, E.; Litvan, I.; Hoglinger, G. U.; Burn, D.; Lees, A.; Andres, M. V., et al. A Phase 2 trial of the GSK-3 inhibitor tideglusib in progressive supranuclear palsy. *Movement Disorders* **2014**, 470–478.

[16] Hoglinger, G. U.; Huppertz, H. -J.; Wagenpfeil, S.; Andres, M. V.; Belloch, V.; Leon, T.; del Ser, T. Tideglusib reduces progression of brain atrophy in progressive supranuclear palsy in a randomized trial. *Movement Disorders* **2014**, 479–487.

[17] Haltiwanger, R. S.; Holt, G. D.; Hart, G. W. Enzymatic addition of O-GlcNAc to nuclear and cytoplasmic proteins. Identification of a uridine diphospho-N-acetylglucosamine:peptide beta-N-acetylglucosaminyltransferase. *J. Biol. Chem.* **1990**, *265*, 2563–2568.

[18] Comtesse, N.; Maldener, E.; Meese, E. Identification of a nuclear variant of MGEA5, a cytoplasmic hyaluronidase and a beta-N-acetylglucosaminidase. *Biochem. Biophys. Res. Commun.* **2001**, *283*, 634–640.

[19] Slawson, C.; Zachara, N. E.; Vosseller, K.; Cheung, W. D.; Lane, M. D.; Hart, G. W. Perturbations in O-linked beta-N-acetylglucosamine protein modification cause severe defects in mitotic progression and cytokinesis. *J. Biol. Chem.* **2005**, *280*, 32944–32956.

[20] O'Donnell, N.; Zachara, N. E.; Hart, G. W.; Marth, J. D. Ogt-dependent X-chromosome linked protein glycosylation is a requisite modification in somatic cell function and embryo viability. *Mol. Cell. Biol.* **2004**, *24*, 1680–1690.

[21] Hanover, J. A.; Forsythe, M. E.; Hennessey, P. T.; Brodigan, T. M.; Love, L. C.; Ashwell, G.; Krause, M. A Caenorhabditis elegans model of insulin resistance: altered macronutrient storage and dauer formation in an OGT-1 knockout. *Proc. Natl. Acad. Sci. U.S.A.* **2005**, *102*, 11266–11271.

[22] Guinez, C.; Morelle, W.; Michalski, J. C.; Lefebvre, T. O-GlcNAc glycosylation: a signal for the nuclear transport of cytosolic proteins? *Int. J. Biochem. Cell. Biol.* **2005**, *37*, 765–774.

[23] Yuzwa, S. A.; Vocadlo, D. J. O-GlcNAc and neurodegeneration: biochemical mechanisms and potential roles in Alzheimer's disease and beyond. *Chem. Soc. Rev.* **2014**, DOI: 10. 1039/C4CS00038B.

[24] Liu, F.; Iqbal, K.; Grundke-Iqbal, I.; Hart, G. W.; Gong, C. X. O-GlcNAcylation regulates phosphorylation of tau: a mechanism involved in Alzheimer's disease. *Proc. Natl. Acad. Sci. U.S.A.* **2004**, *101*, 10804–10809.

[25] Liu, F.; Iqbal, K.; Grundke-Iqbal, I.; Hart, G. W.; Gong, C. X. O-GlcNAcylation regulates phosphorylation of tau: a mechanism involved in Alzheimer's disease. *Proc. Natl. Acad. Sci. U.S.A.* **2004**, *101*, 10804–10809.

[26] Yuzwa, S. A.; Cheung, A. H.; Okon, M.; McIntosh, L. P.; Vocadlo, D. J. O-GlcNAc modification of tau directly inhibits its aggregation without perturbing the conformational properties of tau monomers. *J. Mol. Biol.* **2014**, *426*, 1736–1752.

[27] Macauley, M. S.; Vocadlo, D. J. Increasing O-GlcNAc levels: an overview of small-molecule inhibitors of O-GlcNAcase. *Biochim. Biophys. Acta.* **1800**, *2010*, 74–91.

[28] Vocadlo, D. J.; Davies, G. J. Mechanistic insights into glycosidase chemistry. *Curr. Opin. Chem. Biol.* **2008**, *12*, 539–555.

[29] Schimpl, M.; Borodkin, V. S.; Gray, L. J.; van Aalten, D. M. F. Synergy of peptide and sugar in O-GlcNAcase substrate recognition. *Chem. Biol.* **2012**, *19*, 173–178.

[30] de Alencar, N. A. N.; Sousa, P. R. M.; Silva, J. R. A.; Lameira, J.; Nahum Alves, C.; Martí, S.; Moliner, V. Computational analysis of human OGA structure in complex with PUG-NAc and NAG-thiazoline derivatives. *J. Chem. Inf. Model.* **2012**, *52*, 2775–2783.

[31] Yuzwa, S. A.; Macauley, M. S.; Heinonen, J. E.; Shan, X.; Dennis, R. J.; He, Y., et al. A potent mechanism-inspired O-GlcNAcase inhibitor that blocks phosphorylation of tau in vivo. *Nat. Chem. Biol.* **2008**, *4*, 483–490.

[32] Yuzwa, S. A.; Shan, X.; Macauley, M. S.; Clark, T.; Skorobogatko, Y.; Vosseller, K.; Vocadlo, D. J. Increasing O-GlcNAc slows neurodegeneration and stabilizes tau against aggregation. *Nat. Chem. Biol.* **2012**, *8*, 393–399.

[33] http://www.natap.org/2010/newsUpdates/081210_01.htm.

[34] Morris, M.; Maeda, S.; Vossel, K.; Mucke, L. The many faces of tau. *Neuron.* **2011**, *70*, 410–426.

[35] Butner, K. A.; Kirschner, M. W. Tau protein binds to microtubules through a flexible array of distributed weak sites. *J. Cell Biol.* **1991**, *115*, 717–730.

[36] King, M. E.; Kan, H. M.; Baas, P. W.; Erisir, A.; Glabe, C. G.; Bloom, G. S. Tau-dependent microtubule disassembly initiated by prefibrillar β-amyloid. *J. Cell Biol.* **2006**, *175*, 541–546.

[37] Stamer, K.; Vogel, R.; Thies, E.; Mandelkow, E.; Mandelkow, E. M. Tau blocks traffic of organelles, neurofilaments, and APP vesicles in neurons and enhances oxidative stress. *J. Cell Biol.* **2002**, *156*, 1051–1063.

[38] Ramachandran, G.; Udgaonkar, J. B. Mechanistic studies unravel the complexity inherent in tau aggregation leading to Alzheimer's disease and the tauopathies. *Biochemistry* **2013**, *52*, 4107–4126.

[39] Dawson, H. N.; Cantillana, V.; Jansen, M.; Wang, H.; Vitek, M. P.; Wilcock, D. M., et al. Loss of tau elicits axonal degeneration in a mouse model of Alzheimer's disease. *Neuroscience* **2010**, *169*, 516–531.

[40] Roberson, E. D.; Scearce-Levie, K.; Palop, J. J.; Yan, F.; Cheng, I. H.; Wu, T., et al. Reducing endogenous tau ameliorates amyloid β-induced deficits in an Alzheimer's disease mouse model. *Science* **2007**, *316*, 750–754.

[41] Zhang, B.; Maiti, A.; Shively, S.; Lakhani, F.; McDonald-Jones, G.; Bruce, J., et al. Microtubule-binding drugs offset tau sequestration by stabilizing microtubules and reversing fast axonal transport deficits in a tauopathy model. *Proc. Natl. Acad. Sci. U.S.A.* **2005**, *102*, 227–231.

[42] Peck, A.; Sargin, M. E.; LaPointe, N. E.; Rose, K.; Manjunath, B. S.; Feinstein, S. C.; Wilson, L. Tau isoform-specific modulation of kinesin-driven microtubule gliding rates and trajectories as determined with tau-stabilized microtubules. *Cytoskeleton* **2011**, *68*, 44–55.

[43] Brunden, K. R.; Zhang, B.; Carroll, J.; Yao, Y.; Potuzak, J. S.; Hogan, A. M., et al. Epothilone D improves microtubule density, axonal integrity, and cognition in a transgenic mouse model of tauopathy. *J. Neurosci.* **2010**, *30*, 13861–13866.

[44] Zhang, B.; Carroll, J.; Trojanowski, J. Q.; Yao, Y.; Iba, M.; Potuzak, J. S., et al. The microtubule-stabilizing agent, epothilone d, reduces axonal dysfunction, neurotoxicity,

cognitive deficits, and Alzheimer-like pathology in an interventional study with aged tau transgenic mice. *J. Neurosci.* **2012**, *32*, 3601–3611.

[45] Barten, D. M.; Fanara, P.; Andorfer, C.; Hoque, N.; Wong, P. Y. A.; Husted, K. H., et al. Hyperdynamic microtubules, cognitive deficits, and pathology are improved in tau transgenic mice with low doses of the microtubule-stabilizing agent BMS-241027. *J. Neurosci.* **2012**, *32*, 7137–7145.

[46] Cartelli, D.; Casagrande, F.; Busceti, C. L.; Bucci, D.; Molinaro, G.; Traficante, A., et al. Microtubule alterations occur early in experimental parkinsonism and the microtubule stabilizer epothilone D is neuroprotective. *Sci. Rep.* **1837**, *2013*, 3.

[47] Study to evaluate the safety, tolerability and the effect of BMS-241027 on cerebrospinal fluid biomarkers in subjects with mild Alzheimer's disease. http://clinicaltrials.gov/show/NCT01492374.

[48] Brunden, K. R.; Yao, Y.; Potuzak, J. S.; Ferrer, N. I.; Ballatore, C.; James, M. J., et al. The characterization of microtubule-stabilizing drugs as possible therapeutic agents for Alzheimer's disease and related tauopathies. *Pharmacol. Res.* **2011**, *63*, 341–351.

[49] Lou, K.; Yao, Y.; Hoye, A. T.; James, M. J.; Cornec, A. -S.; Hyde, E., et al. Brain-penetrant, orally bioavailable microtubule-stabilizing small molecules are potential candidate therapeutics for Alzheimer's disease and related tauopathies. *J. Med. Chem.* **2014**, *57*, 6116–6127.

[50] Bassan, M.; Zamostiano, R.; Davidson, A.; Pinhasov, A.; Giladi, E.; Perl, O., et al. Complete sequence of a novel protein containing a femtomolar-activity-dependent neuroprotective peptide. *J. Neurochem.* **1999**, *72*, 1283–1293.

[51] Vulih-Shultzman, I.; Pinhasov, A.; Mandel, S.; Grigoriadis, N.; Touloumi, O.; Pittel, Z.; Gozes, I. Activity-dependent neuroprotective protein snippet NAP reduces tau hyperphosphorylation and enhances learning in a novel transgenic mouse model. *J. Pharm. Exp. Ther.* **2007**, *323*, 438–449.

[52] Oz, S.; Ivashko-Pachima, Y.; Gozes, I. The ADNP derived peptide, NAP modulates the tubulin pool: implication for neurotrophic and neuroprotective activities. *PLoS One.* **2012**, *7*, e51458.

[53] Matsuoka, Y.; Gray, A. J.; Hirata-Fukae, C.; Minami, S. S.; Waterhouse, E. G.; Mattson, M. P., et al. Intranasal NAP administration reduces accumulation of amyloid peptide and tau hyperphosphorylation in a transgenic mouse model of Alzheimer's disease at early pathological stage. *J. Mol. Neurosci.* **2007**, *31*, 165–170.

[54] Matsuoka, Y.; Jouroukhin, Y.; Gray, A. J.; Ma, L.; Hirata-Fukae, C.; Li, H. F., et al. A neuronal microtubule-interacting agent, NAPVSIPQ, reduces tau pathology and enhances cognitive function in a mouse model of Alzheimer's disease. *J. Pharmacol. Exp. Ther.* **2008**, *325*, 146–153.

[55] Shiryaev, N.; Jouroukhin, Y.; Giladi, E.; Polyzoidou, E.; Grigoriadis, N. C.; Rosenmann, H.; Gozes, I. NAP protects memory, increases soluble tau and reduces tau hyperphosphorylation in a tauopathy model. *Neurobiol. Dis.* **2009**, *34*, 381–388.

[56] Jouroukhin, Y.; Ostritsky, R.; Assaf, Y.; Pelled, G.; Giladi, E.; Gozes, I. NAP (davunetide) modifies disease progression in a mouse model of severe neurodegeneration: protection against impairments in axonal transport. *Neurobiol. Dis.* **2013**, *56*, 79–94.

[57] Morimoto, B. H.; De Lannoy, I.; Fox, A. W.; Gozes, I.; Stewart, A. Davunetide pharmacokinetics and distribution to brain after intravenous or intranasal administration to rat. *Chimica Oggi—Chemistry Today.* **2009**, *27*, 16–20.

[58] Gozes, I.; Morimoto, B. H.; Tiong, J.; Fox, A.; Sutherland, K.; Dangoor, D., et al. NAP: research and development of a peptide derived from activity-dependent neuroprotective protein (ADNP). *CNS Drug Rev.* **2005**, *11*, 353–368.

[59] http://clinicaltrials.gov/ct2/results?term=davunetide&Search=Search.

[60] Magen, I.; Gozes, I. Microtubule-stabilizing peptides and small molecules protecting axonal transport and brain function: Focus on davunetide (NAP). *Neuropeptides* **2013**, *47*, 489–495.

[61] Gold, M.; Lorenzl, S.; Stewart, A. J.; Morimoto, B. H.; Williams, D. R.; Gozes, I. Critical appraisal of the role of davunetide in the treatment of progressive supranuclear palsy. *Neuropsychiatr. Dis. Treat.* **2012**, *8*, 85–93.

[62] Boxer, A. L.; Lang, A. E.; Grossman, M.; Knopman, D. S.; Miller, B. L.; Schneider, L. S., et al. Davunetide in patients with progressive supranuclear palsy: a randomised, double-blind, placebo-controlled Phase 2/3 trial. *Lancet Neurol.* **2014**, *13*, 676–685.

[63] Gozes, I.; Schirer, Y.; Idan-Feldman, A.; David, M.; Furman-Assaf, S. NAP alpha-aminoisobutyric acid (IsoNAP). *J. Mol. Neurosci.* **2014**, *52*, 1–9.

[64] Shoji, M.; Golde, T. E.; Ghiso, J.; Cheung, T. T.; Estus, S.; Shaffer, L. M., et al. Production of the Alzheimer amyloid β protein by normal proteolytic processing. *Science* **1992**, *258*, 126–129.

[65] Haapasalo, A.; Kovacs, D. M. The many substrates of presenilin/gamma-secretase. *J. Alzheimer's Dis.* **2011**, *25*, 3–28.

[66] Takami, M.; Funamoto, S. γ-Secretase-dependent proteolysis of transmembrane domain of amyloid precursor protein: successive tri- and tetrapeptide release in amyloid β-protein production. *Int. J. Alzheimer's Dis.* **2012**, *2012* 591392.

[67] Selkoe, D. J. Alzheimer's disease: genes, proteins, and therapy. *Physiol. Rev.* **2001**, *81*, 741–766.

[68] Golde, T. E.; Koo, E. H.; Felsenstein, K. M.; Osborne, B. A.; Miele, L. γ-Secretase inhibitors and modulators. *Biochim. Biophys. Acta.* **1828**, *2013*, 2898–2907.

[69] Hopkins, C. R. ACS Chemical Neuroscience Molecule Spotlight on Semagacestat (LY450139). *ACS Chem. Neurosci.* **2010**, *1*, 533–534.

[70] Gitter, B. D.; Czilli, D. L.; Li, W.; Dieckman, D. K.; Bender, M. H.; Nissen, J. S., et al. Stereoselective inhibition of amyloid beta peptide secretion LY450139, a novel functional gamma secretase inhibitor. *Neurobiol. Aging.* **2004**, *25*, S571.

[71] May, P. C.; Yang, Z.; Li, W.; Li, W. -Y.; Hyslop, P. A.; Siemers, E.; Boggs, L. N. Multicompartmental pharmaco-dynamic assessment of the functional gamma-secretase inhibitor LY450139 in PDAPP transgenic mice and non-transgenic mice. *Neurobiol. Aging.* **2004**, *25*, S65.

[72] Ness, D. K.; Boggs, L. N.; Hepburn, D. L.; Gitter, B. D.; Long, G. G.; May, P. C., et al. Reduced β-amyloid burden, increased C-99 concentrations and evaluation of neuropathology in the brains of PDAPP mice given LY450139 dihydrate daily by gavage for 5 months. *Neurobiol. Aging.* **2004**, *25*, 238–239.

[73] Siemers, E.; Skinner, M.; Dean, R. A.; Gonzales, C.; Satterwhite, J.; Farlow, M., et al. Safety, tolerability and changes in amyloid β concentrations after administration of a γ-secretase inhibitor in volunteers. *Clin. Neuropharmacol.* **2005**, *28*, 126–132.

[74] Siemers, E. R.; Quinn, J. F.; Kaye, J.; Farlow, M. R.; Porsteinsson, A.; Tariot, P., et al. Effects of a gamma-secretase inhibitor in a randomized study of patients with Alzheimer disease. *Neurology* **2006**, *66*, 602–604.

[75] Fleisher, A. S.; Raman, R.; Siemers, E. R.; Becerra, L.; Clark, C. M.; Dean, R. A., et al. Phase 2 safety trial targeting amyloid beta production with a gamma-secretase inhibitor in Alzheimer disease. *Arch. Neurol.* **2008**, *65*, 1031–1038.

[76] Doody, R. S.; Raman, R.; Farlow, M.; Iwatsubo, T.; Vellas, B.; Joffe, S., et al. A Phase 3 trial of semagacestat for treatment of Alzheimer's disease. *N. Engl. J. Med.* **2013**, *369*, 341–350.

[77] Schor, N. F. What the halted Phase III gamma-secretase inhibitor trial may (or may not) be telling us. *Ann. Neurol.* **2011**, *69*, 237–239.

[78] Coric, V.; van Dyck, C. H.; Salloway, S.; Andreasen, N.; Brody, M.; Richter, R. W., et al. Safety and tolerability of the gamma-secretase inhibitor avagacestat in a Phase 2 study of mild to moderate Alzheimer disease. *Arch. Neurol.* **2012**, *69*, 1430–1440.

[79] Murphy, M. P.; Hickman, L. J.; Eckman, C. B.; Uljon, S. N.; Wang, R.; Golde, T. E. Gamma-secretase, evidence for multiple proteolytic activities and influence of

membrane positioning of substrate on generation of amyloid beta peptides of varying length. *J. Biol. Chem.* **1999**, *274*, 11914–11923.

[80] Weggen, S.; Eriksen, J. L.; Sagi, S. A.; Pietrzik, C. U.; Ozols, V.; Fauq, A., et al. Evidence that nonsteroidal anti-inflammatory drugs decrease amyloid beta 42 production by direct modulation of gamma-secretase activity. *J. Biol. Chem.* **2003**, *278*, 31831–31837.

[81] Green, R. C.; Schneider, L. S.; Amato, D. A.; Beelen, A. P.; Wilcock, G.; Swabb, E. A., et al. Effect of tarenflurbil on cognitive decline and activities of daily living in patients with mild Alzheimer disease: a randomized controlled trial. *JAMA.* **2009**, *302*, 2557–2564.

[82] Peretto, I.; Radaelli, S.; Parini, C.; Zandi, M.; Raveglia, L. F.; Dondio, G., et al. Synthesis and biological activity of flurbiprofen analogues as selective inhibitors of b-amyloid1-42 secretion. *J. Med. Chem.* **2005**, *48*, 5705–5720.

[83] Imbimbo, B. P.; Del Giudice, E.; Cenacchi, V.; Volta, R.; Villetti, G.; Facchinetti, F., et al. In vitro and in vivo profiling of CHF5022 nd CHF5074 two beta-amyloid 1-42 lowering agents. *Pharmacol. Res.* **2007**, *55*, 318–328.

[84] Balducci, C.; Mehdawy, B.; Mare, L.; Giuliani, A.; Lorenzini, L.; Sivilia, S., et al. The c-secretase modulator CHF5074 restores memory and hippocampal synaptic plasticity in plaque-free Tg2576 mice. *J. Alzheimer's Dis.* **2011**, *24*, 799–816.

[85] Giuliani, A.; Beggiato, S.; Baldassarro, V. A.; Mangano, C.; Giardino, L.; Imbimbo, B. -P., et al. CHF5074 restores visual memory ability and pre-synaptic cortical acetylcholine release in pre-plaque Tg2576 mice. *J. Neurochem.* **2013**, *124*, 613–620.

[86] Imbimbo, B. P.; Giardino, L.; Sivilia, S.; Giuliani, A.; Gusciglio, M.; Pietrini, V., et al. CHF5074, a novel gamma secretase modulator, restores hippocampal neurogenesis potential and reverses contextual memory deficit in a transgenic mouse model of Alzheimer's disease. *J. Alzheimer's Dis.* **2010**, *20*, 159–173.

[87] Sivilia, S.; Lorenzini, L.; Giuliani, A.; Gusciglio, M.; Fernandez, M.; Baldassarro, V. A., et al. Multi-target action of the novel anti-Alzheimer compound CHF5074: in vivo study of long term treatment in Tg2576 mice. *BMC Neurosci.* **2013**, *14*, 44.

[88] Imbimbo, B. P.; Del Giudice, E.; Colavito, D.; D'Arrigo, A.; Dalle Carbonare, M.; Villetti, G., et al. 1-(39,49-Dichloro-2-fluoro[1,19-biphenyl]-4-yl)-cyclopropanecarboxylic acid (CHF5074), a novel gamma-secretase modulator, reduces brain beta-amyloid pathology in a transgenic mouse model of Alzheimer's disease without causing peripheral toxicity. *J. Pharmacol. Exp. Ther.* **2007**, *323*, 822–830.

[89] Imbimbo, B. P.; Hutter-Paier, B.; Villetti, G.; Facchinetti, F.; Cenacchi, V.; Volta, R., et al. CHF5074, a novel gamma-secretase modulator, attenuates brain beta-amyloid pathology and learning deficit in a mouse model of Alzheimer's disease. *Br. J. Pharmacol.* **2009**, *156*, 982–993.

[90] Lanzillotta, A.; Sarnico, I.; Benarese, M.; Branca, C.; Baiguera, C.; Hutter-Paier, B., et al. The gamma-secretase modulator CHF5074 reduces the accumulation of native hyperphosphorylated tau in a transgenic mouse model of Alzheimer's disease. *J. Mol. Neurosci.* **2011**, *45*, 22–31.

[91] Poli, G.; Corda, E.; Lucchini, B.; Puricelli, M.; Martino, P. A.; Dall'Ara, P., et al. Therapeutic effect of CHF5074, a new γ-secretase modulator, in a mouse model of scrapie. *Prion.* **2012**, *6*, 62–72.

[92] Branca, C.; Sarnico, I.; Ruotolo, R.; Lanzillotta, A.; Viscomi, A. R.; Benarese, M., et al. Pharmacological targeting of the b-amyloid precursor protein intracellular domain. *Sci. Rep.* **2014**, *4*, 4618.

[93] Konietzko, U. AICD nuclear signaling and its possible contribution toAlzheimer's disease. *Curr. Alzheimer Res.* **2012**, *9*, 200–216.

[94] Ghosal, K.; Vogt, D. L.; Liang, M.; Shen, Y.; Lamb, B. T.; Pimplikar, S. W. Alzheimer's disease-like pathological features in transgenic mice expressing the APP intracellular domain. *Proc. Natl. Acad. Sci. U.S.A.* **2009**, *106*, 18367–18372.

[95] Ghosal, K.; Stathopoulos, A.; Pimplikar, S. W. APP intracellular domain impairs adult neurogenesis in transgenic mice by inducing neuroinflammation. *PLoS One.* **2010**, *5* e11866.

[96] Imbimbo, B. P.; Frigerio, E.; Breda, M.; Fiorentini, F.; Fernandez, M.; Sivilia, S., et al. Pharmacokinetics and pharmacodynamics of CHF5074 after short-term administration in healthy subjects. *Alzheimer Dis. Assoc. Disord.* **2013**, *27*, 278–286.

[97] http://www.cerespir.com/c/csp1103/.

[98] Yesuvadian, R.; Krishnamoorthy, J.; Ramamoorthy, A.; Bhunia, A. Potent γ-secretase inhibitors/modulators interact with amyloid-β fibrils but do not inhibit fibrillation: a high-resolution NMR study. *Biochem. Biophys. Res. Commun.* **2014**, *447*, 590–595.

[99] Borgegard, T.; Jureus, A.; Olsson, F.; Rosqvist, S.; Sabirsh, A.; Rotticci, D., et al. First and second generation γ-secretase modulators (GSMs) modulate amyloid-β (Aβ) peptide production through different mechanisms. *J. Biol. Chem.* **2012**, *287*, 11810–11819.

[100] Hashimoto, T.; Ishibashi, A.; Hagiwara, H.; Murata, Y.; Takenaka, O.; Miyagawa, T. E2012: a novel gamma-secretase modulator—pharmacology. *Alzheimers Dement.* **2010**, *6*, S242.

[101] Nakano-Ito, K.; Fujikawa, Y.; Hihara, T.; Shinjo, H.; Kotani, S.; Suganuma, A., et al. E2012-induced cataract and its predictive biomarkers. *Toxicol. Sci.* **2014**, *137*, 249–258.

[102] Imbimbo, B. P.; Giardina, G. M. γ-Secretase inhibitors and modulators for the treatment of Alzheimer's disease: disappointments and hopes. *Curr. Top. Med. Chem.* **2011**, *11*, 1555–1570.

[103] Nagy, C.; Schuck, E.; Ishibashi, A.; Nakatani, Y.; Rege, B.; Logovinsky, V. E2012, a novel gamma secretase modulator, decreases plasma amyloid-beta (Aβ) levels in humans. *Alzheimers Dement.* **2010**, *6*, S574.

[104] Yu, Y.; Logovinsky, V.; Schuck, E.; Kaplow, J.; Chang, M. -K.; Miyagawa, T., et al. Safety, tolerability, pharmacokinetics, and pharmacodynamics of the novel g-secretase modulator, E2212, in healthy human subjects. *J. Clin. Pharmacol.* **2014**, *54*, 528–536.

[105] Vohora, D.; Bhowmik, M. Histamine H3 receptor antagonists/inverse agonists on cognitive and motor processes: relevance to Alzheimer's disease, ADHD, schizophrenia, and drug abuse. *Front. System Neurosci.* **2012**, *6*, 72.

[106] Meneses, A. Role of 5-HT6 receptors in memory formation. *Drug News Perspect.* **2001**, *14*, 396–400.

[107] Roth, B. L.; Craigo, S. C.; Choudhary, M. S.; Uluer, A.; Monsma, F. J., Jr.; Shen, Y., et al. Binding of typical and atypical antipsychotic agents to 5-hydroxytryptamine-6 and 5-hydroxytryptamine-7 receptors. *J. Pharmacol. Exp. Ther.* **1994**, *268*, 1403–1410.

[108] Shadurskii, K. S.; Matveeva, I. A.; Iliuchenok, T. Therapeutic and protective properties of dimebon in burns. *Farmakol. Toksikol.* **1983**, *46*, 90–92.

[109] Cano-Cuenca, N.; Solís-García del Pozo, J. E.; Jordan, J. Evidence for the efficacy of latrepirdine (dimebon) treatment for improvement of cognitive function: a meta-analysis. *J. Alzheimer's Dis.* **2014**, *38*, 155–164.

[110] Wu, J.; Li, Q.; Bezprozvanny, I. Evaluation of Dimebon in cellular model of Huntington's disease. *Mol. Neurodegener.* **2008**, *3*, 15.

[111] Eckert, S. H.; Eckmann, J.; Renner, K.; Eckert, G. P.; Leuner, K.; Muller, W. E. Dimebon ameliorates amyloid-beta induced impairments of mitochondrial form and function. *J. Alzheimer's Dis.* **2012**, *31*, 21–32.

[112] Doody, R. S. Dimebon as a potential therapy for Alzheimer's disease. *CNS Spectr.* **2009**, *14*, 14–18.

[113] Bezprozvanny, I. The rise and fall of dimebon. *Drug News Perspect* **2010**, *23*, 518–523.

[114] Weisova, P.; Alvarez, S. P.; Kilbride, S. M.; Anilkumar, U.; Baumann, B.; Jordan, J., et al. Latrepirdine is a potent activator of AMP-activated protein kinase and reduces neuronal excitability. *Transl. Psychiatry.* **2013**, *3*, e317.

[115] Bharadwaj, P. R.; Verdile, G.; Barr, R. K.; Gupta, V.; Steele, J. W.; Lachenmayer, M. L., et al. Latrepirdine (Dimebon™) enhances autophagy and reduces intracellular GFP-Abeta$_{42}$ levels in yeast. *J. Alzheimer's Dis.* **2012**, *32*, 949–967.

[116] Steele, J. W.; Ju, S.; Lachenmayer, L.; Liken, J.; Stock, A.; Kim, S. H., et al. Latrepirdine stimulates autophagy and reduces accumulation of α-synuclein in cells and in mouse brain. *Mol. Psychiatry.* **2013**, *18*, 882–888.

[117] Yamashita, M.; Nonaka, T.; Arai, T.; Kametani, F.; Buchman, V. L.; Ninkina, N., et al. Methylene blue and dimebon inhibit aggregation of TDP-43 in cellular models. *FEBS Lett.* **2009**, *583*, 2419–2424.

[118] Bharadwaj, P. R.; Bates, K. A.; Porte, T.; Teimouri, E.; Perry, G.; Steele, J. W., et al. Latrepirdine: molecular mechanisms underlying potential therapeutic roles in Alzheimer's and other neurodegenerative diseases. *Transl. Psychiatry.* **2013**, *3*, e332.

[119] Steele, J. W.; Kim, S. H.; Cirrito, J. R.; Verges, D. K.; Restivo, J. L.; Westaway, D., et al. Acute dosing of latrepirdine (Dimebon™), a possible Alzheimer therapeutic, elevates extracellular amyloid-beta levels in vitro and in vivo. *Mol. Neurodegener.* **2009**, *4*, 51.

[120] Perez, S. E.; Nadeem, M.; Sadleir, K. R.; Matras, J.; Kelley, C. M.; Counts, S. E., et al. Dimebon alters hippocampal amyloid pathology in 3xTg-AD mice. *Int. J. Physiol. Pathophysiol. Pharmacol.* **2012**, *4*, 115–127.

[121] Wang, J.; Ferruzzi, M. G.; Varghese, M.; Qian, X.; Cheng, A.; Xie, M., et al. Preclinical study of dimebon on beta-amyloid-mediated neuropathology in Alzheimer's disease. *Mol. Neurodegener.* **2011**, *6*, 7.

[122] Bachurin, S.; Bukatina, E.; Lermontova, N.; Tkachenko, S.; Afanasiev, A.; Grigoriev, V., et al. Antihistamine agent Dimebon as a novel neuroprotector and a cognition enhancer. *Ann. N. Y. Acad. Sci.* **2001**, *939*, 425–435.

[123] Giorgetti, M.; Gibbons, J. A.; Bernales, S.; Alfaro, I. E.; Drieu La Rochelle, C.; Cremers, T., et al. Cognition-enhancing properties of Dimebon in a rat novel object recognition task are unlikely to be associated with acetylcholinesterase inhibition or N-methyl-D-aspartate receptor antagonism. *J. Pharmacol. Exp. Ther.* **2010**, *333*, 748–757.

[124] Steele, J. W.; Lachenmayer, M. L.; Ju, S.; Stock, A.; Liken, J.; Kim, S. H., et al. Latrepirdine improves cognition and arrests progression of neuropathology in an Alzheimer's mouse model. *Mol. Psychiatry.* **2013**, *18*, 889–897.

[125] Peters, O. M.; Shelkovnikova, T.; Tarasova, T.; Springe, S.; Kukharsky, M. S.; Smith, G. A., et al. Chronic administration of dimebon does not ameliorate amyloid-β pathology in 5xFAD transgenic mice. *J. Alzheimer's Dis.* **2013**, *36*, 589–596.

[126] Peters, O. M.; Connor-Robson, N.; Sokolov, V. B.; Aksinenko, A. Y.; Kukharsky, M. S.; Bachurin, S. O., et al. Chronic administration of dimebon ameliorates pathology in tauP301S transgenic mice. *J. Alzheimer's Dis.* **2013**, 1041–1049.

[127] Bachuin, S. O.; Shelkovnikova, T. A.; Ustyugov, A. A.; Peters, O.; Khritankova, I.; Afanasieva, M. A., et al. Dimebon slows progression of proteinopathy in gamma-synuclein transgenic mice. *Neurotoxicol. Res.* **2012**, *22*, 33–42.

[128] Shelkovnikova, T. A.; Ustyugov, A. A.; Millership, S.; Peters, O.; Anichtchik, O.; Spillantini, M. G., et al. Dimebon does not ameliorate pathological changes caused by expression of truncated (1-120) human alpha-synuclein in dopaminergic neurons of transgenic mice. *Neurodegener. Dis.* **2011**, *8*, 430–437.

[129] Doody, R. S.; Gavrilova, S. I.; Sano, M.; Thomas, R. G.; Aisen, P. S.; Bachurin, S. O., et al. Effect of dimebon on cognition, activities of daily living, behaviour, and global function in patients with mild-to-moderate Alzheimer's disease: a randomised, double-blind, placebo-controlled study. *Lancet.* **2008**, *372*, 207–215.

[130] Kieburtz, K.; McDermott, M. P.; Voss, T. S.; Corey-Bloom, J.; Deuel, L. M.; Dorsey, E. R., et al. A randomized, placebo-controlled trial of latrepirdine in Huntington disease. *Arch. Neurol.* **2010**, *67*, 154–160.

[131] HORIZON Investigators of the Huntington Study Group and European Huntington's Disease Network. A randomized, double-blind, placebo-controlled study of latrepirdine in patients with mild to moderate Huntington disease. *JAMA Neurol.* **2013**, *70*, 25–33.

[132] http://investors.medivation.com/releasedetail.cfm?ReleaseID=448818.

[133] http://investors.medivation.com/releasedetail.cfm?ReleaseID=639515.

[134] Pieper, A. A.; McKnight, S. L.; Ready, J. M. P7C3 and an unbiased approach to drug discovery for neurodegenerative diseases. *Chem. Soc. Rev.* **2014**, S60448A DOI: 10. 1039/ C3C.

[135] Pieper, A. A.; Xie, S.; Capota, E.; Estill, S. J.; Zhong, J.; Long, J. M., et al. Discovery of a proneurogenic, neuroprotective chemical. *Cell.* **2010**, *142*, 39–51.

[136] MacMillan, K. S.; Naidoo, J.; Liang, J.; Melito, L.; Williams, N. S.; Morlock, L., et al. Development of proneurogenic, neuroprotective small molecules. *J. Am. Chem. Soc.* **2011**, *133*, 1428–1437.

[137] De Jesus-Cortes, H.; Xu, P.; Drawbridge, J.; Estill, S. J.; Huntington, P.; Tran, S., et al. Neuroprotective efficacy of aminopropyl carbazoles in a mouse model of Parkinson disease. *Proc. Natl. Acad. Sci. U.S.A.* **2012**, *109*, 17010–17015.

[138] Tesla, R.; Wolf, H. P.; Xu, P.; Drawbridge, J.; Estill, S. J.; Huntington, P., et al. Neuroprotective efficacy of aminopropyl carbazoles in a mouse model of amyotrophic lateral sclerosis. *Proc. Natl. Acad. Sci. U.S.A.* **2012**, *109*, 17016–10721.

[139] Blaya, M. O.; Bramlett, H.; Nadoo, J.; Pieper, A. A.; Dietrich, W. D., III. Neuroprotective efficacy of a proneurogenic compound after traumatic brain injury. *J. Neurotrauma.* **2014**, *31*, 476–468.

[140] Walker, A. K.; Rivera, P. D.; Wang, Q.; Chuang, J. -C.; Tran, S.; Osborne-Lawrence, S., et al. The P7C3-class of neuroprotective compounds exerts antidepressant efficacy in mice by increasing hippocampal neurogenesis. *Mol. Psychiatry.* **2014**, DOI:10.1038/ mp.2014.34.

2

Targeting the Protein Quality Control (PQC) Machinery

2.1 MOLECULAR CHAPERONES, PQC, AND NEURODEGENERATION

Misfolded/misdecorated protein monomers in general [1], and tau monomers in particular [2], are processed through several paths in neurons. Such paths may be detrimental—leading to neurotoxic aggregates—or beneficial—either refolding misfolded protein copies, or disposing of them. The former must be antagonized, and the latter must be stimulated to prevent the aggregation of amyloidogenic proteins.

Abnormally decorated, misfolded protein copies are toxic to neurons [3,4]. The cellular protein quality control (PQC) system takes care of misfolded proteins in normal cells. Protein chaperones control the PQC activity network, described in detail in the biology-oriented companion book [5]. Chaperones are intrinsically disordered proteins with broad substrate specificity, with high affinity for misfolded proteins, and reduced affinity for their natively folded, physiologically useful counterparts [6].

Chaperones contribute to preserve the needed level of functional proteins in the cell. Chaperone *holdases* hold aggregation-prone, partially misfolded substrates to prevent their aggregation and to stabilize them [7,8]. Chaperone *(un)foldases* bind partially or totally misfolded proteins, unfold and correctly refold them. The process requires ATP hydrolysis and leads to energy-minimized, physiologically competent client proteins through binding and localized reorganization with disordered chaperone regions [9,10]. Once refolding is complete, (un)foldases and correctly folded client proteins split, due to reduced binding affinity. *Disaggregating chaperones* bind to aggregates containing many copies of misfolded substrates to unfold–refold them, causing their disaggregation and refolding–detoxification [10,11]. Overall, refolding of misfolded proteins

Chemical Modulators of Protein Misfolding and Neurodegenerative Disease. http://dx.doi.org/10.1016/B978-0-12-801944-3.00002-3

reduces the risk of formation of toxic oligomers and aggregation, and provides functionally competent protein copies [12].

If any abnormal splicing/post-translational modification (PTM) pattern of tau is not rapidly reversed, or if pathological tau mutations are expressed, abnormal tau copies adopt non-native conformations. Abundant, misfolded tau copies tend to aggregate, eventually recruiting also wild-type (WT) tau copies.

Chaperones are a prospective point of intervention against tauopathies, due to their interactions with tau [13,14]. Small heat shock proteins (sHsps) counteract protein aggregation, and *Hsp27* in particular shows strong interactions with tau and Aβ [15]. *Hsp70* family members [16] bind to tau [17], decrease its aggregation [18], and increase its microtubule (MT) binding either as such [19], or through complexes with Hsp70 co-chaperones [20]. *Hsp90* [21] has a major impact on tau aggregation and folding *per se* [22], or through complexes with immunophilin FK-binding protein 51 (FKBP51) [23] and FK-binding protein 52 (FKBP52) [24]. Hsp70 and Hsp90 chaperone families may be functionally impaired in tauopathies [12], and are validated targets for their treatment. The same is true for Hsp70– and Hsp90–co-chaperone complexes that promote degradation/elimination of misfolded tau copies or aggregates [12].

Thus, the next three sections describe small molecule modulators acting on Hsp27, Hsp70 and Hsp90 to restore the impairment of PQC in neurodegenerative disease (NDD).

2.2 Hsp27

The ATP-independent nature of Hsp27 prevents the comparatively easy identification of ATP-competitive inhibitors of chaperone activity, witnessed for Hsp90 and Hsp70. Nevertheless, Hsp27 modulators are known either to act directly on Hsp27 (modulators binding to Hsp27), or indirectly (modulation of Hsp27 expression or phosphorylation). Hsp27 inhibitors could be of use in stressed-PQC impaired conditions that are common in NDDs and, in particular, in tauopathies. Conversely, compounds activating Hsp27 (either alone or with complexed Hsp70) are reported as neuroprotective in animal models of *Parkinson's disease* (PD) and *Alzheimer's disease* (AD) (see below). Thus, Hsp27 modulators should be useful to study the transition between a neuroprotective and a neurotoxic role for Hsp27 and chaperones in cellular and *in vivo* models of proteinopathies/NDDs.

Five small molecule inhibitors directly bind to Hsp27. The isoxazole KRIBB3 (**2.1**, Figure 2.1) is the result of a high-throughput screening (HTS)/structural optimization campaign looking for inhibitors of cell migration [25]. A biotinylated derivative of KRIBB3 pulls down Hsp27

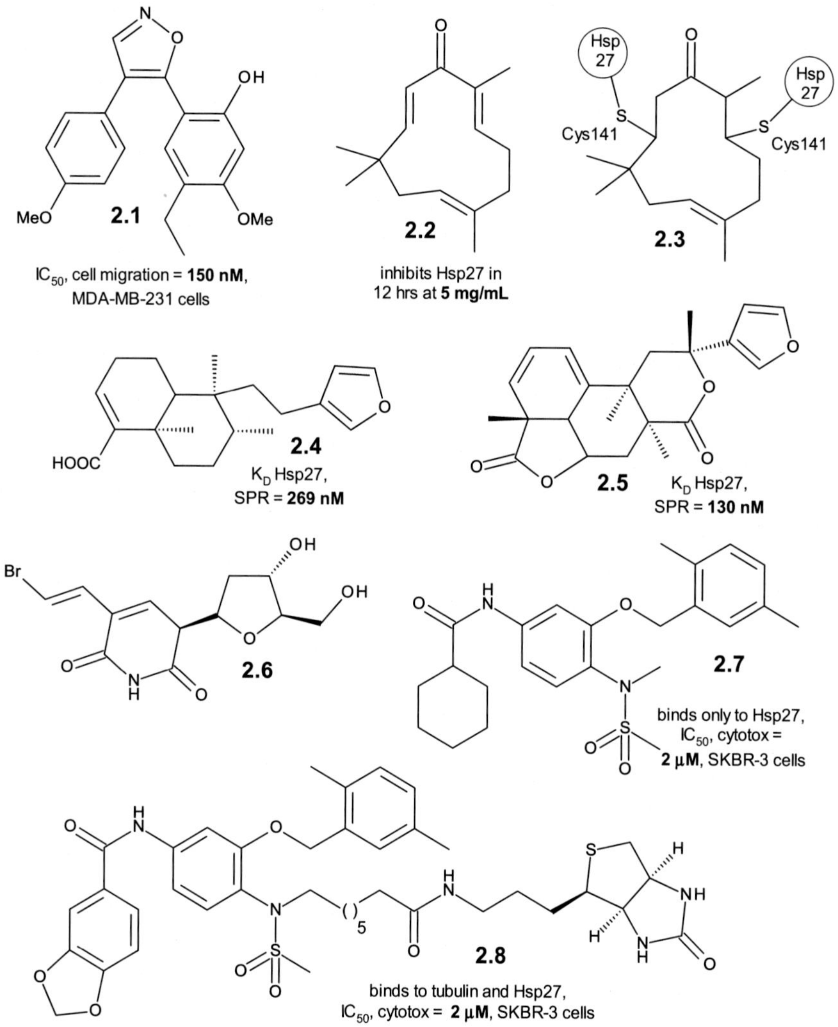

FIGURE 2.1   Small molecule inhibitors of Hsp27: chemical structures, **2.1–2.8**.

in a cellular lysate, while small interfering RNA (siRNA) knockdown of Hsp27 also inhibits cell migration. The KRIBB3 binding site on Hsp27 is not identified, but an inhibitory effect on the phosphorylation of Ser78 (either through physical shielding of the site, or due to conformational changes) is observed [25]. Subsequent reports [26] question the relevance of Hsp27 inhibition in the antitumoral–proapoptotic action of KRIBB3, and show that it is also a potent inhibitor of tubulin polymerization. The relevance of Hsp27 inhibition for KRIBB3 in various cellular assays may vary [27–29], but its structure (providing that its exact binding site on

Hsp27 will be determined) could be of use in the rational design of Hsp27 inhibitors.

Naturally occurring, α,β-unsaturated ketone/sesquiterpenone zerumbone (**2.2**) inhibits Hsp27 in Hsp27-overexpressing lung adenocarcinoma cells [30]. Treatment with zerumbone causes the dimerization of Hsp27 through a Michael reaction between the two α,β-unsaturated moieties in **2.2**, and the Cys141 residue in two Hsp27 copies. The stable dimer **2.3** is specifically formed by 8-carbonyl-bearing zerumbone, as its 8-OH and 8-CH_2 congeners do not promote covalent, stable dimerization of Hsp27. Zerumbone-promoted dimerization does not happen aspecifically with Cys residues in other proteins [30], although Michael reactions between zerumbone and, *inter alia*, kelch-like ECH-associated protein 1 (Keap1) and human antigen R (HuR) to provide covalent zerumbone protein adduct are reported [31]. Interestingly, zerumbone induces a heat shock response in mouse hepatoma cells [32], and activates two main mechanisms for misfolded protein disposal, the ubiquitin–proteasome system (UPS, see Chapter 3) and autophagy (see Chapters 4 and 5) [33].

Hardwickiic acid (HAA, **2.4**) is a naturally occurring neoclerodane diterpene, whose binding with Hsp27 is determined through a chemical proteomic experiment [34]. Surface plasmon resonance (SPR) determines a nM K_D for the HAA–Hsp27 interaction, and identifies other terpenoid structures (i.e., the tetracyclic bis-lactone **2.5**) as Hsp27 inhibitors. Hsp27 incubation with HAA *in vitro* protects it from proteolytic cleavage in the N-terminal region, precisely between amino acids (AAs) 26 and 36 (probably where the HAA–Hsp27 interaction takes place) [34].

Brivudine (RP101, **2.6**) is a vinyluridine marketed against varicella zoster virus [35]. Clinical testing of brivudine in combination with gemcitabine and cisplatin in pancreatic cancer patients improves the efficiency of chemotherapy [36]. The effect of brivudine may be due to down-regulation of Hsp27 activity in tumor cells [37], although other molecular targets are proposed [38]. Affinity chromatography with brivudine-loaded magnetic beads pulls down only Hsp27. A computational model of Hsp27 suggests a putative RP101 binding site in the N-terminal Hsp27 domain, where Phe29 and Phe33 residues are essential to establish the RP101–Hsp27 interaction. Partial validation of this binding site is obtained with double Phe29E/Phe33E mutant Hsp27 (no binding with RP101), although single Phe mutations do not affect brivudine–Hsp27 binding [37].

The arylsulfonamide JCC76 (**2.7**) is an Hsp27 inhibitor identified during the optimization of anticancer agents with unknown mechanism of action [39]. The biotinylated arylsulfonamide **2.8** (Figure 2.1) strongly binds to α- and β-tubulin, and to Hsp27 in breast cancer cells [40]. While the interaction with tubulin is clarified (**2.8** binds to the colchicine binding site), the binding site with Hsp27 is not known [40]. Synthesis and biological characterization of a larger set of analogs provides compounds with stronger tubulin binding, and compounds with preponderant Hsp27 inhibition [41]. The arylsulfonamide **2.7** is the most potent and selective among them.

Structural characterization of Hsp27—in particular the N-terminal WDPF motif and the α-crystallin domain (ACD)—should facilitate the rational design of Hsp27 modulators acting on the dynamic organization of Hsp27 oligomers. The high flexibility/low secondary structure content of the N- and C-terminal domains of Hsp27 prevents the crystallization of full length (FL) Hsp27. The structure of a double Glu125Ala/Glu126Ala mutant in the 90-171 fragment of Hsp27, containing most of the ACD domain, is published [42,43] and could be useful in the rational drug design of Hsp27. One should consider, though, that the double mutation probably influences the oligomerization equilibrium (i.e., the functionality as a chaperone) of the ACD domain, and of Hsp27 in general. The crystal structure of a hybrid construct where a 14-mer sequence between the N-terminal WDPF sequence and the ACD domain of human Hsp27 is inserted into Hsp16.5 from *Methanocaldococcus jannaschii* (a crystallized Hsp27 congener) is known [44]. It partially elucidates the mechanism of Hsp27 activation, and the influence of oligomer size on the accessibility of Hsp27 domains to client protein binding [44]. This knowledge should at some point be converted into rationally designed small molecule modulators of the Hsp27 chaperone cycle.

The arylethynyltriazolyl ribonucleoside **2.9** (Figure 2.2) inhibits the growth of pancreatic tumor cells *in vitro* and *in vivo* [45]. Its effect is in part

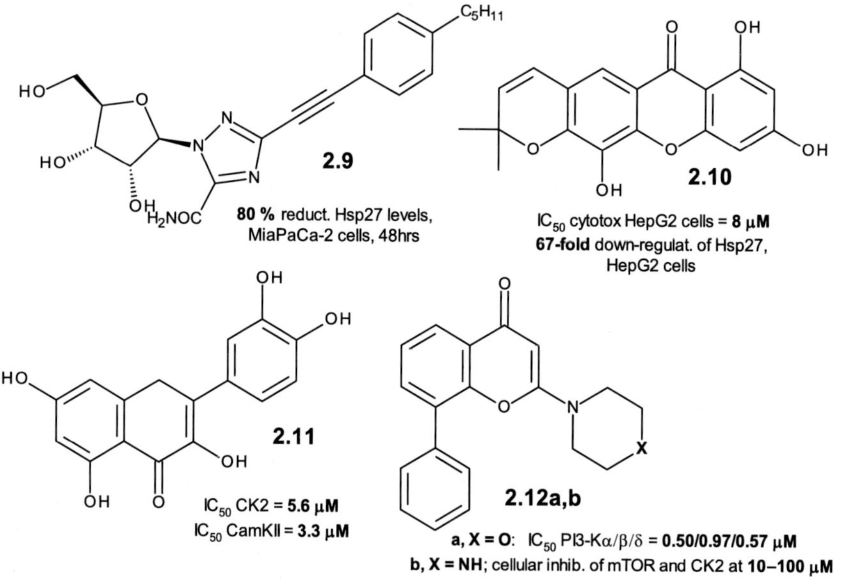

FIGURE 2.2   Small molecule down-regulators of Hsp27: chemical structures, **2.9–2.12b**.

due to strong down-regulation of Hsp27, through an unknown mechanism. A preliminary structure–activity relationship (SAR) identifies the 3-ethynyl moiety as essential for anticancer activity [45]. The naturally occurring xanthone TDP (**2.10**) inhibits the growth of hepatocellular carcinoma (HCC) by inducing caspase-dependent apoptosis [46]. Proteomic profiling identifies nine down-regulated proteins, among which Hsp27 is the most affected. The role of Hsp27 in the activity of TDP is validated by experiments with Hsp27 siRNA silencing or overexpression, and by the observed up-regulation of Hsp27 in HCC. The mechanism of action by which TDP down-regulates Hsp27 is not disclosed [46].

The naturally occurring flavone quercetin (**2.11**) shows biological effects related to its antioxidant and free radical scavenging properties [47,48]. Quercetin inhibits several kinases [49], and inhibits Hsp70 induction *via* suppression of heat shock factor 1 (HSF1) activation/inhibition of HSF1 phosphorylation by casein kinase 2 (CK2) and Ca^{2+}/calmodulin-dependent protein kinase II (CamKII) [49]. Quercetin causes an increase of Ser78 phosphorylation on Hsp27 (potentially activating Hsp27 chaperone functions), but its suppression of HSF1 induction causes an overall down-regulation of Hsp27 expression [50]. Down-regulation of Hsp27 levels by quercetin is confirmed in breast cancer stem cells [51,52] and in oral squamous carcinoma cells [53].

The benzopyranmorpholine LY294002 (**2.12a**) is a kinase inhibitor that down-regulates Hsp27 and Hsp70 by suppressing HSF1 binding to DNA [54]. Inhibition of phosphatidylinositol-3-kinase (PI3-K), a known LY294002 target, causes protein kinase B (Akt) inactivation and may eventually lead to inhibition of HSF1 phosphorylation/ activation [54]. LY294002 has limited kinase specificity, and other targets may influence its cellular effects. Its benzopyranpiperazine analog LY303511 (**2.12b**, Figure 2.2) does not inhibit PI3-K, but blocks phosphorylation by mammalian target of rapamycin (mTOR) kinase and CK2 [55]. LY303511 affects the Hsp27 chaperone cycle in cervical cancer HeLa cells causing a slow, permanent nuclear translocation of Hsp27 that lowers the cytosolic Hsp27 pool [56]. Namely, rapid, transient nuclear translocation coupled with phosphorylation-dependent oligomeric restructuring of Hsp27 supports the PQC machinery in stress conditions [57]. LY303511 induces a sustained phosphorylation of Hsp27 by rapid activation of p38 kinase (promoting MAPK-activated protein kinase 2 (MK2) driven phosphorylation of Hsp27), and inhibits Hsp27 dephosphorylation by protein phosphatase 2A (PP2A) [56].

Indirect inhibition of Hsp27 activation/chaperoning should be granted by MK2 inhibition/prevention of Hsp27 phosphorylation. ATP-competitive MK2 inhibitors are abundantly reported in the literature as anti-inflammatory agents, although their therapeutic potential in NDDs has not been evaluated yet. Benzopyranopyridines such as **2.13** (Figure 2.3) [58] are HTS-derived, structurally optimized MK2 inhibitors with good,

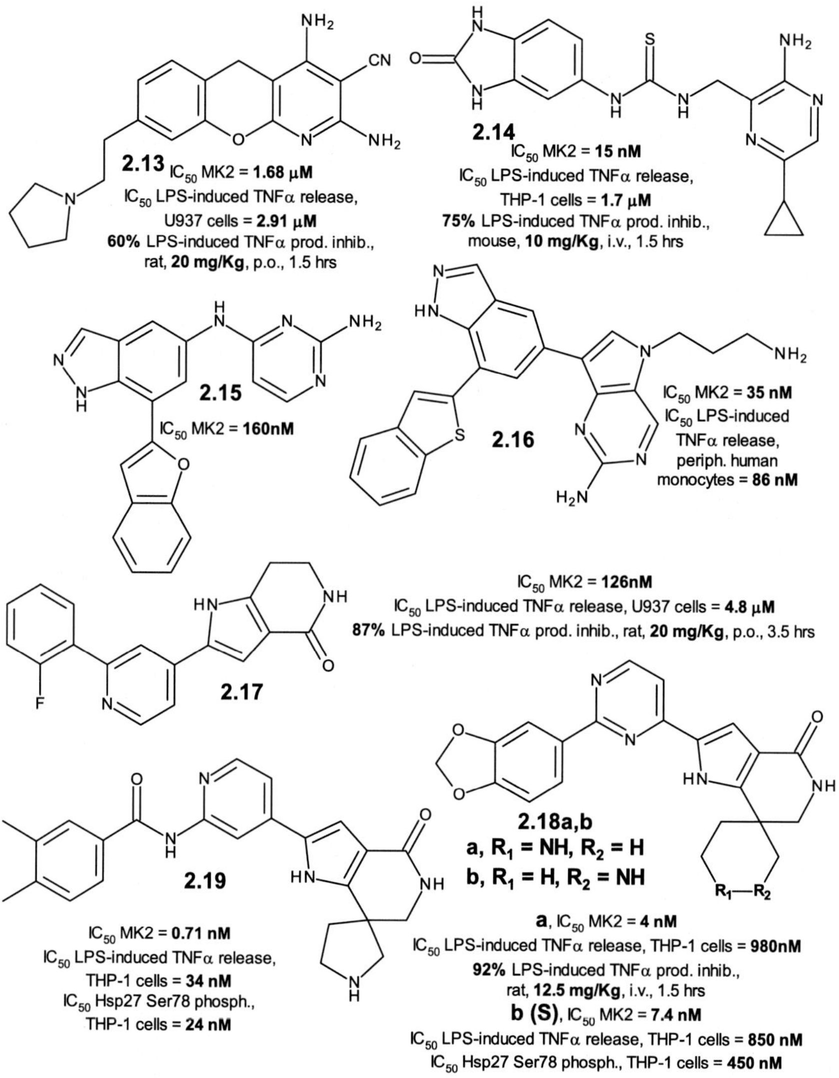

2.13 IC_{50} MK2 = **1.68 μM**
IC_{50} LPS-induced TNFα release,
U937 cells = **2.91 μM**
60% LPS-induced TNFα prod. inhib.,
rat, **20 mg/Kg**, p.o., 1.5 hrs

2.14
IC_{50} MK2 = **15 nM**
IC_{50} LPS-induced TNFα release,
THP-1 cells = **1.7 μM**
75% LPS-induced TNFα prod. inhib.,
mouse, **10 mg/Kg**, i.v., 1.5 hrs

2.15
IC_{50} MK2 = **160nM**

2.16
IC_{50} MK2 = **35 nM**
IC_{50} LPS-induced
TNFα release,
periph. human
monocytes = **86 nM**

2.17
IC_{50} MK2 = **126nM**
IC_{50} LPS-induced TNFα release, U937 cells = **4.8 μM**
87% LPS-induced TNFα prod. inhib., rat, **20 mg/Kg**, p.o., 3.5 hrs

2.18a,b
a, R$_1$ = NH, R$_2$ = H
b, R$_1$ = H, R$_2$ = NH

2.19
IC_{50} MK2 = **0.71 nM**
IC_{50} LPS-induced TNFα release,
THP-1 cells = **34 nM**
IC_{50} Hsp27 Ser78 phosph.,
THP-1 cells = **24 nM**

a, IC_{50} MK2 = **4 nM**
IC_{50} LPS-induced TNFα release, THP-1 cells = **980nM**
92% LPS-induced TNFα prod. inhib.,
rat, **12.5 mg/Kg**, i.v., 1.5 hrs
b (S), IC_{50} MK2 = **7.4 nM**
IC_{50} LPS-induced TNFα release, THP-1 cells = **850 nM**
IC_{50} Hsp27 Ser78 phosph., THP-1 cells = **450 nM**

FIGURE 2.3   Small molecule inhibitors of MK2 and other Hsp27-phosphorylating kinases: chemical structures, **2.13–2.19**.

nM potency in cell-free and cellular assays, such as lipopolysaccharide (LPS)-stimulated tumor necrosis factor α (TNFα) release. Their selectivity *vs.* other kinases is not disclosed. Compound **2.13** is orally active *in vivo* in an acute LPS rat model of inflammation.

Aminopyrazinthioureas such as **2.14** [59] arise from the structural optimization of HTS hits. They inhibit Hsp27 phosphorylation *in vitro* at nM concentration and show low μM potency in cellular assays of

LPS-stimulated tumor TNFα production. Their selectivity *vs.* other kinases is not disclosed. Compound **2.14** is active by i.v. administration *in vivo* in an LPS mouse model of inflammation.

Structure-based drug design and crystallography, using two catalytically inactive fragments of MK2 and starting from the kinase-privileged 2,4-diaminopyrimidine scaffold, is used to design the ATP-competitive indazole-pyrimidine compound **2.15**, and to structurally characterize its interaction with MK2 [60]. Structural optimization of **2.15** leads to indazole-pyrrolo[3,2-*d*] pyrimidines such as **2.16** [61], endowed with strong cell-free potency on MK2, and on LPS-stimulated TNFα release. The selectivity of **2.16** and of its analogs *vs.* other kinases is good. Compound **2.16** has a poor pharmacokinetic (PK) profile in rats by oral administration [61].

Structural optimization of an HTS hit leads to 2-pyridyl-tetrahydro-4H-pyrrolo[3,2-*c*]pyridine-4-ones such as **2.17** [62]. Compound **2.17** shows excellent cell-free inhibition of Hsp27 phosphorylation, moderate activity on LPS-stimulated TNFα release and on cellular phosphorylation of Hsp27 Ser78, and good oral potency *in vivo* in an LPS mouse model of inflammation. It has good selectivity *vs.* kinases closely related to MK2, most likely due to steric clashes of the arylpyridine substituent with their bulky active site residues corresponding to the small Cys140 residue in MK2 [62]. The pyrrolopyridine scaffold occurs also in 2-pyrimidinyl substituted spiro compounds such as **2.18a,b** and **2.19**, obtained through modifications suggested by X-ray crystal complexes of MK2 with **2.18**-like inhibitors [63]. Compound **2.18a** is highly potent in cell-free MK2 inhibition, and moderately active in cellular LPS-stimulated TNFα release and in Hsp27 phosphorylation assays. It is selective *vs.* other kinases, and active by s.c. and i.v. administration in an LPS rat model of inflammation [63]. The (S)-isomer of compound **2.18b** is even more selective *vs.* other kinases than **2.18a**, is orally bioavailable, and has a better PK profile, although its chiral nature could hinder its development [64]. Introduction of a benzamide substituent and replacement of the pyrimidine ring with a pyridine, as in compound **2.19** (Figure 2.3), provides an extremely potent subfamily of *in vitro* and *in vivo* active MK2 inhibitors [65].

Constraining 2-pyridyl pyrrolopiridinones into pyrrole-based tetracycles, such as **2.20** (Figure 2.4) [66], provides some of the most potent ATP-competitive MK2 inhibitors. Compound **2.20** is endowed with nM inhibition of cellular Hsp27 phosphorylation, and of LPS-stimulated TNFα release. It also shows limited efficacy as an oral treatment in mouse and rat models of rheumatoid arthritis [66].

Closely related 2-pyridyl-dihydropyrrolopyrimidinones, such as **2.21**, are potent *in vitro* inhibitors of MK2-mediated Hsp27 phosphorylation, and of LPS-stimulated TNFα release [67]. Their selectivity *vs.* other kinases is limited, and no data are available about their *in vivo* potency [67].

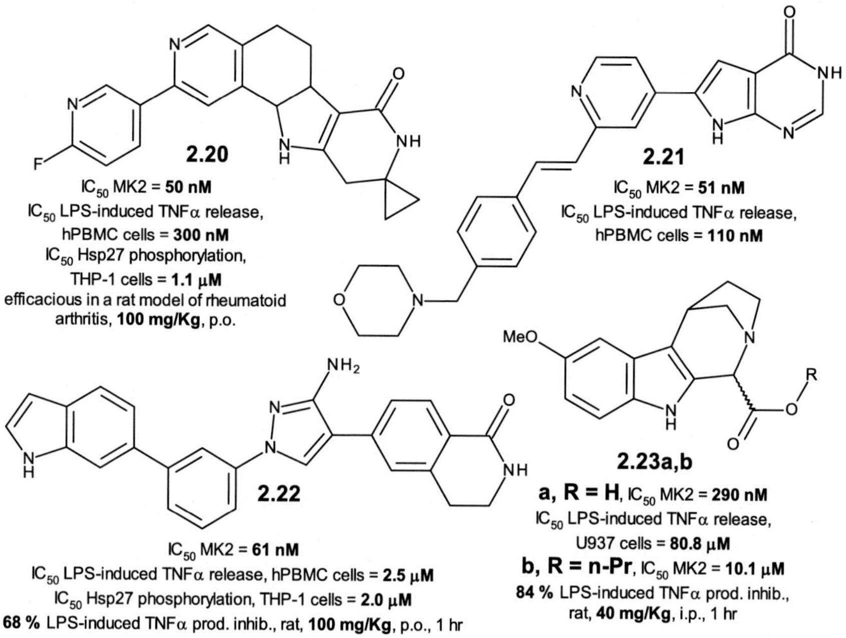

FIGURE 2.4 Small molecule inhibitors of MK2 and other Hsp27-phosphorylating kinases: chemical structures, **2.20–2.23b**.

1,4-Diaryl-3-aminopyrazoles are the result of scaffold hopping [68] on the pyrrolopyrimidinone lead structure. Rational drug design leads to the synthesis and characterization of 1,4-bisaryl-3-aminopyrazoles such as **2.22** [69]. Compound **2.22** is a potent nM inhibitor of Hsp27 phosphorylation in cell-free assays, and inhibits LPS-stimulated TNFα release and anysomicin-stimulated Hsp27 phosphorylation in cells. The selectivity of this compound class *vs.* other kinases is not disclosed. Compound **2.22** is orally active *in vivo* in an acute LPS rat model of inflammation [69].

Carbolines, such as compounds **2.23a,b**, are the result of the optimization of hits from an HTS campaign [70]. Among ≈100 synthesized analogs, the acidic compound **2.23a** shows nM potency in cell-free MK2 assays but no cellular activity, due to poor membrane permeability. The ester **2.23b** (Figure 2.4), a pro-drug of **2.23a**, shows reasonable cellular potency against LPS-stimulated TNFα release, and limited *in vivo* efficacy by intraperitoneal (i.p.) administration in the LPS rat model of inflammation [70].

Carbolines from another HTS campaign lead to the identification of thiazolamide-substituted carbolines, such as **2.24** (Figure 2.5) [71]. Although their cellular activity is limited, due to poor cell permeability, they possess cell-free potency and selectivity [71]. An improvement in terms of cellular potency, solubility, lipophilicity, and *in vivo* PK profile

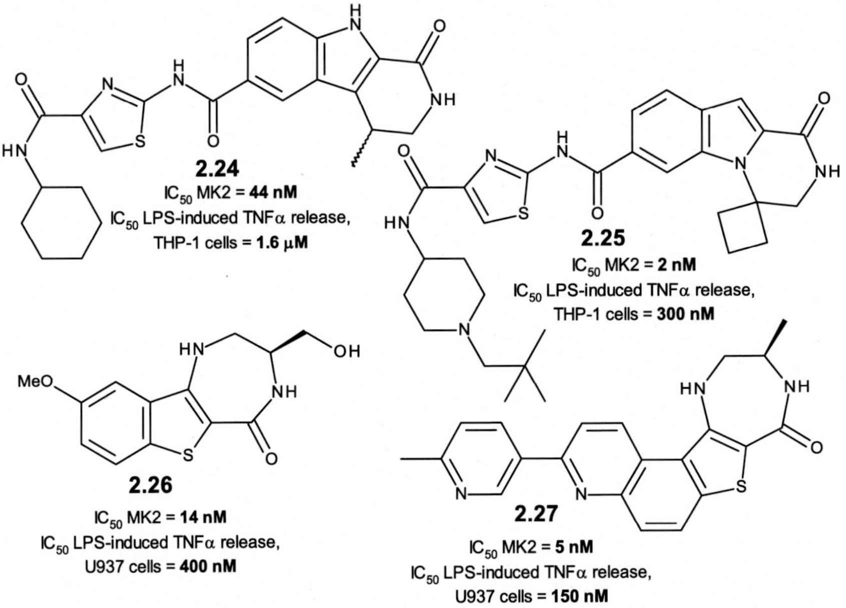

FIGURE 2.5 Small molecule inhibitors of MK2 and other Hsp27-phosphorylating kinases: chemical structures, **2.24–2.27**.

is provided by piperidine-containing spiroderivatives such as **2.25** [72], which do not show efficacy in a mouse model of LPS-stimulated TNFα production.

Benzothiophene lactams, such as **2.26**, show potent inhibition in cell-free MK2 assays, and in cellular assays against LPS-stimulated TNFα release and Hsp27 phosphorylation on Ser78 [73]. Compound **2.26** is poorly selective *vs.* other kinases, while second-generation tetracyclic benzothiophenes such as **2.27** (Figure 2.5) are more potent in cellular assays, and more kinase-selective [74]. *In vivo* data for these compounds are not disclosed.

Increased chaperone activity by Hsp27 may be the result of several causes, that sometimes happen simultaneously. The expression of Hsp27, and of other Hsp family members, is induced by HSF1, so that HSF1 activators cause an increase in Hsp27 levels. The bisarylester bicyclol (**2.28**, Figure 2.6) induces increased Hsp27 and Hsp70 expression through dose-dependent promotion of the nuclear translocation of HSF1, and its binding to the heat shock gene promoter [75]. A protective effect of bicyclol is observed in Hsp27-dependent mice models of liver injury by acetaminophen [75] and concanavalin A [76]. Bicyclol-dependent increase of Hsp27 and Hsp70 expression suppresses nuclear factor kappa-light-chain-enhancer of activated B cells (NF-κB) activation caused by concanavalin A, and reduces inflammation [76]. The effect of bicyclol in both animal

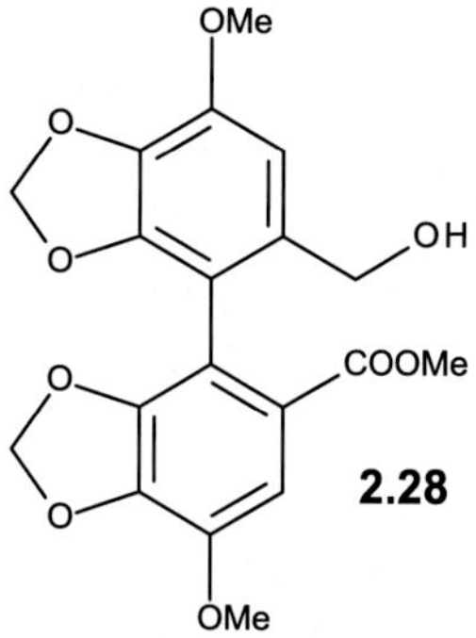

2.28

1,23-fold Hsp27 increase,
50 μM, HepG2 cells
2.6-fold Hsp27 increase,
mice, **200 mg/Kg**, p.o., 6 hrs

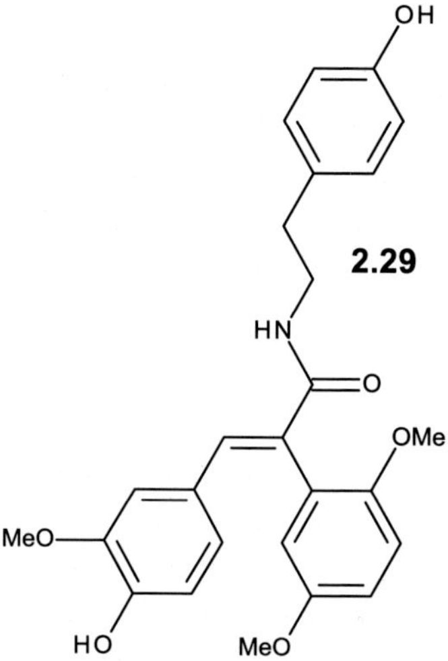

2.29

5-fold Hsp27 increase,
5 μM, SH-SY5Y cells
2-fold Akt phosphorylation
increase, **10 μM**, 2 hrs

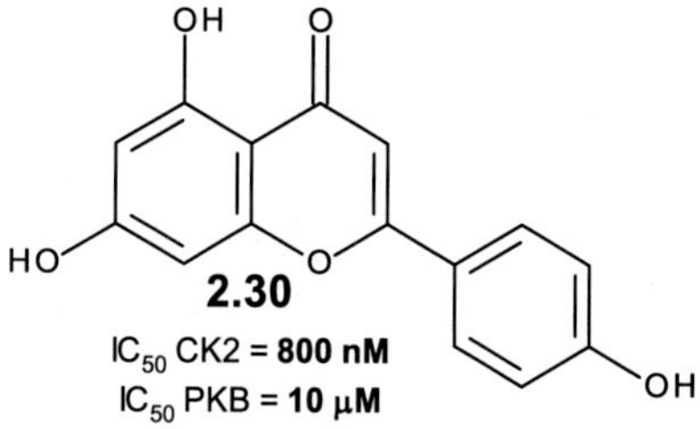

2.30

IC_{50} CK2 = **800 nM**
IC_{50} PKB = **10 μM**

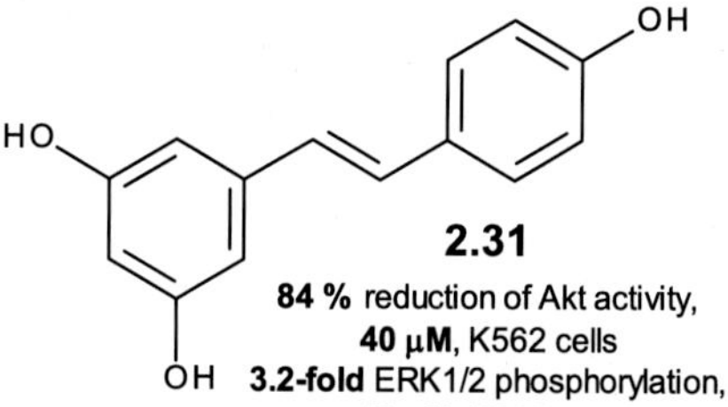

2.31

84 % reduction of Akt activity,
40 μM, K562 cells
3.2-fold ERK1/2 phosphorylation,
40 μM, K562 cells

FIGURE 2.6 Small molecule regulators of multiple events influencing Hsp27 activity: chemical structures, **2.28–2.31**.

models is antagonized by quercetin, validating its dependence on Hsp27. Bicyclol is clinically tested as a treatment for chronic viral hepatitis B and C, and is currently approved for medical use in China [77].

The naturally occurring squamosamide analog FLZ (**2.29**) facilitates HSF1 activation and translocation, leading to an increase in Hsp27 and Hsp70 expression [78]. Its effect is most likely due to the dose-dependent activation of Akt, causing HSF1 phosphorylation/activation and its nuclear translocation/heat shock gene promoter binding [78]. Interestingly, FLZ shows neuroprotective effects in a 1-methyl-4-phenylpyridinium (MPP^+) induced model of PD (effects prevented by co-administration of quercetin) [78], and in a β-amyloid-induced mouse model of hippocampal injury and memory/cognitive deficits [79].

The naturally occurring flavone apigenin (**2.30**) induces higher Hsp27 chaperone activity, but does not increase Hsp27 levels [80]. Rather, it increases phosphorylation of Hsp27 and causes its activation. Apigenin promotes an early phosphorylation on Ser15 and Ser82 (15 minutes' treatment), and a late phosphorylation of Ser15, Ser78, and Ser82 (6 hours' treatment). p38/MAPK appears to phosphorylate all three Ser residues, while the protein kinase C delta (PKCδ) isoform phosphorylates Ser15 and Ser82 [80]. Apigenin shows multiple PQC-related effects, including modulation of the MAPK and PI3-K–Akt pathways [81], CK2 inhibition, dissociation of cell division cycle 37 homolog (cdc37) Hsp90–kinase tertiary complexes [82] (see also section 2.4), and inhibition of proteasome activity [83] (see also Chapter 3).

The polyphenolic natural product resveratrol (**2.31**, Figure 2.6) has a wide range of effects related to Hsps. It is reported to inhibit Akt and to stimulate extracellular signal-regulated kinases 1/2 (ERK1/2), causing HSF1 down-regulation and Hsp expression in human chronic myeloid leukemia (CML) K562 cells [84]; to suppress the inflammatory cascade during rat liver carcinogenesis by antagonizing Hsp70 overexpression and NF-κB nuclear translocation [85]; and to induce Hsp27 down-regulation in human breast adenocarcinoma MCF-7 cells, inducing apoptosis [86]. Conversely, resveratrol induces the heat shock response by activating HSF1-dependent transcription of Hsps in severe heat stress conditions in COS-7 cells [87]; and it extends the survival time of the transgenic (TG) mouse model of amyotrophic lateral sclerosis (ALS) caused by the G93A-Cu^{2+}/Zn^{2+} superoxide dismutase 1 (SOD1) mutation, possibly by promoting HSF1 activation, and subsequent up-regulation of Hsps, through silent mating type information regulation 2 homolog 1 (sirtuin 1, SIRT-1)-dependent HSF1 deacetylation [88]. Depending on cellular/tissue conditions, one could hypothesize that resveratrol counteracts pathological imbalances (i.e., Hsp overexpression in cancer, or Hsp overwhelming protein misfolding in some proteinopathies) by inducing contrasting effects. A better understanding of such a pathology-adaptable therapeutic behavior could assist in the rational design of more effective PQC-directed small molecules against NDDs at various development stages.

2.3 Hsp70

The Hsp70 machinery has a broad client specificity. Although ATPase inhibition seems to cause tau-specific effects [89], one cannot exclude side effects on other client proteins. Specific Hsp70 family member–co-chaperone complexes are therapeutically validated as tauopathy-related targets. Among them, more is known about the Hsp70–BAG-1 (Bcl-2-associated athanogene 1) [90] complex. BAG-1 associates with the Hsp70–tau complex, but not with tau alone, in a tau isoform-independent manner [20]. Thus, selective inhibitors of the BAG-1–Hsp70–tau ternary complex should have a therapeutic role against tauopathies.

2.3.1 Hsp70 Inhibitors

Hsp70 inhibitors are known in literature [13,91], although as a smaller pool of compounds and at an earlier developmental stage than their Hsp90-targeted counterparts. The latter inhibitors benefit from earlier validation as an oncology target for Hsp90, and by a clearer pathology-Hsp90 connection. Such a connection is still being established for Hsp70 in oncology [92,93]. Conversely, the Hsp70 machinery may have a stronger impact in other therapeutic areas [13].

Spergualin derivatives are known antibiotic and antitumoral compounds [94]. 15-Deoxyspergualin (gusperimus, 15-DSG, **2.32**, Figure 2.7) [95] interacts with Hsp70 family members [96]. 15-DSG shows a suboptimal bioavailability profile [97]. It is used by i.v. infusion in several clinical trials in oncology [98] and allograft rejection [99] (the latter being more promising than the former). A second generation, inverted amide analog (**2.33**, tresperimus) [100] shows better stability and bioavailability, with good clinical results in allograft rejection [101].

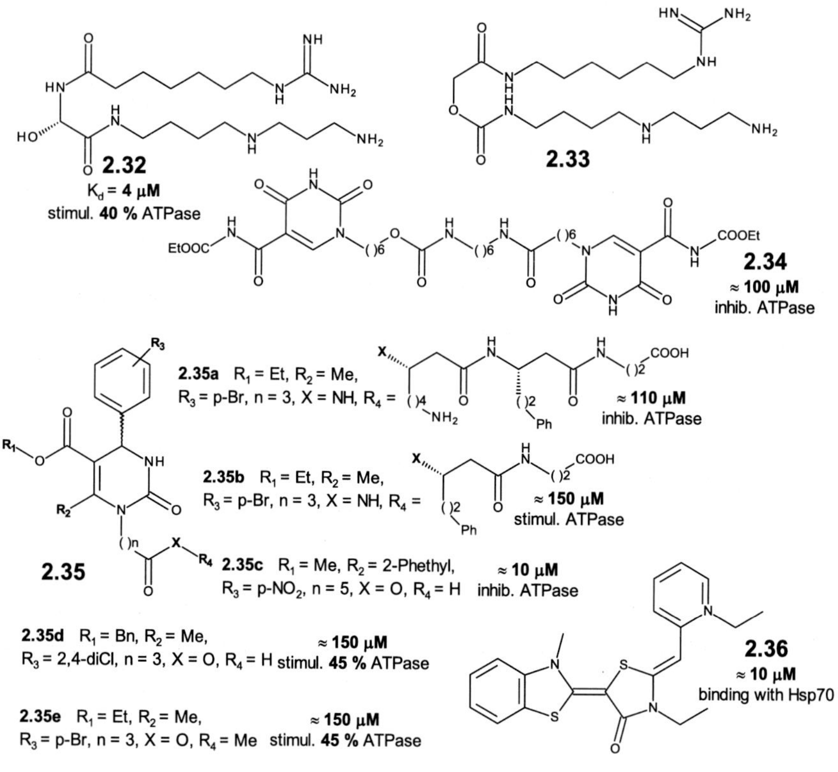

FIGURE 2.7 Small molecules binding to Hsp70 and regulating its chaperone activity: chemical structures, **2.32–2.36**.

Hsp70 binding, and the resulting ATPase stimulation [102], is important for the pharmacological action of spergualin derivatives [103]. Their effects are due also to multiple, non-Hsp70-related effects [13].

The dihydropyrimidine (DHP)-containing spergualin **2.34** is an Hsp70 ATPase inhibitor [104]. Consequently, the DHP nucleus was targeted as a source of Hsp70 modulators. A first set of ≈50 DHPs includes compounds **2.35a** and **2.35b** [105], respectively a weak stimulator and an inhibitor of Hsp70 ATPase activity selective for eukaryotic *vs.* bacterial Hsp70 family members [106]. A larger set of ≈180 DHPs is represented by compound **2.35c**, showing low μM inhibition of Hsp70 ATPase [107]. The molecular interaction between the ATP stimulator DHP **2.35d** and Hsp70 is elucidated through nuclear magnetic resonance (NMR), computational models, and biochemical assays [108]. The compound binds to a previously unknown allosteric site adjacent to, but not coincident with, the J-domain co-chaperone binding site on Hsp70. This explains the switch between DHP activators—smaller, no hindrance between J-domain co-chaperones and DHPs, synergistic effect on ATPase activity—and DHP inhibitors—bulkier, steric clashes with J-domain co-chaperones, inhibition of ATPase activity [108].

Some ATPase-inhibitory DHPs show potent, J-domain co-chaperone-dependent cytotoxic activity on tumor cells [109]. Cytotoxicity stems from Hsp70 ATPase activity impairment and subsequent degradation of the Hsp70 client Akt kinase [110]. DHP-centered activators of Hsp70 ATPase show interesting *in vitro* and *in vivo* effects in AD models. Compound **2.35e** [105] displays a modest but measurable Hsp70- (but not Hsp90- or Hsp40-) dependent anti-aggregation activity on Aβ, while an inhibitor shows an aggregation-stimulating behavior [111]. DHP stimulators **2.35d** and **2.35e** (50 μM) cause a neurotoxic ≈250–300% increase of total and hyperphosphorylated (HP) tau levels in cells through Hsp70 ATPase activation, while inhibitors lead to depletion of tau levels [89]. A similar pattern (ATPase stimulators–tau aggregation inhibitors–increased neurotoxicity; ATPase inhibitors–tau aggregation promoters–decreased neurotoxicity) is observed when compound **2.35e** and DHP-based ATPase inhibitors are tested in cellular models of Huntington disease (HD) and other polyQ repeat-expressing NDDs [112].

The stabilizing activity of Hsp70 on tau and polyQ proteins, through ATPase-promoted events, becomes detrimental in a gain-of-function (GOF) neurotoxic scenario, where misfolded/HP tau is stabilized in tauopathy environments. It would be interesting to check if Hsp70 ATPase stimulators also stabilize soluble, neurotoxic Aβ oligomers. Conversely, ATPase inhibitors switch misfolded/HP tau and polyQ proteins towards aggregation–degradation–clearance [89,112]. Thus, they should have an application against NDDs.

The rhodacyanine dye MKT-077 (**2.36**, Figure 2.7) [113] binds Hsp70 family members with varying affinities, the highest being for mitochondrial mtHsp70 [114]. It shows potent antitumor effects in various models of cancer [115], and a reasonable preclinical activity–toxicity profile [116]. Its clinical use is prevented by renal toxicity [117]. MKT-077 binds on an allosteric site nearby the ADP-bound nucleotide-binding domain (NBD)–ATPase site of Hsp70, as determined through NMR and molecular modeling [118]. Structural modifications likely to increase the binding strength for MKT-077, and possibly to decrease its nephrotoxicity, are known [118].

Sulfogalactosylceramides (SCGs) such as **2.37** (adaSGC, Figure 2.8) bind to Hsp70 family members [119] with varying affinities for eukaryotic

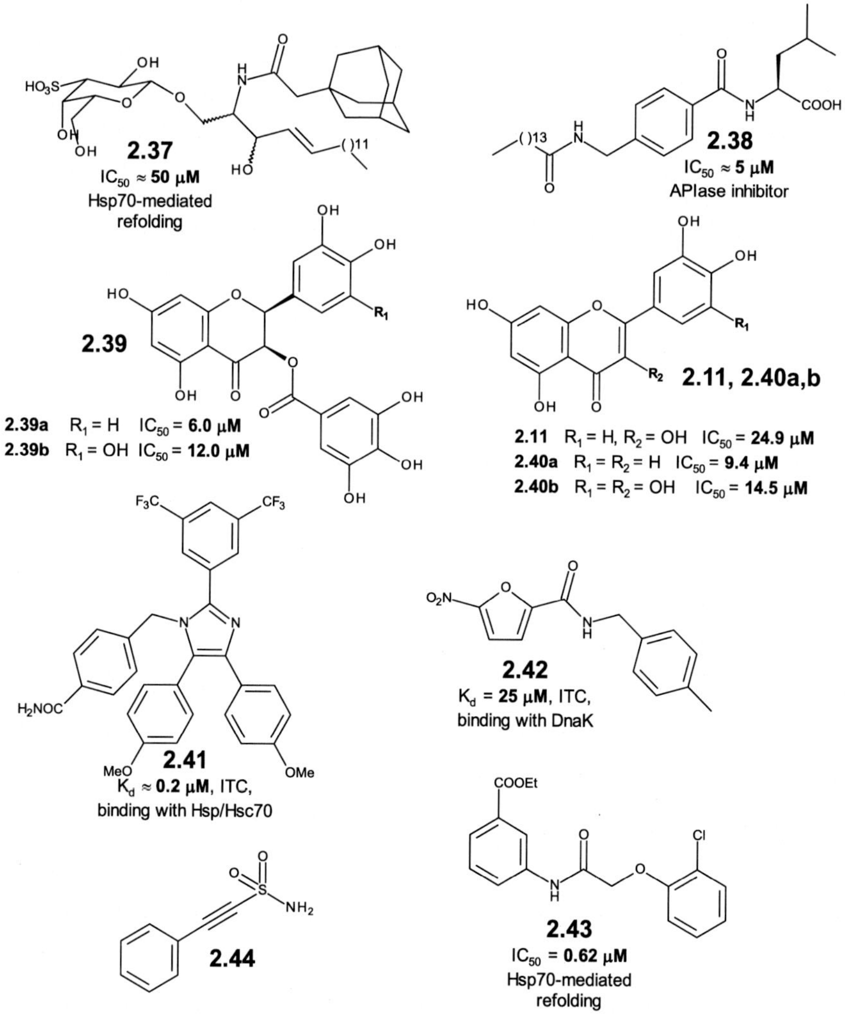

FIGURE 2.8 Small molecules binding to Hsp70 and regulating its chaperone activity: chemical structures, **2.11, 2.37–2.44**.

and bacterial Hsp70s [120]. An SGC-specific binding site on Hsp90 is defined on the NBD [119], but the cellular mechanism of action for these compounds is not fully clarified. adaSGC prevents misfolding and degradation of a mutant protein through an Hsp70- and J-domain-dependent mechanism in a cellular cystic fibrosis model [121].

Acylbenzamide-containing fatty acids, such as **2.38**, inhibit the bacterial Hsp70 chaperone DnaK [122]. They inhibit the aminopeptidyl isomerase (APIase) activity of DnaK, and show strong antibacterial activity. APIase inhibition and minimal inhibitory concentration (MIC) values for a limited number of acylbenzamide-containing fatty acids do not correlate, hinting to other factors than DnaK APIase inhibition to determine their antibacterial potency [122]. Their potency on eukaryotic Hsp70 is not disclosed.

Natural products are a valuable source of compounds directed against molecular targets and diseases [123]. Six natural extracts show activity in a phenotypic screening designed to identify allosteric inhibitors of the ATPase activity of the bacterial DnaK (Hsp70 family)–DnaJ (J-domain co-chaperone)–GroEL (NEF co-chaperone) ternary complex [124]. The identified active principle from white tea is the flavonoid epicatechin-3-gallate **2.39a** (ECG). Epigallocatechin-3-gallate (EGCG) **2.39b**, quercetin **2.11** (see also section 2.2 and Figure 2.2), luteolin **2.40a,** and myricetin **2.40b** are similarly active (μM potency, allosteric ATPase inhibitors [124]).

An Hsp70-targeted activity of EGCG, quercetin, and myricetin is supported by other studies. EGCG competes with ATP in the binding site of Hsp70, and lowers Hsp70 and Hsp90 levels *in vitro* and *in vivo* [125]. Quercetin, in addition to Hsp27-targeted activities (see also section 2.2), causes Hsp70 inhibition, likely through inhibition of CK2 [50]. Biotinylated, active quercetin analogs prove that a direct Hsp70–quercetin binding is established in Hsp70-overexpressing cells [126]. Myricetin is identified as an Hsp70 ATPase inhibitor in an HTS campaign [89].

A multi-targeted action of flavonoids on Hsp chaperone systems, including ATP-competitive and allosteric inhibition of ATPase activities, and CK2-mediated inhibition of Hsp70–co-chaperone complexes, is plausible. Conversely, an allosteric binding site for myricetin on the NBD of DnaK, identified through NMR and mutagenesis studies, provides rational elements for the design of ATPase inhibitors preventing the Hsp70–J-domain co-chaperone interaction [124]. Flavonoids can at least partially be targeted against specific mechanisms *via* chemical modifications [124]. A pleiotropic, moderate activity against a number of tauopathy-relevant targets may be a sensible, although complex therapeutic option. Positive results obtained by treating animals with natural mixtures containing flavonoids indirectly support this theory [127–129].

Apoptozole (**2.41**) is a tetrasubstituted imidazole, named after its apoptosis-inducing properties in cells [130]. Its activity is Hsc/Hsp binding-dependent, as proven by affinity chromatography with a supported apoptozole derivative [130]. The compound restores the defective cellular processing of ΔF508-CFTR (cystic fibrosis transmembrane conductance regulator), a mutant, misfolded protein involved in cystic fibrosis [131]. Apoptozole binds to the ATP binding domain of Hsp70 and blocks its ATPase activity, as measured by isothermal calorimetry (ITC). Its rescuing effect is likely due to the disruption of the tertiary Hsp70–C-terminus of Hsc70 interacting protein (CHIP)-mutant protein complex, and to the prevention of CHIP-mediated ubiquitination and degradation of ΔF508-CFTR [131]. The effect of apoptozole on Hsp-mediated tau events is not reported.

Heterocyclic 2-carboxamides such as **2.42** bind to DnaK in an allosteric pocket, which regulates communication between NBD and substrate-binding domain (SBD) [132]. ITC experiments show that they inhibit the DnaK ATPase activity. They also show antibiotic activity on several bacterial strands [132]. Phenoxy-N-arylacetamides such as **2.43** interfere with the Hsp70–DnaJ complex, and inhibit the ATPase activity of Hsp70 [133].

The only SBD-binding Hsp70 inhibitor is pifithrin-μ (PES, **2.44**, Figure 2.8), discovered in a screening targeted towards inhibiting p53-mediated apoptosis [134]. PES has multiple actions on the Hsp70 machinery [135]. Complexes between Hsp70 and co-chaperones CHIP, BAG-1, and, in general, J-domain proteins are affected by PES. PES modulates binding of Hsp70 with tumor suppressor protein p53, apoptosis protease activating factor 1 (APAF-1) and the autophagy marker sequestosome-1/p62. Its multiple actions—probably not limited to Hsp70-related processes—are mirrored by a strong anticancer action against, *inter alia*, lymphoma [135] and leukemias [136]. Chemical reactivity of the triple bond in PES justifies an unspecific activity profile, surprisingly contrasting with its peculiar mechanism of action and *inter*-Hsp70 selectivity. PES binds to inducible Hsp70 and to its bacterial ortholog DnaK, but does not bind to constitutive Hsc70 or to endoplasmic reticulum-located Hsp70 chaperone Grp78. Its selectivity is due to PES interaction with the C-terminus/SBD domain of Hsp70, as confirmed by deletion studies [135]. Such *inter*-chaperone selectivity should be studied with more drug-like analogs of PES to dissect the roles of Hsp70 family members, to understand their co-chaperone interactions and—most important—to fully appreciate their druggability.

2.3.2 Hsp70-BAG-1 Inhibitors

Thioflavin S (**2.45a,b**, Figure 2.9) [137] is a mixture of ammonium salts showing sub-μM potency in preventing the Hsc/Hsp70–BAG-1 interaction [138]. It blocks the Hsc/Hsp70–BAG-1-dependent activation of ERK kinases, and the potentiation of vitamin V3 receptor activity. It is cytotoxic on several tumor cell lines, while close structural analogs are devoid of Hsc/Hsp70–BAG-1 affinity and cytotoxicity [138]. Unfortunately, the potency of thioflavin S is batch-dependent, its cellular activity is at least partially due to Hsc/Hsp70–BAG-1-unrelated

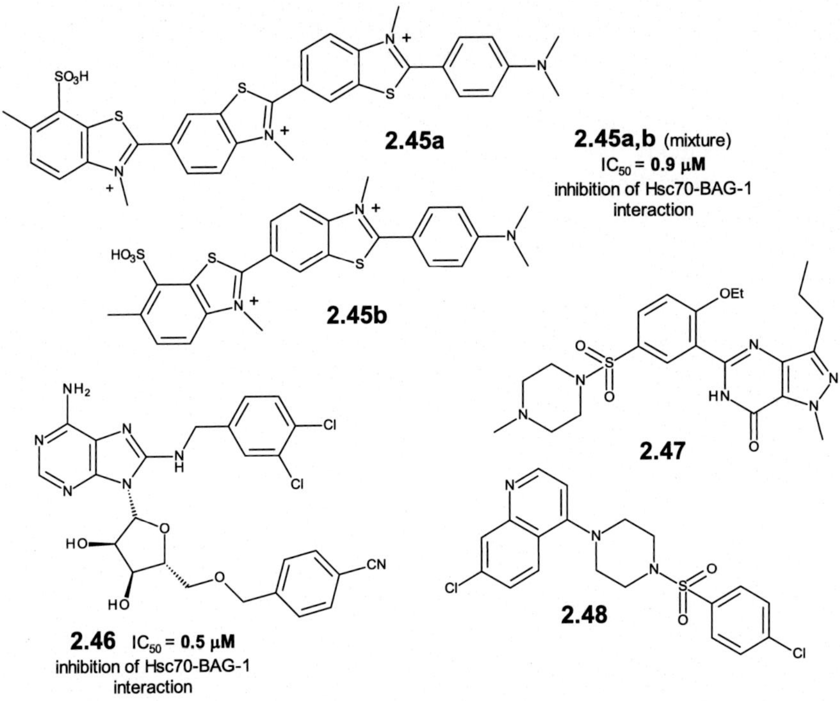

FIGURE 2.9 Small molecules Hsp70–BAG-1 complex inhibitors: chemical structures, 2.45a–2.48.

mechanisms, and its cationic structure hinders its optimization to a drug-like candidate.

Selected adenosine mimics, such as **2.46** (VER-155008), are ATP-competitive NBD binders. Their rational design, synthesis, and structural optimization is based on the X-ray structure of the Hsc70–BAG-1 complex [139]. The pan-Hsp70 family inhibition shown by VER-155008 can be partially focused against Gpr78, a validated chaperone target in oncology, by rational design and chemical modifications [140]. VER-155008 shows efficacy *in vitro* on colon and breast cancer cell lines, but is not orally available and has a poor PK profile [141].

Known phosphodiesterase type 5 (PDE5) inhibitors sildenafil and KM11060 (**2.47** and **2.48**, Figure 2.9) are respectively a weaker and a stronger Hsc70–BAG-1 inhibitor [142]. A computational model built on thioflavin S is used to calculate their affinities. The observed potency of both compounds as mutant CFTR protein correctors in cystic fibrosis is ascribed to their Hsc70 chaperone-inhibiting properties. SAR indications to optimize their CFTR corrector activity are provided [142].

These chemotypes could be of use in neurodegeneration, especially if chaperone complex-selective effects of such Hsc/Hsp70–BAG-1 inhibitors would be confirmed *in vitro* and *in vivo*. A more favorable adverse event profile, compared with Hsp70 inhibitors, could be expected from selective Hsc/Hsp70–BAG-1 inhibitors. The available X-ray structure of the Hsc70–BAG-1 complex [139] should assist in the search of novel, complex-selective inhibitors.

2.4 Hsp90

Hsp90 family proteins are mostly targeted in oncology [21,143] but their relevance against NDDs is evident [144]. Hsp90 co-chaperones provide substrate specificity, selective client-targeted efficiency, and dynamic turnover to direct the fate of Hsp90 substrate proteins [14]. Targeting Hsp90–co-chaperone complexes relevant for tau pathologies, either directly (small molecules binding to the Hsp90–co-chaperone interface) or indirectly (small molecules regulating co-chaperone activation) should represent a safer therapeutic approach than targeting Hsp90 family members.

2.4.1 Hsp90 Inhibitors

The first clinical trial employing an Hsp90 inhibitor against cancer dates back to 1998 [145]. Since then, several clinical candidates were selected, and hundreds of oncology trials—either as standalone agents, or in combination with cytotoxic agents—were started. Some excellent reviews [143,146,147] describe in detail their status and perspectives. Here, we briefly review clinically tested compounds—mostly related to three main chemotypes—with a disclosed chemical structure. Compound classes at an earlier development stage can be accessed elsewhere [148,149].

Geldanamycin (**2.49a**, Figure 2.10) [150] is the first characterized, naturally occurring Hsp90 inhibitor. It competes with ATP for binding to the NBD–ATP binding site of Hsp90 [151]. Toxicity, limited solubility and poor stability prevent clinical testing of geldanamycin [152]. Four semisynthetic geldanamycin derivatives (**2.49b–d, 2.50**), replacing the nonessential C-17 methoxy group with solubilizing-detoxifying amines, are being evaluated in the clinics. 17-Allylamino-17-demethoxy geldanamycin (17-AAG, tanespimycin, **2.49b**) is the most advanced Hsp90 inhibitor in development as an i.v. treatment [153], having reached Phase III in combination with the proteasome inhibitor bortezomib against multiple myelomas [154]. Its development is on hold [155], although nanoparticle-bound 17-AAG (ABI-1010) is under evaluation in a Phase I trial [156]. 17-N,N-dimethylaminoethyl-17-demethoxy geldanamycin (17-DMAG, alvespimycin, **2.49c**) has similar efficacy to 17-AAG, and has long been considered a second priority compound [157]. Recent Phase I studies on advanced solid tumors include 17-DMAG as a single treatment [157] or in combination

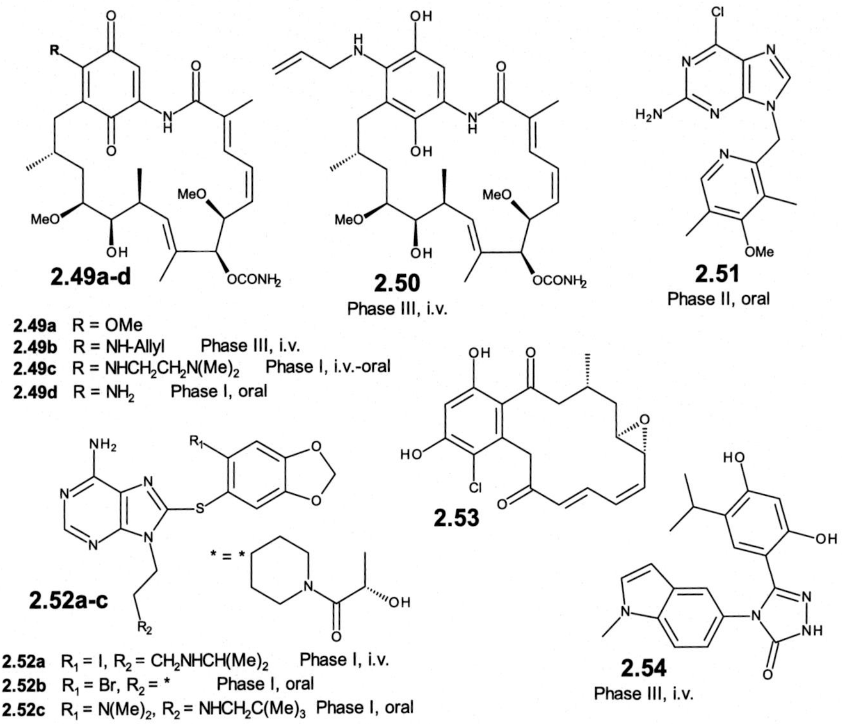

2.49a-d

2.49a R = OMe
2.49b R = NH-Allyl   Phase III, i.v.
2.49c R = NHCH$_2$CH$_2$N(Me)$_2$ Phase I, i.v.-oral
2.49d R = NH$_2$ Phase I, oral

2.50
Phase III, i.v.

2.51
Phase II, oral

2.52a-c

* = *

2.52a R$_1$ = I, R$_2$ = CH$_2$NHCH(Me)$_2$ Phase I, i.v.
2.52b R$_1$ = Br, R$_2$ = *   Phase I, oral
2.52c R$_1$ = N(Me)$_2$, R$_2$ = NHCH$_2$C(Me)$_3$ Phase I, oral

2.53

2.54
Phase III, i.v.

FIGURE 2.10   Clinically tested Hsp90 inhibitors: chemical structures, **2.49a–2.54**.

with trastuzumab [158]. The simplest 17-amino-17-demethoxy geldana-mycin (IPI-493, **2.49d**) is the orally bioavailable, active metabolite of 17-AAG. Its overall poor PK profile is observed in a Phase I study [159]. Conversely, the hydroquinone-hydrochloride form of 17-AAG (IPI-504, retaspimycin, **2.50**) [160] has reached the Phase III developmental stage. Such Phase III trial has been recently terminated, due to higher mortality rate for treated patients [161]. Although geldanamycin derivatives were the first Hsp90 inhibitors, it is unclear if any of them will ever reach the market.

Purine-like, ATP-competitive synthetic Hsp90 inhibitors received atten-tion, due to structural peculiarities of the ATPase domain of Hsp90 com-plexed with nucleotides [162]. Orally active BIIB021 (CNF 2024, **2.51**) [163] contains a 6-amine and a 2-Cl group, and bears a short, pyridyl-containing substituent in position 9. Its Phase I–II development status on breast and gastrointestinal tumors [143,164] is hampered by the recent de-prioritiza-tion of BIIB021 by its developing company [147]. I.v.-administered PU-H71 (**2.52a**) [165] bears an 8-arylsufide moiety and a linear, basic chain in position 9. It is undergoing Phase I studies against advanced solid tumors

[166]. MPC-3100 (**2.52b**) [167] has a Br/I replacement on the substituent in position 8 with respect to PU-H71, and bears an acylpiperidine-containing alkyl chain in position 9. It causes supraventricular tachycardia in patients [168], and investigational new drug (IND) filing of a pro-drug is expected [169]. Debio 932 (**2.52c**) [170] has a dimethylamino group on the substituent in position 8 and a basic, neopentyl-substituted alkyl chain in position 9. It crosses the blood–brain barrier (BBB) [171], and is currently being tested in a Phase I trial against advanced solid tumors [172].

Radicicol (**2.53**) [173] is a naturally occurring Hsp90 inhibitor that inspired the design of four clinically tested resorcinol-containing synthetic analogs. Among them, three are developed as i.v. treatments. Triazole-containing ganetespib (STA-9090, **2.54**, Figure 2.10) [174] shows a good clinical efficacy profile in Phase I and Phase II studies [175]. An open label Phase II trial on non-small cell lung cancer (NSCLC) patients is ongoing [176], and a Phase IIb/III combination with docetaxel is being tested on second-line NSCLC patients [177]. Rationally designed isoxazole-containing NVP-AUY922 (VER-52296, **2.55**, Figure 2.11) [178] shows efficacy in Phase I [179] and Phase II [180] studies. Its evaluation as a stand-alone treatment, and in combination with bortezomib in a Phase II study against relapsed or refractory multiple myeloma, should be completed soon [181]. KW-2748 (**2.56**) [182] is endowed with a good toxicity–PK profile, determined in a Phase I study [183]. Its evaluation in combination with bortezomib in a Phase I/II study against relapsed or refractory multiple myeloma is ongoing [184]. AT13387 (**2.57**) has been rationally designed and synthesized using fragment-based drug optimization [185,186]. It is clinically developed as an oral treatment, and is undergoing several Phase I [187] and Phase II [188] studies.

Three structurally unrelated small molecules have also reached clinical development. HSP990 (**2.58**) [189] is a tetrahydropyrimidinpyrazone undergoing Phase I evaluation [190]. SNX-5422 (PF-04929113, **2.59**) [191] is a 2-aminobenzamide tested in a Phase I trial on patients suffering from solid tumors and lymphomas [192]. Its development is halted, due to severe ocular toxicity [193]. XL888 (**2.60**, Figure 2.11) [194] is a 3-aminobenzamide whose clinical development [195] seems to be discontinued [146].

2.4.2 Hsp90–Co-chaperone Complexes: Direct Inhibition

Inhibitors of the specific functions of Hsp90 complexed with co-chaperones should be more selective than ATPase-targeted Hsp90 inhibitors [196]. Clinical use of ATPase-targeted Hsp90 inhibitors in acute, life-threatening oncology diseases may be granted in future, but their use as chronic treatments in NDDs is questionable. Unfortunately, targeting of Hsp90 co-chaperones, either *per se* or complexed with Hsp90 family members, has proven to be difficult.

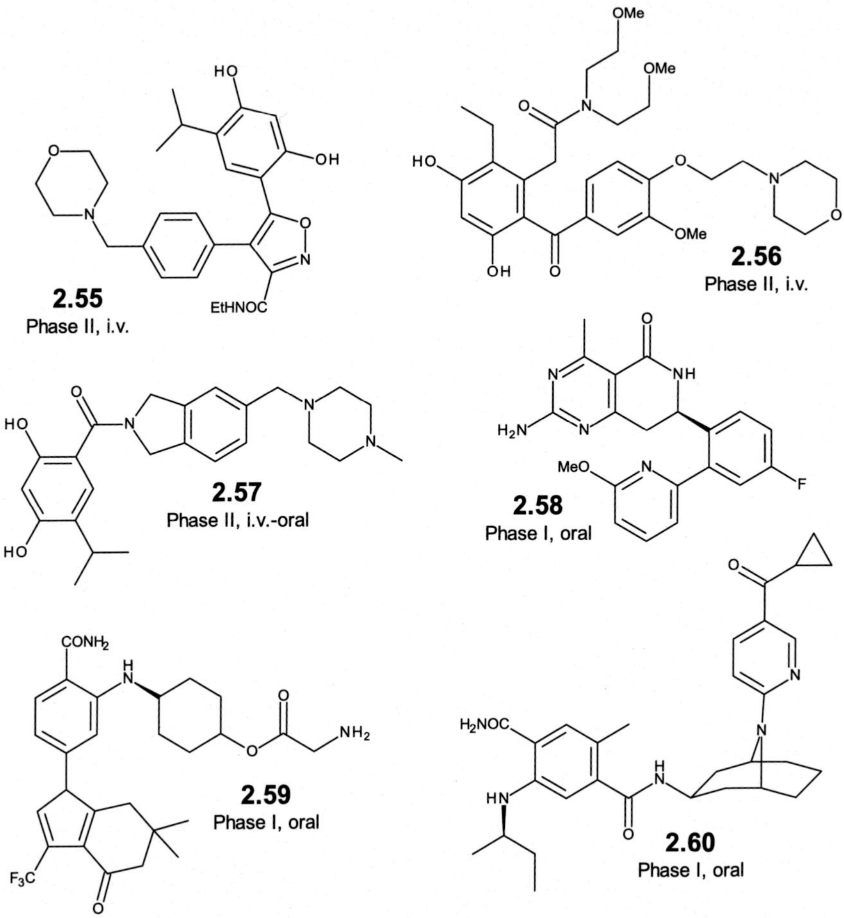

FIGURE 2.11   Clinically tested Hsp90 inhibitors: chemical structures, **2.55–2.60**.

Immunophilin *FKBP* inhibitors stem from FK506 (**2.61**, Figure 2.12) [197], a naturally occurring immunosuppressive compound endowed with neurotrophic and neuroprotective activity [198]. Non-immunosuppressive, neuroprotective smaller ligands such as **2.62** (GPI-1046) [199] and its optimized analog V-10367 (**2.63a**) [200] mimic the FKBP-binding domain of FK506. Their neurotrophic activity is observed in clinical trials [201]. The neuroprotective action of pipecoline ketoamides **2.62** and **2.63a** *in vivo* is likely mediated by larger FKBP52 immunophilins [202].

Fourteen FKBP proteins are known [201], and a clarification of their role in the nervous system is still awaited. A systematic evaluation of binding potencies and affinity trends for FK506-related, small immunophilin ligands on FKBP proteins (including tau-relevant FKBP51 and FKBP52)

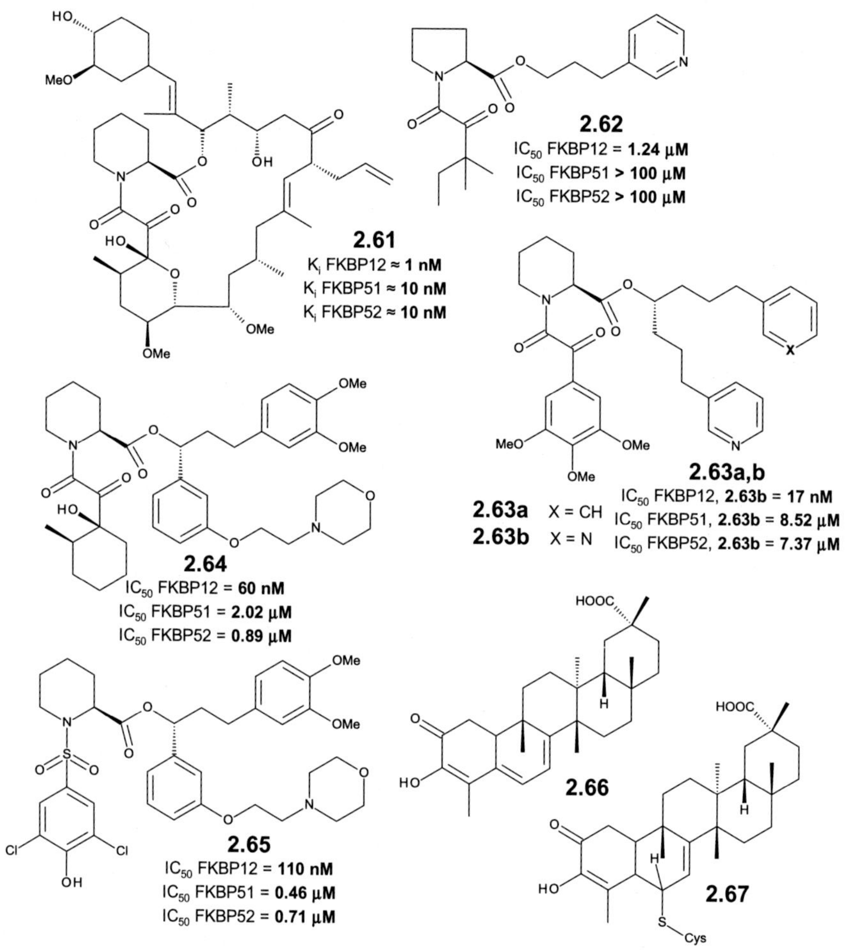

FIGURE 2.12 Direct inhibitors of Hsp90–co-chaperone complexes: chemical structures, **2.61–2.67**.

is now available [203,204]. Compounds bind more strongly—low nM potency—to the small FKBP12 protein, while they show low μM potency on both FKBP51 and FKBP52. The pipecolyl ketoamide **2.64** [203] is the only sub-μM FKBP52 binder, with a slight preference over FKBP51. Its binding to FKBP12 remains stronger, although not as much as for its analogs. The pipecolyl sulfonamide **2.65** [204] is the only ligand showing almost equivalent binding to FKBP12, FKBP51, and FKBP52.

FKBP51 inhibitors [205] and FKBP52 stimulators [206] should have beneficial effects in tauopathies. FKBP12 is also related to AD and tau [207]. Thus, neuroprotective small immunophilin ligands should be tested in models of tauopathies to observe their effects. The comparison of compounds similarly active on small and large FKBPs, such as **2.65**, with FKBP12-targeted compounds could clarify the role of FKBPs in tauopathies. To this regard, FKBP12-selective pipecolyl ketoamide VX-710 (**2.63b**) is a multidrug resistance inhibitor clinically tested in oncology [208] for which no neuroprotective action is reported.

The only known inhibitor of the cdc37–Hsp90 interaction is the triterpene natural product celastrol (**2.66**) [209]. An NMR study indicates that celastrol binds to cdc37 (spectral changes are observed by incubating cdc37 with celastrol), while the NMR spectrum of Hsp90 is not modified by celastrol [210]. Molecular interactions between celastrol and cdc37 include the reaction/addition of one or more Cys residues in the N-terminal, client kinase-binding domain of cdc37 onto the quinone methide/Michael acceptor moiety of celastrol to yield the physiologically reversible Michael adduct **2.67** (Figure 2.12) [210]. A mechanistic study confirms the inhibitory effect of celastrol on the formation of the Hsp90–cdc37 complex, favoring a direct interaction between celastrol and the C-terminus region of Hsp90 [211]. A Michael reaction of celastrol with Cys residues on both cdc37 and Hsp90 could explain these observations [211].

Celastrol shows multiple biological actions, and the cdc37–Hsp90 complex is not its only target. As to AD and neurodegeneration, celastrol reduces the severity of Aβ pathology in a TG mouse model [212]. Its action involves NF-κB-mediated reduction of the expression of β-secretase 1 (BACE-1), leading to lower Aβ levels. This effect should be cdc37–Hsp90-independent, although I kappa B kinase (IKK) activation (in itself an NF-κB regulator) may be cdc37–Hsp90-dependent [212]. Celastrol shows *in vivo* neuroprotection in models of PD and HD [213], and of ALS [214]. Unfortunately, it also shows anti-inflammatory and anticancer activities [215] that may make it suitable for other therapeutic applications, but should cause problems during chronic treatment. For example, celastrol hits several transcription factors [216] through direct Hsp90-transcription factor, Hsp90–Hsp70-Hsp90 organizing protein (Hop), and/or Hsp90–p23 interactions. Identifying cdc37–Hsp90-directed celastrol effects, checking its activity in tauopathy models, and gathering structural information on the cdc37–Hsp90–client kinase complexes are needed steps towards the validation of a therapeutic Hsp90-co-chaperone target against tauopathies.

2.4.3 Hsp90–Co-chaperone Complexes: Indirect Inhibition

CK2 inhibitors are extensively studied as treatments for a number of diseases [217]. Recently, their role in neurodegeneration has been reviewed [218]. Their structure and subdivision into main structural classes has also been reported [217–219].

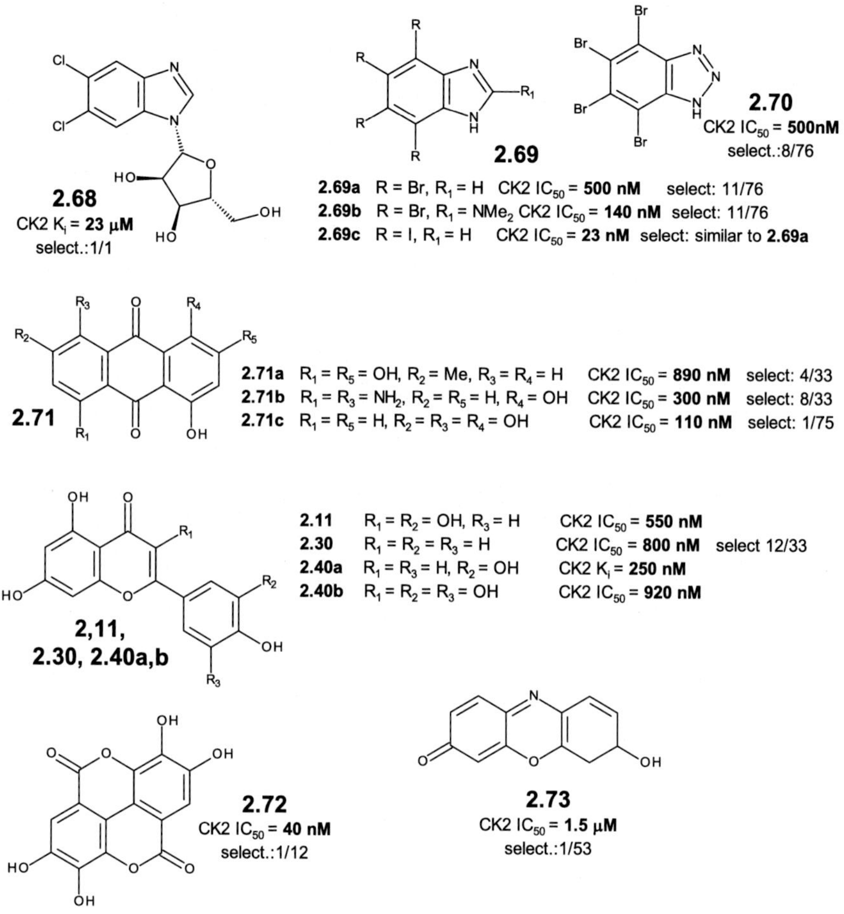

2.68
CK2 K_i = 23 μM
select.:1/1

2.69

2.69a   R = Br, R_1 = H   CK2 IC_{50} = **500 nM**   select: 11/76
2.69b   R = Br, R_1 = NMe$_2$ CK2 IC_{50} = **140 nM**   select: 11/76
2.69c   R = I, R_1 = H   CK2 IC_{50} = **23 nM**   select: similar to **2.69a**

2.70
CK2 IC_{50} = 500nM
select.:8/76

2.71

2.71a R_1 = R_5 = OH, R_2 = Me, R_3 = R_4 = H   CK2 IC_{50} = **890 nM**   select: 4/33
2.71b R_1 = R_3 = NH$_2$, R_2 = R_5 = H, R_4 = OH   CK2 IC_{50} = **300 nM**   select: 8/33
2.71c R_1 = R_5 = H, R_2 = R_3 = R_4 = OH   CK2 IC_{50} = **110 nM**   select: 1/75

2,11,
2.30, 2.40a,b

2.11 R_1 = R_2 = OH, R_3 = H   CK2 IC_{50} = **550 nM**
2.30 R_1 = R_2 = R_3 = H   CK2 IC_{50} = **800 nM**   select 12/33
2.40a R_1 = R_3 = H, R_2 = OH   CK2 K_i = **250 nM**
2.40b R_1 = R_2 = R_3 = OH   CK2 IC_{50} = **920 nM**

2.72
CK2 IC_{50} = 40 nM
select.:1/12

2.73
CK2 IC_{50} = 1.5 μM
select.:1/53

FIGURE 2.13   Indirect inhibitors of Hsp90–co-chaperone complexes *via* CK2 inhibition: chemical structures, **2.11, 2.30, 2.40a,b, 2.68–2.73**.

Poly-halogenated benzimidazoles and benzotriazoles derive from the dichlorobenzimidazole-based nucleoside DRB (**2.68**, Figure 2.13) [220], an aspecific kinase inhibitor. Its structural optimization leads to a tetrabromo-benzimidazole (TBBz, **2.69a**) [221], its 2-dimethylamino substituted analog (DMAT, **2.69b**) [222], a tetraiodobenzimidazole (TIBI, **2.69c**) [223], and a tetrabromobenzotriazole (TBBt, **2.70**) [224]. Compound **2.68** is a weak CK2 inhibitor, while compounds **2.69a,b** and **2.70** are potent and relatively selective, as measured on a panel of 76 kinases [225]. Interestingly, dual specificity tyrosine-phosphorylation-regulated kinase 1A (DYRK1A)—a target kinase related to tau alternative splicing [226]—is inhibited by this class of compounds. Tetraiodo derivatives, such as **2.69c**, are more potent and similarly selective, but may present cell penetration-solubility issues [224].

Compounds **2.69b** and **2.70** show Hsp90–co-chaperone complexes-mediated activity on models of NDDs. Compound **2.69b** (5 μM) prevents the acute cognitive decline-correlated inhibition of synaptic transmission caused by oligomeric $A\beta_{42}$ peptides in squid ganglia [227]. It could synergize with Hsp90–cdc37-directed activities of CK2 inhibitors. Compound **2.69b** modulates HD-relevant, site-specific phosphorylation of huntingtin in cellular assays [228]. It also shows *in vivo* activity—attributed to the inhibition of the ternary Hsp90–cdc37– inositol-requiring protein 1 alpha (IRE1α) complex—at 50 mg/kg in a multiple myeloma mouse model [229]. Compound **2.70** inhibits phosphorylation of the Ser13 epitope of cdc37 in COS7 cells in a dose-dependent manner [230]. It does not affect cdc37 levels, and does not significantly reduce the levels of Hsp90–cdc37 client kinases MAPK/MAK/MRK-overlapping kinase (MOK) and rapid accelerated fibrosarcoma 1 (Raf-1) [230].

The naturally occurring quinone emodin (**2.71a**) [231] is a sub-μM CK2 inhibitor, with good selectivity *vs.* other kinases. The X-ray complex between emodin and CK2 is available, and the synthetic analog **2.71b**—slightly improved potency on CK2, slightly decreased kinase selectivity, lower risk of DNA intercalation—results from structural optimization of emodin [232]. Quinalizarin (**2.71c**) is an extremely potent and kinase-selective quinone. Multiple interactions are observed in the X-ray complex between quinalizarin and CK2 [233]. Quinalizarin modulates site-specific phosphorylation of huntingtin in cellular assays [228].

Naturally occurring flavonoids are kinase inhibitors, and CK2 is among their targets. Hsp70 inhibitors quercetin, myricetin, and luteolin (**2.11, 2.40a,b**, see respectively section 2.3.1 and Figures 2.2 and 2.8) [234] may represent multi-targeted agents against tauopathies. Apigenin (**2.30**, see also section 2.2 and Figure 2.6) [235] has moderate potency and limited kinase selectivity—a common feature to other flavonoids [234]. Its biological effects include modulation of kinase activity [81], NF-κB inhibition [83], proteasome inhibition [236], and apoptosis induction [237]. A mechanistic study shows that CK2 inhibition-mediated dephosphorylation of the Ser13 epitope of cdc37 is responsible for most, if not all the effects of apigenin [82]. In fact, disassociation of upstream client kinase–Hsp90–cdc37 ternary complexes, followed by destabilization/degradation of cdc37-free kinases, may be a common mechanism of action for other flavonoids, such as **2.11** and **2.40a,b** [82].

Ellagic acid (**2.72**) [238] is a potent and selective coumarin. It shows cytotoxicity on several tumor cell lines, but suffers from poor bioavailability [239]. Resorufin (**2.73**, Figure 2.13) [240] is a resorcinol-related synthetic derivative endowed with moderate CK2 inhibition. Its kinase selectivity, its cytotoxicity on various cell lines [219], and the availability of the X-ray structure of a resorufin-CK2 complex [241] may prompt its chemical modification to find drug-like analogs.

Pyrazolo[1,5-*a*][1,3,5]triazine-based compounds, such as **2.74** (Figure 2.14) [242], are sub-nM CK2 inhibitors. Their cellular activity is

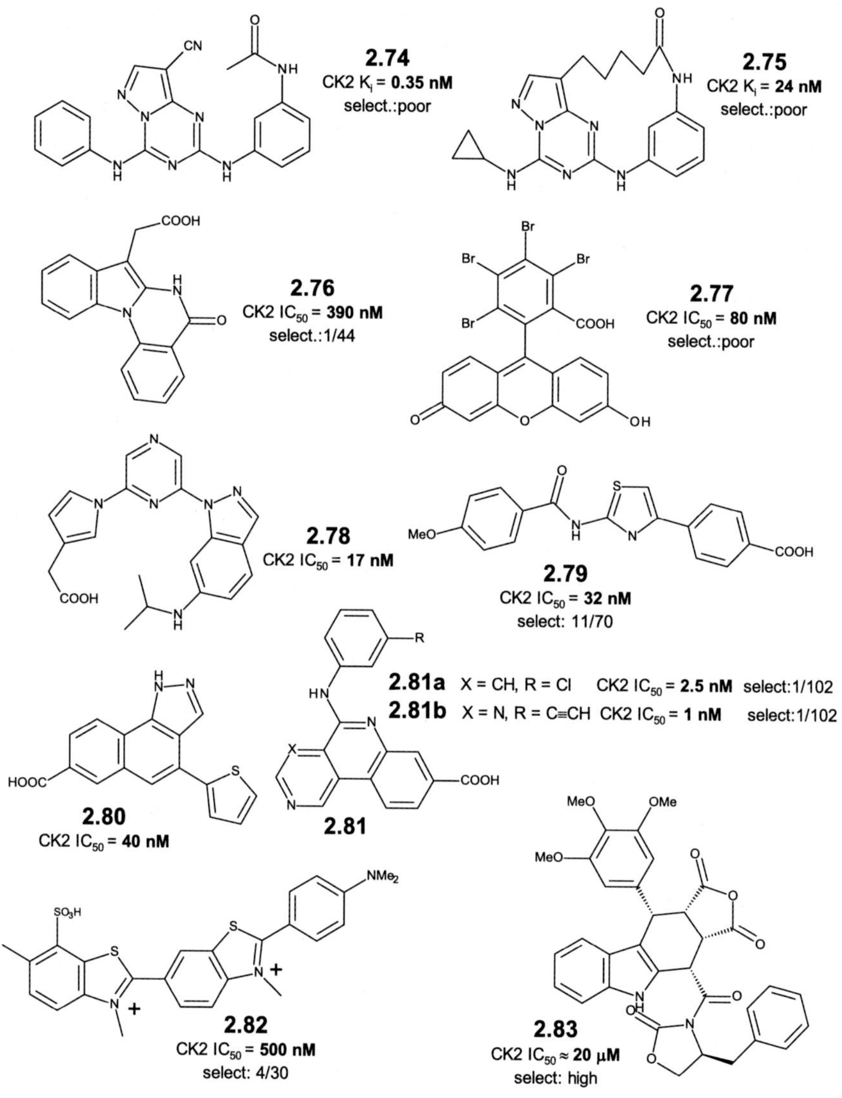

FIGURE 2.14 Indirect inhibitors of Hsp9–co-chaperone complexes *via* CK2 inhibition: chemical structures, **2.74–2.83**.

much weaker, likely due to poor permeability, and they have limited kinase selectivity [243]. The macrocyclic analog **2.75** [244] loses cell-free potency, but gains in terms of cellular penetration.

Several carboxylates inhibit CK2. The indoloquinazoline acetic acid IQA (CGP029482, **2.76**) is an nM CK2 inhibitor with good selectivity *vs.* other kinases, including DYRK1A. It shows a cleaner CK2-related activity spectrum than TBBt/**2.70** in cellular models [245], but has a limited aqueous stability [246]. The xanthenyl benzoic acid derivative **2.77** [247] is extremely potent, but it interferes with the cellular translation process [248]. Other promising carboxylic CK2 inhibitors include the substituted pyrazine pyrrole acetic acid CC04820 (**2.78**) [249], the thiazolylbenzoic acid **2.79** [250] and the benzo[g]indazole carboxylate **2.80** [251].

Tricyclic carboxylate CK2 inhibitors include CX-4945 (**2.81a**) and CX-5011 (**2.81b**). The former is a potent, selective, and orally active CK2 inhibitor [252] that acts on multiple pathology-related targets [253]. Its effects on the Hsp90 chaperone machinery have not been reported yet. The preclinical profile of CX-4945 is remarkable [254], and the candidate is being clinically tested in oncology. Preliminary PK and pharmacodynamic results from a Phase I study are available [255]. The clinical development of CX-4945 supports CK2 inhibitors as chronic treatments against NDDs. Second generation compounds, such as CX-5011, show an even higher selectivity than CX-4945 [256] and represent viable, further options for clinical development.

The described structural classes represent ATP-competitive CK2 inhibitors. Two ATP-noncompetitive inhibitors can be mentioned, although far from being drug-like. The benzothiazolium sulfonate **2.82** shows potency on the isolated enzyme and good selectivity [247]. Its binding site/mode is unknown and it is inactive in cellular assays, likely due to cell permeability issues. Podophyllotoxin-inspired compound W16 (**2.83**, Figure 2.14) binds at the interface between the α and β subunits of CK2, preventing their association and inhibiting kinase activity with moderate potency and high selectivity [257]. Its complexity hinders its use as a scaffold for lead optimization.

The CK2 inhibitory landscape is rich and varied, but proof of concept data (cellular and especially *in vivo* data) validating their application against tauopathy-inducing/progressing molecular events are missing. The potential multi-targeting profile of such inhibitors against tauopathies (CK2 directly phosphorylates tau, negatively regulates cdc37 and FKBP52 among Hsp90 co-chaperones, and reduces amyloid neurotoxicity) grants further investigation, maybe using clinically compliant CX-4945.

2.5 RECAP

This chapter deals with small molecule modulators of neuropathological alterations caused by early-mid steps in the pathway to protein misfolding and aggregation in general, and to tau and/or tau-connected events in particular. Three potential therapeutic mechanisms were examined in detail in the biology-oriented companion book [5], and at least one target was arbitrarily chosen for each mechanism. Eighty-three scaffolds/molecules shown in Figures 2.1 to 2.14, acting on the three selected targets, are described in detail in this chapter, and are briefly summarized in Table 2.1. The chemical core of each scaffold/compound is structurally

TABLE 2.1　Compounds **2.1–2.83:** Chemical Class, Target, Developing Organization, Development Status

Number	Chemical cpd./class	Target	Organization	Dev. status
2.1	KRIBB3	Hsp27 inhibition	Korea Research Institute of Bioscience and Biotechnology	LO
2.2	Zerumbone	Hsp27 inhibition	Korea Institute of Radiological and Medical Science	LO
2.4, 2.5	Clerodanes, hardwickiic acid	Hsp27 inhibition	Salerno University, Italy	DD
2.6	Brivudine, RP-101	Hsp27 inhibition	RESprotect, Dresden, Germany	Ph II
2.7	Arylsulfonamides, JCC76	Hsp27 inhibition	Cleveland State University	LO
2.9	Arylethynyltriazolyl ribonucleoside	Hsp27 down-regulation	Wuhan University, China	LO
2.10	Xanthones, TDP	Hsp27 down-regulation	The Chinese University of Hong Kong	LO
2.11	Quercetin	CK2, CamKII	Chung San Medical University, Taiwan	PE
2.12	LY303511	mTOR, CK2	National University of Singapore	DD
2.13	Benzopyranopyridines	MK2	Pfizer	LO
2.14	Aminopyrazinthio-ureas	MK2	Merck	LO
2.15	Indazolepyrimidines	MK2	Abbott	DD
2.16	Indazole-pyrrolo[3,2-*d*] pyrimidines	MK2	Abbott	LO
2.17	2-pyridyl-tetrahydro-4H-pyrrolo [3,2-*c*]pyridine-4-ones	MK2	Pfizer	LO
2.18a,b	2-pyrimidinyl-tetrahydro-4H-pyrrolo[3,2-*c*] pyridine-4-one spiro compounds	MK2	Merck	LO
2.19	Benzamide-substituted tetrahydro-4H-pyrrolo[3,2-*c*] pyridine-4-one spiro compounds	MK2	Merck	DD

(Continued)

TABLE 2.1   Compounds **2.1–2.83**: Chemical Class, Target, Developing Organization, Development Status (*cont.*)

Number	Chemical cpd./class	Target	Organization	Dev. status
2.20	Pyrrole-based tetracycles	MK2	Novartis	LO
2.21	2-Pyridyl-dihydropyr-rolopyrimidinones	MK2	Novartis	DD
2.22	1,4-Bisaryl-3-aminopyrazoles	MK2	Novartis	LO
2.23a,b	Carbolines	MK2	Pfizer	LO
2.24	Thiazolamide-substituted carbolines	MK2	Boehringer Ingelheim	DD
2.25	Piperidine-containing spiroderivatives	MK2	Boehringer Ingelheim	LO
2.26, 2.27	Benzothiophene lactams	MK2	Pfizer	DD
2.28	Bicyclol	Hsp27 inducer	Chinese Academy of Medical Sciences	MKTD
2.29	Squamosamide analogs, FLZ	HSF1 activa-tor, Akt up-regulation	Chinese Academy of Medical Sciences	DD
2.30	Apigenin	CK2, cdc37 inhibitor	Cognitive Res. Center, Bejing, China	LO
2.31	Resveratrol	Hsp27 inducer/down-regulator	Several	PE
2.32	Gusperimus	Hsp70	Nordic	Ph III
2.33	Tresperimus	Hsp70	Laboratoires Fournier	PE
2.34	DHP-spergualin	Hsp70, ATPase inhib.	University of Pitts-burgh	DD
2.35a,d,e	DHPs	Hsp70, ATPase stimul.	Alzheimer's Institute, Florida	LO
2.35b,c	DHPs	Hsp70, ATPase inhib.	Alzheimer's Institute, Florida	LO
2.36	MKT-077	Hsp70	Harvard Med. School	Ph I
2.37	Sulfogalactosyl ce-ramides	Hsp70	University of Toronto	Ph I
2.38	Acylbenzamide fatty acids	Hsp70	Max Planck, Halle, D	DD

TABLE 2.1 Compounds **2.1–2.83:** Chemical Class, Target, Developing Organization, Development Status (*cont.*)

Number	Chemical cpd./class	Target	Organization	Dev. status
2.39a,b	Gallates – ECG, EGCG	Hsp70, plus others	Several	PE
2.40a,b	Luteolin, myricetin - flavonoids	Hsp70, plus others	Several	PE
2.41	Apoptozole	Hsp70, ATPase inhib.	Yonsei University, South Korea	LO
2.42	Heterocyclic carbox-amides	Hsp70, ATPase inhib.	Burnham Institute, Calif.	DD
2.43	Phenoxy-N-arylacet-amides	Hsp70, ATPase inhib.	ALS Biopharma	DD
2.44	Pifithrin	Hsp70, plus others	University of Pennsylvania	LO
2.45a,b	Thioflavin S	Hsp70–BAG-1	Cancer Research, UK	DD
2.46	VER-155008	Hsp70–BAG-1	Vernalis	LO
2.47	Sildenafil	Hsp70–BAG-1	University of Genova, I	DD
2.48	KM11060	Hsp70–BAG-1	University of Genova, I	DD
2.49a–d	Geldanamycin derivatives	Hsp90	BMS, Infinity	Ph III
2.50	Retaspimycin	Hsp90	Infinity	Ph III
2.51	BIIB021	Hsp90	Biogen Idec	Ph II
2.52a–c	Purine-like compounds	Hsp90	Sloan-Kettering NY, Myrexis, DebioPharm	Ph I
2.53	Radicicol	Hsp90	–	NP
2.54	Ganetespib	Hsp90	Synta	Ph III
2.55	AUY922	Hsp90	Vernalis, Novartis	Ph II
2.56	KW-2478	Hsp90	Kiowa Hakko Kirin	Ph II
2.57	AT13387	Hsp90	Astex	Ph II
2.58	HSP990	Hsp90	Novartis	Ph I
2.59	SNX-5422	Hsp90	Esanex	Ph I
2.60	XL888	Hsp90	Exelixis	Ph I
2.61	FK506	Hsp90–FKBPs	Astellas	MKTD

(Continued)

TABLE 2.1   Compounds 2.1–2.83: Chemical Class, Target, Developing Organization, Development Status (cont.)

Number	Chemical cpd./class	Target	Organization	Dev. status
2.62	GPI-1046	Hsp90–FKBPs	Guilford, Amgen	Ph I
2.63a,b	V1-0367, VX-710	Hsp90–FKBPs	Vertex	Ph II
2.64	Pipecolyl ketoamide	Hsp90–FKBPs	Max Planck, Munich, D	DD
2.65	Pipecolyl sulfonamide	Hsp90–FKBPs	Max Planck, Munich, D	DD
2.66	Celastrol	Hsp90–cdc37, plus others	–	NP
2.68	DRB	CK2	University of Pennsylvania	LO
2.69a–c	Tetrahalo benzimidazoles	CK2	University of Padova	LO
2.70	Tetrabromo benzotriazole	CK2	University of Padova	LO
2.71a-c	Quinones	CK2	University of Padova	DD
2.72	Ellagic acid	CK2	–	NP
2.73	Resorufin	CK2	Southern Denmark University	DD
2.74	Pyrazolotriazines	CK2	Polaris	DD
2.75	Macrocyclic pyrazolotriazines	CK2	Polaris	DD
2.76	IQA	CK2	University of Padova	LO
2.77	Xanthenyl benzoic acids	CK2	INSERM, Grenoble	DD
2.78	CC04820	CK2	Kyoto University, Jp	DD
2.79	Thiazolylbenzoic acids	CK2	Kyoto University, Jp	DD
2.80	Benzo[g]indazole carboxylate	CK2	Kyoto University, Jp	DD
2.81a,b	Tricyclic carboxylates	CK2	Cylene Pharmaceuticals	Ph I
2.82	Benzothiazolium sulfonates	CK2	INSERM, Grenoble	DD
2.83	W16	CK2	INSERM, Grenoble	DD

Not progressed, NP; early discovery, DD; lead optimization, LO; preclinical evaluation, PE; clinical Phase I-II-III, Ph I–Ph III; marketed, MKTD.

defined; its molecular target is mentioned; the developing laboratory (either public or private) is listed; and the development status—according to publicly available information—is finally provided.

References

[1] Friedman, R. Aggregation of amyloids in a cellular context: modeling and experiment. *Biochem. J.* **2011**, *438*, 415–426.

[2] Badiola, N.; Suarez-Calvet, M.; Lleo, A. Tau phosphorylation and aggregation as a therapeutic target in tauopathies. *CNS Neurol. Dis. Drug Targets* **2010**, *9*, 727–740.

[3] Warren, J. D.; Rohrer, J. D.; Schott, J. M.; Fox, N. C.; Hardy, J.; Rossor, M. N. Molecular nexopathies: a new paradigm of neurodegenerative disease. *Tr. Neurosci.* **2013**, *36*, 561–569.

[4] Martin, L.; Latypova, X.; Terro, F. Post-translational modifications of tau protein: Implications for Alzheimer's disease. *Neurochem. Int.* **2011**, *58*, 458–471.

[5] Seneci, P. *Molecular targets in protein misfolding and neurodegenerative disease.* Elsevier, **2014**, 278 pages.

[6] Bukau, B.; Weissman, J.; Horwich, A. Molecular chaperones and protein quality control. *Cell* **2006**, *125*, 443–451.

[7] Kim, R.; Lai, L.; Lee, H. H.; Cheong, G. W.; Kim, K. K., et al. On the mechanism of chaperone activity of the small heat-shock protein of Methanococcus jannaschii. *Proc. Natl. Acad. Sci. U.S.A.* **2003**, *100*, 8151–8155.

[8] Haslbeck, M.; Miess, A.; Stromer, T.; Walter, S.; Buchner, J. Disassembling protein aggregates in the yeast cytosol: the cooperation of HSP26 with Ssa1 and Hsp104. *J. Biol. Chem.* **2005**, *280*, 23861–23868.

[9] Slepenkov, S. V.; Witt, S. N. The unfolding story of the Escherichia coli Hsp70 DnaK: is DnaK a holdase or an unfoldase? *Mol. Microbiol.* **2002**, *45*, 1197–1206.

[10] Sharma, S. K.; Christen, P.; Goloubinoff, P. Disaggregating chaperones: an unfolding story. *Curr. Protein Pept. Sci.* **2009**, *10*, 432–446.

[11] De Los Rios, P.; Ben-Zvi, A.; Slutsky, O.; Azem, A.; Goloubinoff, P. Hsp70 chaperones accelerate protein translocation and the unfolding of stable protein aggregates by entropic pulling. *Proc. Natl. Acad. Sci. U.S.A.* **2006**, *103*, 6166–6171.

[12] Uversky, V. N. Flexible nets of malleable guardians: intrinsically disordered chaperones in neurodegenerative diseases. *Chem. Rev.* **2011**, *111*, 1134–1166.

[13] Evans, C. G.; Chang, L.; Gestwicki, J. E. Heat shock protein 70 (Hsp70) as an emerging drug target. *J. Med. Chem.* **2010**, *53*, 4585–4602.

[14] Salminen, A.; Ojala, J.; Kaarniranta, K.; Hiltunen, M.; Soininen, H. Hsp90 regulates tau pathology through co-chaperone complexes in Alzheimer's disease. *Progr. Neurobiol.* **2011**, *93*, 99–110.

[15] Abisambra, J. F.; Blair, L. J.; Hill, S. E.; Jones, J.; Kraft, C.; Rogers, J., et al. Phosphorylation dynamics regulate Hsp27-mediated rescue of neuronal plasticity deficits in tau transgenic mice. *J. Neurosci.* **2010**, *30*, 15374–15382.

[16] Mayer, M. P.; Bukau, B. Hsp70 chaperones: cellular functions and molecular mechanisms. *Cell. Mol. Life Sci.* **2005**, *62*, 670–684.

[17] Sarkar, M.; Kuret, J.; Lee, G. Two motifs within the tau-microtubule-binding domain mediate its association with the hsc70 molecular chaperone. *J. Neurosci. Res.* **2008**, *86*, 2763–2773.

[18] Petrucelli, L.; Dickson, D.; Kehoe, K.; Taylor, J.; Snyder, H.; Grover, A., et al. CHIP and Hsp70 regulate tau ubiquitination, degradation and aggregation. *Hum. Mol. Genet.* **2004**, *13*, 703–714.

[19] Dou, F.; Netzer, W. J.; Tanemura, K.; Li, F.; Hartl, F. U.; Takashima, A., et al. Chaperones increase association of tau protein with microtubules. *Proc. Natl. Acad. Sci. U.S.A.* **2003**, *100*, 721–726.

[20] Elliott, E.; Tsvetkov, P.; Ginzburg, I. BAG-1 associates with Hsc70-tau complex and regulates the proteasomal degradation of tau protein. *J. Biol. Chem.* **2007**, *282*, 37276–37284.

[21] Pearl, L. H.; Prodromou, C. Structure and mechanism of the Hsp90 molecular chaperone machinery. *Annu. Rev. Biochem.* **2006**, *75*, 271–294.

[22] Luo, W.; Dou, F.; Rodina, A.; Chip, S.; Kim, J.; Zhao, Q., et al. Roles of heat-shock protein 90 in maintaining and facilitating the neurodegenerative phenotype in tauopathies. *Proc. Natl. Acad. Sci. U.S.A.* **2007**, *104*, 9511–9516.

[23] Jinwai, U. K.; Koren, J., III.; Borysov, S. I.; Schmid, A. B.; Abisambra, J. F.; Blair, L. J., et al. The Hsp90 cochaperone, FKBP51, increases tau stability and polymerizes microtubules. *J. Neurosci.* **2010**, *30*, 591–599.

[24] Chambraud, B.; Sardin, E.; Giustiniani, J.; Dounane, O.; Schumacher, M.; Goedert, M.; Baulieu, E. -E. A role for FKBP52 in tau protein function. *Proc. Natl. Acad. Sci. U.S.A.* **2010**, *107*, 2658–2663.

[25] Shin, K. D.; Lee, M. Y.; Shin, D. S.; Lee, S.; Son, K. H.; Koh, S., et al. Blocking tumor cell migration and invasion with biphenyl isoxazole derivative KRIBB3, a synthetic molecule that inhibits Hsp27 phosphorylation. *J. Biol. Chem.* **2005**, *280*, 41439–41448.

[26] Shin, K. D.; Yoon, Y. J.; Kang, Y. -R.; Son, K. -H.; Kim, H. M.; Kwon, B. -M.; Han, D. C. KRIBB3, a novel microtubule inhibitor, induces mitotic arrest and apoptosis in human cancer cells. *Biochem. Pharmacol.* **2008**, *7*, 383–394.

[27] Li, J.; Hu, W.; Lan, Q. The apoptosis-resistance in t-AUCB-treated glioblastoma cells depends on activation of Hsp27. *J. Neurooncol* **2012**, *110*, 187–194.

[28] Kaigorodova, E. V.; Ryazantseva, N. V.; Novitskii, V. V.; Maroshkina, A. N.; Belkina, M. V. Effects of HSP27 chaperone on THP-1 tumor cell apoptosis. *Bull. Exp. Biol. Med.* **2012**, *154*, 77–79.

[29] Kaigorodova, E. V.; Litvinova, L. S.; Konovalova, E. V.; Klimova, S. V.; Tashireva, L. A.; Nosareva, O. L.; Novitskiy, V. V. The inhibition of Hsp27 chaperone affects the level of p53 protein in tumor cells. *Int. J. Biol.* **2013**, *5*, 13–18.

[30] Choi, S. -H.; Lee, Y. -J.; Seo, W. D.; Lee, H. -J.; Nam, J. -W.; Lee, Y. J., et al. Altered crosslinking of Hsp27 by zerumbone as a novel strategy for overcoming Hsp27-mediated radioresistance. *Int. J. Radiation Oncol. Biol. Phys.* **2011**, *79*, 1196–1205.

[31] Ohnishi, K.; Irie, K.; Murakami, A. In vitro covalent binding proteins of zerumbone, a chemopreventive food factor. *Biosci. Biotechnol. Biochem.* **2009**, *73*, 1905–1907.

[32] Ohnishi, K.; Ohkura, S.; Nakahata, E.; Ishisaka, A.; Kawai, Y.; Terao, J., et al. Nonspecific protein modifications by a phytochemical induce heat shock response for self-defense. *PLoS ONE* **2013**, *8*, e58641.

[33] Ohnishi, K.; Nakahata, E.; Irie, K.; Murakami, A. Zerumbone, an electrophilic sesquiterpene, induces cellular proteo-stress leading to activation of ubiquitin–proteasome system and autophagy. *Bioch. Biophys. Res. Commun.* **2013**, *430*, 616–622.

[34] Faiella, L.; Dal Piaz, F.; Bisio, A.; Tosco, A.; De Tommasi, N. A chemical proteomics approach reveals Hsp27 as a target for proapoptotic clerodane diterpenes. *Mol. BioSyst.* **2012**, *8*, 2637–2644.

[35] De Clercq, E. Dancing with chemical formulae of antivirals: A personal account. *Biochem. Pharmacol.* **2013**, *86*, 711–725.

[36] Fahrig, R.; Quietzsch, D.; Heinrich, J. -C.; Heinemann, V.; Boeck, S.; Schmid, R. M., et al. RP101 improves the efficacy of chemotherapy in pancreas carcinoma cell lines and pancreatic cancer patients. *Anti-Cancer Drugs* **2006**, *17*, 1045–1056.

[37] Heinrich, J. -C.; Tuukkanen, A.; Schroeder, M.; Fahrig, T.; Fahrig, R. RP101 (brivudine) binds to heat shock protein HSP27 (HSPB1) and enhances survival in animals and pancreatic cancer patients. *J. Cancer Res. Clin. Oncol.* **2011**, *137*, 1349–1361.

[38] Dawelbait, G.; Winter, C.; Zhang, Y.; Pilarsky, C.; Grutzmann, R.; Heinrich, J. -C.; Schroeder, M. Structural templates predict novel protein interactions and targets from pancreas tumour gene expression data. *Bioinformatics* **2007**, *23*, i115–i124.

[39] Zhong, B.; Cai, X.; Yi, X.; Zhou, A.; Chen, S.; Su, B. In vitro and in vivo effects of a cyclooxygenase-2 inhibitor nimesulide analog JCC76 in aromatase inhibitors-insensitive breast cancer cells. *J. Steroid Biochem. Mol. Biol.* **2011**, *126* (10), 18.

[40] Yi, X.; Zhong, B.; Smith, K. M.; Geldenhuys, W. J.; Feng, Y.; Pink, J. J., et al. Identification of a class of novel tubulin inhibitors. *J. Med. Chem.* **2012**, *55*, 3425–3435.

[41] Zhong, B.; Chennamaneni, S.; Lama, R.; Yi, X.; Geldenhuys, W. J.; Pink, J. J., et al. Synthesis and mechanism investigation of dual Hsp27 and tubulin inhibitors. *J. Med. Chem.* **2013**, *56*, 5306–5320.

[42] Baranova, E. V.; Beelen, S.; Gusev, N. B.; Strelkov, S. V. The taming of small heat-shock proteins: crystallization of the alpha-crystallin domain from human Hsp27. *Acta Crystallogr., Sect. F: Struct. Biol. Cryst. Commun.* **2009**, *65*, 1277–1281.

[43] Baranova, E. V.; Weeks, S. D.; Beelen, S.; Bukach, O. V.; Gusev, N. B.; Strelkov, S. V. Three-dimensional structure of α-crystallin domain dimers of human small heat shock proteins HSPB1 and HSPB6. *J. Mol. Biol.* **2011**, *411*, 110–122.

[44] Mchaourab, H. S.; Lin, Y. -L.; Spiller, B. W. Crystal structure of an activated variant of small heat shock protein Hsp16.5. *Biochemistry* **2012**, *51*, 5105–5112.

[45] Xia, Y.; Liu, Y.; Wan, J.; Wang, M.; Rocchi, P.; Qu, F., et al. Novel triazole ribonucleoside down-regulates heat shock protein 27 and induces potent anticancer activity on drug-resistant pancreatic cancer. *J. Med. Chem.* **2009**, *52*, 6083–6096.

[46] Fu, W. -m.; Zhang, J. -f.; Wang, H.; Xi, Z. -c.; Wang, W. -m.; Zhuang, P., et al. Heat shock protein 27 mediates the effect of 1,3,5-trihydroxy-13, 13-dimethyl-2H-pyran [7,6-b] xanthone on mitochondrial apoptosis in hepatocellular carcinoma. *J. Proteomics* **2012**, *75*, 4833–4843.

[47] Pietta, P. G. Flavonoids as antioxidants. *J. Nat. Prod.* **2000**, *63*, 1035–1042.

[48] Nair, H. B.; Sung, B.; Yadav, V. R.; Kannappan, R.; Chaturvedi, M. M.; Agarwal, B. B. Delivery of antiinflammatory nutraceuticals by nanoparticles for the prevention and treatment of cancer. *Biochem. Pharmacol.* **2010**, *80*, 1833–1843.

[49] Davies, S. P.; Reddy, H.; Caivano, M.; Cohen, P. Specificity and mechanism of action of some commonly used protein kinase inhibitors. *Biochem. J.* **2000**, *351*, 95–105.

[50] Wang, R. E.; Kao, J. L. -F.; Hilliard, C. A.; Pandita, R. K.; Roti Roti, J. L.; Hunt, C. R.; Taylor, J. S. Inhibition of heat shock induction of heat shock protein 70 and enhancement of heat shock protein 27 phosphorylation by quercetin derivatives. *J. Med. Chem.* **2009**, *52*, 1912–1921.

[51] Wei, L.; Liu, T. T.; Wang, H. H.; Hong, H. M.; Yu, A. L.; Feng, H. P.; Chang, W. W. Hsp27 participates in the maintenance of breast cancer stem cells through regulation of epithelialemesenchymal transition and nuclear factor-kappa B. *Breast Cancer Res.* **2011**, *13*, R101.

[52] Lee, C. H.; Hong, H. -M.; Chang, Y. -Y.; Chang, W. W. Inhibition of heat shock protein (Hsp) 27 potentiates the suppressive effect of Hsp90 inhibitors in targeting breast cancer stem-like cells. *Biochimie* **2012**, *94*, 1382–1389.

[53] Chen, S. -F.; Nieh, S.; Jao, S. -W.; Liu, C. -L.; Wu, C. -H.; Chang, Y. -C., et al. Quercetin suppresses drug-resistant spheres via the p38 MAPK–Hsp27 apoptotic pathway in oral cancer cells. *PLoS ONE* **2012**, *7* e49275.

[54] Ohnishi, K.; Yasumoto, J.; Takahashi, A.; Ohnishi, T. LY294002, an inhibitor of PI-3K, enhances heat sensitivity independently of p53 status in human lung cancer cells. *Int. J. Oncol.* **2006**, *29*, 249–253.

[55] Kristof, A. S.; Pacheco-Rodriguez, G.; Schremmer, B.; Moss, J. LY303511 (2-piperazinyl-8-phenyl-4H-1-benzopyran-4-one) acts via phosphatidylinositol 3-kinase-independent pathways to inhibit cell proliferation via mammalian target of rapamycin (mTOR)- and non-mTOR-dependent mechanisms. *J. Pharmacol. Exp. Ther.* **2005**, *314*, 1134–1143.

[56] Mellier, G.; Liu, D.; Bellot, G.; Holme, A. L.; Pervaiz, S. Small molecule sensitization to TRAIL is mediated via nuclear localization, phosphorylation and inhibition of chaperone activity of Hsp27. *Cell Death Dis.* **2013**, *4* e890.

[57] Bryantsev, A. L.; Loktionova, S. A.; Ilyinskaya, O. P.; Tararak, E. M.; Kampinga, H. H.; Kabakov, A. E. Distribution, phosphorylation, and activities of Hsp25 in heat-stressed H9c2 myoblasts: a functional link to cytoprotection. *Cell Stress Chaperones* **2002**, *7*, 146–155.

[58] Anderson, D. R.; Hegde, S.; Reinhard, E.; Gomez, L.; Vernier, W. F.; Lee, L., et al. Aminocyanopyridine inhibitors of mitogen activated protein kinase-activated protein kinase 2 (MK-2). *Bioorg. Med. Chem. Lett.* **2005**, *15*, 1587–1590.

[59] Lin, S.; Lombardo, M.; Malkani, S.; Hale, J. J.; Mills, S. G.; Chapman, K., et al. Novel 1-(2-aminopyrazin-3-yl)methyl-2-thioureas as potent inhibitors of mitogen-activated protein kinase-activated protein kinase 2 (MK-2). *Bioorg. Med. Chem. Lett.* **2009**, *19*, 3238–3242.

[60] Argiriadi, M. A.; Ericsson, A. M.; Harris, C. M.; Banach, D. L.; Borhani, D. W.; Calderwood, D. J., et al. 2,4-Diaminopyrimidine MK2 inhibitors. Part I: Observation of an unexpected inhibitor binding mode. Bioorg. *Med. Chem. Lett.* **2010**, *20*, 330–333.

[61] Harris, C. M.; Ericsson, A. M.; Argiriadi, M. A.; Barberis, C.; Borhani, D. W.; Burchat, A., et al. 2,4-Diaminopyrimidine MK2 inhibitors. Part II: Structure-based inhibitor optimization. Bioorg. *Med. Chem. Lett.* **2010**, *20*, 334–337.

[62] Anderson, D. R.; Meyers, M. J.; Vernier, W. F.; Mahoney, M. W.; Kurumbail, R. G.; Caspers, N., et al. Pyrrolopyridine inhibitors of mitogen-activated protein kinase-activated protein kinase 2 (MK-2). *J. Med. Chem.* **2007**, *50*, 2647–2654.

[63] Barf, T.; Kaptein, A.; de Wilde, S.; van der Heijden, R.; van Someren, R.; Demont, D., et al. Structure-based lead identification of ATP-competitive MK2 inhibitors. *Bioorg. Med. Chem. Lett.* **2011**, *21*, 3818–3822.

[64] Kaptein, A.; Oubrie, A.; de Zwart, E.; Hoogenboom, N.; de Wit, J.; van de Kar, B., et al. Discovery of selective and orally available spiro-3-piperidyl ATP-competitive MK2 inhibitors. *Bioorg. Med. Chem. Lett.* **2011**, *21*, 3823–3827.

[65] Oubrie, A.; Kaptein, A.; de Zwart, E.; Hoogenboom, N.; Goorden, R.; van de Kar, B., et al. Novel ATP competitive MK2 inhibitors with potent biochemical and cell-based activity throughout the series. *Bioorg. Med. Chem. Lett.* **2012**, *22*, 613–618.

[66] Revesz, L.; Schlapbach, A.; Aichholz, R.; Dawson, J.; Feifel, R.; Hawtin, S., et al. In vivo and in vitro SAR of tetracyclic MAPKAP-K2 (MK2) inhibitors. Part II. *Bioorg. Med. Chem. Lett.* **2010**, *20*, 4719–4723.

[67] Schlapbach, A.; Feifel, R.; Hawtin, S.; Heng, R.; Koch, G.; Moebitz, H., et al. Pyrrolopyrimidones: a novel class of MK2 inhibitors with potent cellular activity. *Bioorg. Med. Chem. Lett.* **2008**, *18*, 6142–6146.

[68] Sun, H.; Tawa, G.; Wallqvist, A. Classification of scaffold-hopping approaches. *Drug Discov. Today* **2012**, *17*, 310–324.

[69] Velcicky, J.; Feifel, R.; Hawtin, S.; Heng, R.; Huppertz, C.; Koch, G., et al. Novel 3-aminopyrazole inhibitors of MK-2 discovered by scaffold hopping strategy. *Bioorg. Med. Chem. Lett.* **2010**, *20*, 1293–1297.

[70] Trujillo, J. I.; Meyers, M. J.; Anderson, D. R.; Hegde, S.; Mahoney, M. W.; Vernier, W. F., et al. Novel tetrahydro-β-carboline-1-carboxylic acids as inhibitors of mitogen activated protein kinase-activated protein kinase 2 (MK-2). *Bioorg. Med. Chem. Lett.* **2007**, *17*, 4657–4663.

[71] Wu, J. -P.; Wang, J.; Abeywardane, A.; Andersen, D.; Emmanuel, M.; Gautschi, E., et al. The discovery of carboline analogs as potent MAPKAP-K2 inhibitors. *Bioorg. Med. Chem. Lett.* **2007**, *17*, 4664–4669.

[72] Goldberg, D. R.; Choi, Y.; Cogan, D.; Corson, M.; DeLeon, R.; Gao, A., et al. Pyrazinoindolone inhibitors of MAPKAP-K2. *Bioorg. Med. Chem. Lett.* **2008**, *18*, 938–941.

[73] Anderson, D. R.; Meyers, M. J.; Kurumbail, R. G.; Caspers, N.; Poda, G. I.; Long, S. A., et al. Benzothiophene inhibitors of MK2. Part 1: Structure–activity relationships, assessments of selectivity and cellular potency. *Bioorg. Med. Chem. Lett.* **2009**, *19*, 4878–4881.

[74] Anderson, D. R.; Meyers, M. J.; Kurumbail, R. G.; Caspers, N.; Poda, G. I.; Long, S. A., et al. Benzothiophene inhibitors of MK2. Part 2: Improvements in kinase selectivity and cell potency. *Bioorg. Med. Chem. Lett.* **2009**, *19*, 4882–4884.

[75] Bao, X. Q.; Liu, G. T. Bicyclol: a novel anti-hepatitis drug with hepatic heat shock protein27/70 inducing activity and cytoprotective effects in mice. *Cell Stress Chaperones* **2008**, *13*, 347–355.

[76] Bao, X. Q.; Liu, G. T. Induction of overexpression of the 27- and 70-kDa heat shock proteins by bicyclol attenuates concanavalin A-induced liver injury through suppression of nuclear factor-kB in mice. *Mol. Pharmacol.* **2009**, *75*, 1180–1188.

[77] Liu, G. T. Bicyclol: a novel drug for treating chronic viral hepatitis B and C. *Med. Chemistry* **2009**, *5*, 29–43.

[78] Kong, X. -c.; Zhang, D.; Qian, C.; Liu, G. -t.; Bao, X. -q. FLZ, a novel HSP27 and HSP70 inducer, protects SH-SY5Y cells from apoptosis caused by MPP+. *Brain Res.* **2011**, *1383*, 99–107.

[79] Fang, F.; Liu, G. -t. Protective effects of compound FLZ on β-amyloid peptide-(25-35)-induced mouse hippocampal injury and learning and memory impairment. *Acta Pharmacol. Sinica* **2006**, *27*, 651–658.

[80] Gonzalez-Mejia, M. E.; Voss, O. H.; Murnan, E. J.; Doseff, A. I. Apigenin-induced apoptosis of leukemia cells is mediated by a bimodal and differentially regulated residue-specific phosphorylation of heat-shock protein 27. *Cell Death Dis.* **2010**, *1*, e64.

[81] Shukla, S.; Gupta, S. Apigenin-induced cell cycle arrest is mediated by modulation of MAPK, PI3K-Akt, and loss of cyclin D1 associated retinoblastoma dephosphorylation in human prostate cancer cells. *Cell Cycle* **2007**, *6*, 1102–1114.

[82] Zhao, M.; Ma, J.; Zhu, H. -Y.; Zhan, X. -H.; Du, Z. -Y.; Xu, Y. -J.; Yu, X. -D. Apigenin inhibits proliferation and induces apoptosis in human multiple myeloma cells through targeting the trinity of CK2. *Cdc37 and Hsp90. Mol. Cancer* **2011**, *10*, 104.

[83] Chen, D.; Landis-Piwowar, K. R.; Chen, M. S.; Dou, Q. P. Inhibition of proteasome activity by the dietary flavonoid apigenin is associated with growth inhibition in cultured breast cancer cells and xenografts. *Breast Cancer Res* **2007**, *9*, R80.

[84] Mustafi, S. B.; Chakraborty, P. K.; Raha, S. Modulation of Akt and ERK1/2 pathways by resveratrol in chronic myelogenous leukemia (CML) cells results in the downregulation of Hsp70. *PLoS ONE* **2010**, *5* e8719.

[85] Bishayee, A.; Waghray, A.; Barnes, K. F.; Mbimba, T.; Bhatia, D.; Chatterjee, M.; Darvesh, A. S. Suppression of the inflammatory cascade is implicated in resveratrol chemoprevention of experimental hepatocarcinogenesis. *Pharm. Res.* **2010**, *27*, 1080–1091.

[86] Díaz-Chavez, J.; Fonseca-Sanchez, M. A.; Arechaga-Ocampo, E.; Flores-Perez, A.; Palacios-Rodríguez, Y.; Domínguez-Gomez, G., et al. Proteomic profiling reveals that resveratrol inhibits HSP27 expression and sensitizes breast cancer cells to doxorubicin therapy. *PLoS ONE* **2013**, *8* e64378.

[87] Putics, A.; Vegh, E. M.; Csermely, P.; Soti, C. Resveratrol induces the heat-shock response and protects human cells from severe heat stress. *Antiox. Redox Signal* **2008**, *10*, 65–75.

[88] Han, S.; Choi, J. -R.; Shin, K. S.; Kang, S. J. Resveratrol upregulated heat shock proteins and extended the survival of G93A-SOD1 mice. *Brain Res.* **1483**, **2012**, 112–117.

[89] Jinwal, U. K.; Miyata, Y.; Koren, J.; Jones, I. I. I.; Trotter, J. R.; Chang, J. H., et al. Chemical manipulation of Hsp70 ATPase activity regulates Tau stability. *J. Neurosci.* **2009**, *29*, 12079–12088.

[90] Takayama, S.; Sato, T.; Krajewski, S.; Kochel, K.; Irie, S.; Millan, J. A.; Reed, J. C. Cloning and functional analysis of BAG-1: a novel Bcl-2-binding protein with anti-cell death activity. *Cell* **1995**, *80*, 279–284.

[91] Powers, M. V.; Jones, K.; Barillari, C.; Westwood, I.; van Montfort, R. L. M.; Workman, P. Targeting Hsp70—the second potentially druggable heat shock protein and molecular chaperone? *Cell Cycle* **2010**, *9*, 1542–1550.

[92] Powers, M. V.; Clarke, A. P.; Workman, P. Death by chaperone—HSP90, HSP70 or both? *Cell Cycle* **2009**, *8*, 518–526.

[93] Goloudina, A. R.; Demidov, O. N.; Garrido, C. Inhibition of HSP70: a challenging anticancer strategy. *Cancer Lett.* **2012**, *325*, 117–124.

[94] Takeuchi, T.; Iinuma, H.; Kunimoto, S.; Masuda, T.; Ishizuka, M.; Takeuchi, M., et al. A new antitumor antibiotic, spergualin: isolation and antitumor activity. *J. Antibiot.* **1981**, *34*, 1619–1621.

[95] Umeda, Y.; Moriguchi, M.; Kuroda, H.; Nakamura, T.; Iinuma, H.; Takeuchi, T.; Umezawa, H. Synthesis and antitumor activity of spergualin analogues. I. Chemical modification of 7-guanidino-3-hydroxyacyl moiety. *J. Antibiot.* **1985**, *38*, 886–898.

[96] Mazzucco, C. E.; Nadler, S. G. A member of the Hsp70 family of heat-shock proteins is a putative target for the immunosuppressant 15-deoxyspergualin. *Ann. N.Y. Acad. Sci.* **1993**, *685*, 202–204.

[97] Thomas, F. T.; Tepper, M. A.; Thomas, J. M.; Haisch, C. E. 15- Deoxyspergualin: a novel immunosuppressive drug with clinical potential. *Ann. N.Y. Acad. Sci.* **1993**, *685*, 175–192.

[98] Dhingra, K.; Valero, V.; Gutierrez, L.; Theriault, R.; Booser, D.; Holmes, F., et al. Phase II study of deoxyspergualin in metastatic breast cancer. *Invest. New Drugs* **1994**, *12*, 235–241.

[99] Kaufman, D. B.; Gores, P. F.; Kelley, S.; Grasela, D. M.; Nadler, S. G.; Ramos, E. 15-Deoxyspergualin: immunotherapy in solid organ and cellular transplantation. *Transplant. Rev.* **1996**, *10*, 160–174.

[100] Lebreton, L.; Annat, J.; Derrepas, P.; Dutartre, P.; Renaut, P. Structure-immunosuppressive activity relationships of new analogues of 15-deoxyspergualin. 1. Structural modifications of the hydroxyglycine moiety. *J. Med. Chem.* **1999**, *42*, 277–290.

[101] Elices, M. J. Laboratoires Fournier. *Curr. Opin. Invest. Drugs* **2001**, *2*, 372–374.

[102] Brodsky, J. L. Selectivity of the molecular chaperone-specific immunosuppressive agent 15-deoxyspergual: modulation of Hsc70 ATPase activity without compromising DnaJ. chaperone interactions. *Biochem Pharmacol.* **1999**, *57*, 877–880.

[103] Simpson, D. Tresperimus: a new agent for transplant tolerance induction. *Exp. Opin. Invest. Drugs* **2001**, *10*, 1381–1386.

[104] Fewell, S. W.; Day, B. W.; Brodsky, J. L. Identification of an inhibitor of hsc70-mediated protein translocation and ATP hydrolysis. *J. Biol. Chem.* **2001**, *276*, 910–914.

[105] Wisen, S.; Androsavich, J.; Evans, C. G.; Chang, L.; Gestwicki, J. E. Chemical modulators of heat shock protein 70 (Hsp70) by sequential, microwave-accelerated reactions on solid phase. *Bioorg. Med. Chem. Lett.* **2008**, *18*, 60–65.

[106] Fewell, S. W.; Smith, C. M.; Lyon, M. A.; Dumitrescu, T. P.; Wipf, P.; Day, B. W.; Brodsky, J. L. Small molecule modulators of endogenous and co-chaperone-stimulated Hsp70 ATPase activity. *J. Biol. Chem.* **2004**, *279*, 51131–51140.

[107] Wisen, S.; Gestwicki, J. E. Identification of small molecules that modify the protein folding activity of heat shock protein 70. *Anal. Biochem.* **2008**, *374*, 371–377.

[108] Wisen, S.; Bertelsen, E. B.; Thompson, A. D.; Patury, S.; Ung, P.; Chang, L., et al. Binding of a small molecule at a protein–protein interface regulates the chaperone activity of Hsp70–Hsp40. *ACS Chem. Biol.* **2010**, *5*, 611–622.

[109] Wright, C. M.; Chovatiya, R. J.; Jameson, N. E.; Turner, D. M.; Zhu, G.; Werner, S., et al. Pyrimidinone-peptoid hybrid molecules with distinct effects on molecular chaperone function and cell proliferation. *Bioorg. Med. Chem.* **2008**, *16*, 3291–3301.

[110] Koren, J.; Jinwal, U. K.; Jin, Y.; O'Leary, J.; Jones, J. R.; Johnson, A. G., et al. Facilitating Akt clearance via manipulation of Hsp70 activity and levels. *J. Biol. Chem.* **2010**, *285*, 2498–2505.

[111] Evans, C. G.; Wisen, S.; Gestwicki, J. E. Heat shock proteins 70 and 90 inhibit early stages of amyloid beta-(1-42) aggregation in vitro. *J. Biol. Chem.* **2006**, *281*, 33182–33191.

[112] Chafekar, S. M.; Wisén, S.; Thompson, A. D.; Echeverria, A. L.; Walter, G. M.; Evans, C. G., et al. Pharmacological tuning of heat shock protein 70 modulates polyglutamine toxicity and aggregation. *ACS Chem. Biol.* **2012**, *7*, 1556–1564.

[113] Cen, L. B.; Shishido, T. Composition containing rhodcyanine dies for treating cancer. *EP 527494 A1* **1993**, .

[114] Wadhwa, R.; Sugihara, T.; Yoshida, A.; Nomura, H.; Reddel, R. R.; Simpson, R., et al. Selective toxicity of MKT-077 to cancer cells is mediated by its binding to the hsp70 family protein mot-2 and reactivation of p53 function. *Cancer Res.* **2000**, *60*, 6818–6821.

[115] Chiba, Y.; Kubota, T.; Watanabe, M.; Matsuzaki, S. W.; Otani, Y.; Teramoto, T., et al. MKT-077, localized lipophilic cation: antitumor activity against human tumor xenografts serially transplanted into nude mice. *Anticancer Res.* **1998**, *18*, 1047–1052.

[116] Koya, K.; Li, Y.; Wang, H.; Ukai, T.; Tatsuta, N.; Kawakami, M.; Shishido, T.; Chen, L. B. MKT-077, a novel rhodacyanine dye in clinical trials, exhibits anticarcinoma activity in preclinical studies based on selective mitochondrial accumulation. *Cancer Res.* **1996**, *56*, 538–543.

[117] Propper, D. J.; Braybrooke, J. P.; Taylor, D. J.; Lodi, R.; Styles, P.; Cramer, J. A., et al. Phase I trial of the selective mitochondrial toxin MKT077 in chemo-resistant solid tumours. *Ann. Oncol.* **1999**, *10*, 923–927.

[118] Rousaki, A.; Miyata, Y.; Jinwal, U. K.; Dickey, C. A.; Gestwicki, J. E.; Zuiderweg, E. R. P. Allosteric drugs: the interaction of antitumor compound MKT-077 with human Hsp70 chaperones. *J. Mol. Biol.* **2011**, *411*, 614–632.

[119] Mamelak, D.; Lingwood, C. The ATPase domain of hsp70 possesses a unique binding specificity for 30-sulfogalactolipids. *J. Biol. Chem.* **2001**, *276*, 449–456.

[120] Mamelak, D.; Mylvaganam, M.; Whetstone, H.; Hartmann, E.; Lennarz, W.; Wyrick, P. B., et al. Hsp70s contain a specific sulfogalactolipid binding site. Differential aglycone influence on sulfogalactosylceramide binding by recombinant prokaryotic and eukaryotic hsp70 family members. *Biochemistry* **2001**, *40*, 3572–3582.

[121] Park, H. J.; Mylvaganum, M.; McPherson, A.; Fewell, S. W.; Brodsky, J. L.; Lingwood, C. A. A soluble sulfogalactosyl ceramide mimic promotes delta F508 CFTR escape from endoplasmic reticulum associated degradation. *Chem. Biol.* **2009**, *16*, 461–470.

[122] Liebscher, M.; Jahreis, G.; Lucke, C.; Grabley, S.; Raina, S.; Schiene-Fischer, C. Fatty acyl benzamido antibacterials based on inhibition of DnaK-catalyzed protein folding. *J. Biol. Chem.* **2007**, *282*, 4437–4446.

[123] Lachance, H., Wetzel, S., Kumar, K., Waldmann, H., Eds. Charting, navigating, and populating natural product chemical space for drug discovery. *Angew. Chem. Int.* **2012**, *55*, 5989–6001.

[124] Chang, L.; Miyata, Y.; Ung, P. M. U.; Bertelsen, E. B.; McQuade, T. J.; Carlson, H. A., et al. Chemical screens against a reconstituted multiprotein complex: Myricetin blocks DnaJ regulation of DnaK through an allosteric mechanism. *Chem. Biol.* **2011**, *18*, 210–221.

[125] Tran, P. L. C. H. B.; Kim, S. A.; Choi, H. S.; Yoon, J. H.; Ahn, S. G. Epigallocatechin-3-gallate suppresses the expression of HSP70 and HSP90 and exhibits anti-tumor activity in vitro and in vivo. *BMC Cancer* **2010**, *10*, 276.

[126] Wang, R. E.; Hunt, C. R.; Chen, J.; Taylor, J. S. Biotinylated quercetin as an intrinsic photoaffinity proteomics probe for the identification of quercetin target proteins. *Bioorg. Med. Chem.* **2011**, *19*, 4710–4720.

[127] Rezai-Zadeh, K.; Shytle, D.; Sun, N.; Mori, T.; Hou, H.; Jeanniton, D., et al. Green tea epigallocatechin-3-gallate (EGCG) modulates amyloid precursor protein cleavage and reduces cerebral amyloidosis in Alzheimer transgenic mice. *J. Neurosci.* **2005**, *25*, 8807–8814.

[128] Guedj, F.; Sebrie, C.; Rivals, I.; Ledru, A.; Paly, E.; Bizot, J. C., et al. Green tea polyphenols rescue of brain deficits induced by overexpression of DYRK1A. *PLoS One* **2009**, *4* e4606.

[129] Santa-Maria, I.; Diaz-Ruiz, C.; Ksiezak-Reding, H.; Chen, A.; Ho, L.; Wang, J.; Pasinetti, G. M. GSPE interferes with tau aggregation in vivo: implication for treating tauopathy. *Neurobiol. Aging* **2012**, *33*, 2072–2081.

[130] Williams, D. R., Ko, S. -K., Park, S., Lee, M. -R., Shin, I., Eds. An apoptosis-inducing small molecule that binds to heat shock protein 70. *Angew. Chem., Int.* **2008**, *47*, 7466–7469.

[131] Cho, H. J.; Gee, H. Y.; Baek, K. -H.; Ko, S. -K.; Park, J. -M.; Lee, H., et al. A small molecule that binds to an ATPase domain of Hsc70 promotes membrane trafficking of mutant cystic fibrosis transmembrane conductance regulator. *J. Am. Chem. Soc.* **2011**, *133*, 20267–20276.

[132] Cellitti, J.; Zhang, Z.; Wang, S.; Wu, B.; Yuan, H.; Hasegawa, P., et al. Small molecule DnaK modulators targeting the beta-domain. *Chem. Biol. Drug Des.* **2009**, *74*, 349–357.

[133] Cassel, J. A.; Ilyin, S.; McDonnell, M. E.; Reitz, A. B. Novel inhibitors of heat shock protein Hsp70-mediated luciferase refolding that bind to DnaJ. *Bioorg. Med. Chem.* **2012**, *20*, 3609–3614.

[134] Strom, E.; Sathe, S.; Komarov, P. G.; Chernova, O. B.; Pavlovska, I.; Shyshynova, I., et al. Small-molecule inhibitor of p53 binding to mitochondria protects mice from gamma radiation. *Nat. Chem. Biol.* **2006**, *2*, 474–479.

[135] Leu, J. I.; Pimkina, J.; Frank, A.; Murphy, M. E.; George, D. L. A small molecule inhibitor of inducible heat shock protein 70. *Mol. Cell* **2009**, *36*, 15–27.

[136] Kaiser, M.; Kühnl, A.; Reins, J.; Fischer, S.; Ortiz-Tanchez, J.; Schlee, C., et al. Antileukemic activity of the HSP70 inhibitor pifithrin-μ in acute leukemia. *Blood Cancer J* **2011**, *1*, e28.

[137] Kelenyi, G. On the histochemistry of azo group-free thiazole dyes. *J. Histochem. Cytochem.* **1967**, *15*, 172–180.

[138] Sharp, A.; Crabb, S. J.; Johnson, P. W. M.; Hague, A.; Cutress, R.; Townsend, P. A., et al. Thioflavin S (NSC71948) interferes with Bcl-2-associated athanogene (BAG-1)-mediated protein-protein interactions. *J. Pharmacol. Exp. Ther.* **2009**, *331*, 680–689.

[139] Williamson, D. S.; Borgognoni, J.; Clay, A.; Daniels, Z.; Dokurno, P.; Drysdale, M. J., et al. Novel adenosine-derived inhibitors of 70 kDa heat shock protein, discovered through structure-based design. *J. Med. Chem.* **2009**, *52*, 1510–1513.

[140] Macias, A. T.; Williamson, D. S.; Allen, N.; Borgognoni, J.; Clay, A.; Daniels, Z., et al. Adenosine-derived inhibitors of 78 kDa glucose regulated protein (Grp78) ATPase: Insights into isoform selectivity. *J. Med. Chem.* **2011**, *54*, 4034–4041.

[141] Massey, A. J.; Williamson, D. S.; Browne, H.; Murray, J. B.; Dokurno, P.; Shaw, T., et al. A novel, small molecule inhibitor of Hsc70/Hsp70 potentiates Hsp90 inhibitor induced apoptosis in HCT116 colon carcinoma cells. *Cancer Chemother. Pharmacol.* **2010**, *66*, 535–545.

[142] Cichero, E.; Basile, A.; Turco, M. C.; Mazzei, M.; Fossa, P. Scouting new molecular targets for CFTR therapy: the HSC70/BAG-1 complex. A computational study. *Med. Chem. Res.* **2012**, *21*, 4430–4436.

[143] Travers, J.; Sharp, S.; Workman, P. HSP90 inhibition: two-pronged exploitation of cancer dependencies. *Drug Discov. Today* **2012**, *17*, 242–252.

[144] Luo, W.; Sun, W.; Taldone, T.; Rodina, A.; Chiosis, G. Heat shock protein 90 in neurodegenerative diseases. *Mol. Neurodeg.* **2010**, *5*, 24.

[145] NTC00003969, http://www.clinicaltrials.gov/ct2/show/NCT00003969?term=17-aag &rank=34.

[146] Jhaveri, K.; Taldone, T.; Modi, S.; Chiosis, G. Advances in the clinical development of heat shock protein 90 (Hsp90) inhibitors in cancers. *Biochim. Biophys. Acta.* **1823**, *2012*, 742–755.

[147] Hong, D. S.; Banerji, U.; Tavana, B.; George, G. C.; Aaron, J.; Kurzrock, R. Targeting the molecular chaperone heat shock protein 90 (HSP90): lessons learned and future directions. *Cancer Treatment Rev.* **2013**, *39*, 375–387.

[148] Roughley, S.; Wright, L.; Brough, P.; Massey, A.; Hubbard, R. E. HSP90 inhibitors and drugs from fragment and virtual screening. *Top. Curr. Chem.* **2012**, *317*, 61–82.

[149] Messaoudi, S.; Peyrat, J. -F.; Brion, J. -D.; Alami, M. Heat-shock protein 90 inhibitors as antitumor agents: a survey of the literature from 2005 to 2010. *Exp. Opin. Ther. Pat.* **2011**, *21*, 1501–1542.

[150] DeBoer, C.; Meulman, P. A.; Wnuk, R. J.; Peterson, D. H. Geldanamycin, a new antibiotic. *J. Antibiot.* **1970**, *23*, 442–447.

[151] Whitesell, L.; Mimnaugh, E. G.; De Costa, B.; Myers, C. E.; Neckers, L. M. Inhibition of heat shock protein HSP90-pp 60v-src heteroprotein complex formation by benzoquinone ansamycins: essential role for stress proteins in oncogenic transformation. *Proc. Natl. Acad. Sci. U.S.A.* **1994**, *91*, 8324–8328.

[152] Supko, J. G.; Hickman, R. L.; Grever, M. R.; Malspeis, L. Preclinical pharmacologic evaluation of geldanamycin as an antitumor agent. *Cancer Chemother. Pharmacol.* **1995**, *36*, 305–315.

[153] Usmani Saad, Z.; Bona, Robert; Zihai, Li 17 AAG for HSP90 inhibition in cancer—from bench to bedside. *Curr. Mol. Med.* **2009**, *9*, 654–664.

[154] NCT00514371, http://www.clinicaltrials.gov/ct2/show/NCT00514371?term=NCT00 514371&rank=1.

[155] http://www.myelomabeacon.com/news/2010/07/22/tanespimycin-development-halted/.

[156] H.C. Pitot, A Trial of ABI-010 & ABI-007 in patients with advanced non- hematologic malignancies. http://clinicaltrials.gov/ct2/show/NCT00820768?term=ABI-010&rank=1.

[157] Pacey, S.; Wilson, R. H.; Walton, M.; Eatock, M. M.; Hardcastle, A.; Zetterlund, A., et al. A phase I study of the heat shock protein 90 inhibitor alvespimycin (17-DMAG) given intravenously to patients with advanced solid tumours. *Clin. Cancer Res.* **2011**, *17*, 1561–1570.

[158] Jhaveri, K.; Miller, K.; Rosen, L.; Schneider, B.; Chap, L.; Hannah, A., et al. A phase 1 dose-escalation trial of trastuzumab and alvespimycin hydrochloride (KOS-1022; 17 DMAG) in the treatment of advanced solid tumors. *Clin. Cancer Res.* **2012**, *18*, 5090–5098.

[159] Infinity. A Phase 1 dose escalation study of IPI-493. http://clinicaltrials.gov/ct2/show/NCT00724425?term=IPI-493&rank=1.

[160] Porter, J. R.; Adams, J.; Ahn, R.; Ammoscato, V.; Arsenault, B.; Austad, B. C., et al. Pharmaceutical development of IPI-504, an Hsp90 inhibitor and clinical candidate for the treatment of cancer. *Drug Devel. Res.* **2010**, *71*, 429–438.

[161] Johnston, A.; Allaire, M. Infinity halts RING trial in advanced gastrointestinal stromal tumors. http://investor.ipi.com/releasedetail.cfm?ReleaseID=377328, 2009.

[162] Chene, P. ATPases as drug targets: learning from their structure. *Nat. Rev. Drug Discov.* **2002**, *1*, 665–673.

[163] Zhang, H.; Neely, L.; Lundgren, K.; Yang, Y. C.; Lough, R.; Timple, N.; Burrows, F. BIIB021, a synthetic Hsp90 inhibitor, has broad application against tumors with acquired multidrug resistance. *Int. J. Cancer* **2010**, *126*, 1226–1234.

[164] Elfiky, A.; Saif, M. W.; Beeram, W.; O'Brien, S.; Lammanna, N.; Castro, J. E., et al. BIIB021, an oral, synthetic non-ansamycin Hsp90 inhibitor: phase I experience. *J. Clin. Oncol.* **2008**, *26*, 2503 abstract.

[165] Caldas-Lopes, E.; Cerchietti, L.; Ahn, J. H.; Clement, C. C.; Robles, A. I.; Rodina, A., et al. Hsp90 inhibitor PU-H71, a multimodal inhibitor of malignancy, induces complete responses in triple-negative breast cancer models. *Proc. Natl. Acad. Sci. U.S.A.* **2009**, *106*, 8368–8373.

[166] S. Kummar,. A Phase I study of the Hsp90 inhibitor, PU-H71, in patients with refractory solid tumors and low-grade non-Hodgkin's lymphoma. http://bethesdatrials.cancer.gov/clinical-research/search_detail.aspx?ProtocolID=NCI-11-C-0150.

[167] Kim, S. -H.; Bajji, A.; Tangallapally, R.; Markovitz, B.; Trovato, R.; Shenderovich, M., et al. Discovery of (2S)-1-[4-(2-{6-Amino-8-[(6-bromo-1,3-benzodioxol-5-yl)sulfanyl]-9H-purin-9-yl}ethyl)piperidin-1-yl]-2-hydroxypropan-1-one (MPC-3100), a purine-based Hsp90 inhibitor. J. Med. Chem. **2012**, 55, 7480–7501.

[168] Yu, M.; Samlowski, W. E.; Baichwal, V.; Brown, B.; Evans, B. A.; Woodland, D., et al. MPC-3100, a fully synthetic, orally bioavailable Hsp90 inhibitor, in cancer patients. J. Clin. Oncol. **2010**, 28 abstract e13112.

[169] Myrexis. http://www.thepharmaletter.com/file/107269/myrexis-to-drop-development-of-azixa-4th-qtr-loss-widens-names-new-ceo.html.

[170] Bao, R.; Lai, C. -J.; Qu, H.; Wang, D.; Yin, L.; Zifcak, B., et al. CUDC-305, a novel synthetic HSP90 inhibitor with unique pharmacologic properties for cancer therapy. Clin. Cancer Res. **2009**, 15, 4046–4057.

[171] Bao, R.; Lai, C. J.; Qu, H.; Wang, D.; Yin, L.; Zifcak, B., et al. CUDC-305, a novel synthetic HSP90 inhibitor with unique pharmacologic properties for cancer therapy. Clin. Cancer Res. **2009**, 15, 4046–4057.

[172] van Ingen, H.; Sanlaville, Y. Study of Debio 0932 in patients with advanced solid tumours or lymphoma. http://clinicaltrials.gov/ct2/show/NCT01168752?term=debio&rank=3.

[173] Schulte, T. W.; Akinaga, S.; Murakata, T.; Agatsuma, T.; Sugimoto, S.; Nakano, H., et al. Interaction of radicicol with members of the heat shock protein 90 family of molecular chaperones. Mol. Endocrinol. **1999**, 13, 1435–1448.

[174] Lin, T. -Y.; Bear, M.; Du, Z.; Foley, K. P.; Ying, W.; Barsoum, J.; London, C. The novel HSP90 inhibitor STA-9090 exhibits activity against Kit-dependent and -independent malignant mast cell tumors. Exp. Hematol. **2008**, 36, 1266–1277.

[175] Choi, H. K.; Lee, H. K. Recent updates on the development of ganetespib as a Hsp90 inhibitor. Arch. Pharmac. Res. **2012**, 35, 1855–1859.

[176] Wong, K.; Koczywas, M.; Goldman, J. W.; Paschold, E. H.; Horn, L.; Lufkin, J. M., et al. An open-label phase II study of the Hsp90 inhibitor ganetespib (STA-9090) as monotherapy in patients with advanced non-small cell lung cancer (NSCLC). J. Clin. Oncol. **2011**, 29, 7500 abstract.

[177] S.P. Corp. Study of ganetespib (STA-9090){ts}+{ts}docetaxel in advanced non small cell lung cancer. http://clinicaltrials.gov/ct2/show/NCT01348126?term=sta9090+and+docetaxel&rank=3.

[178] Eccles, S. A.; Massey, A.; Raynaud, F. I.; Sharp, S. Y.; Box, G.; Valenti, M., et al. NVP-AUY922: a novel heat shock protein 90 inhibitor active against xenograft tumor growth, angiogenesis, and metastasis. Cancer Res. **2008**, 68, 2850–2860.

[179] Samuel, T.; Sessa, C.; Britten, C.; Milligan, K. S.; Mita, M. M.; Banerji, U., et al. AUY922, a novel HSP90 inhibitor: final results of a first-in-human study in patients with advanced solid malignancies. J. Clin. Oncol. **2010**, 28, 2528 abstract.

[180] Garon, E. B.; Moran, T.; Barlesi, F.; Gandhi, L.; Sequist, L. V.; Kim, S. W., et al. Phase II study of the HSP90 inhibitor AUY922 in patients with previously treated, advanced non-small cell lung cancer (NSCLC). J. Clin. Oncol. **2012**, 30, 7543 abstract.

[181] Novartis. A Phase I-Ib/II study to determine the maximum tolerated dose (MTD) of AUY922 alone and in combination with bortezomib, with or without dexamethasone, in patients with relapsed or refractory multiple myeloma. http://clinicaltrials.gov/ct2/show/NCT00708292?term=AUY922&rank=2.

[182] Nakashima, T.; Ishii, T.; Tagaya, H.; Seike, T.; Nakagawa, H.; Kanda, Y., et al. New molecular and biological mechanism of antitumor activities of KW-2478, a novel non-ansamycin heat shock protein 90 inhibitor, in multiple myeloma cells. Clin. Cancer Res. **2010**, 16, 2792–2802.

[183] Cavenagh, J.; Yong, K.; Byrne, J.; Cavet, J.; Johnson, P.; Morgan, G., et al. The safety, pharmacokinetics and pharmacodynamics of KW-2478, a novel Hsp90 antagonist, in patients with B-cell malignancies: a first-in-man, phase I, multicentre, open-label, dose escalation study. *50th ASH Ann. Meet. Exposit.* **2008**, *112*, 958–959.

[184] I. Kyowa Hakko Kirin Pharma. A study of KW-2478 in combination with bortezomib in subjects with relapsed and/or refractory multiple myeloma. http://clinicaltrials. gov/ct2/show/NCT01063907.

[185] Murray, C. W.; Carr, M. G.; Callaghan, O.; Chessari, G.; Congreve, M.; Cowan, S., et al. Fragment-based drug discovery applied to Hsp90. Discovery of two lead series with high ligand efficiency. *J. Med. Chem.* **2010**, *53*, 5942–5955.

[186] Woodhead, A. J.; Angove, H.; Carr, M. G.; Chessari, G.; Congreve, M.; Coyle, J. E., et al. Discovery of (2,4-dihydroxy-5-isopropylphenyl)-[5-(4-methylpiperazin-1-ylmethyl)-1,3-dihydroisoindol-2-yl]methanone (AT13387), a novel inhibitor of the molecular chaperone Hsp90 by fragment based drug design. *J. Med. Chem.* **2010**, *53*, 5956–5969.

[187] Wolanski, A. A study to assess the safety of escalating doses of AT13387 in patients with metastatic solid tumors. http://clinicaltrials.gov/ct2/show/NCT00878423?term=AT-13387&rank=4.

[188] G. Demetri. A study to investigate the safety and efficacy of AT13387, alone or in combination with imatinib, in patients with GIST. http://clinicaltrials.gov/ct2/show/NCT01294202?term=AT-13387&rank=1.

[189] Menezes, D. L.; Taverna, P.; Jensen, M. R.; Abrams, T.; Stuart, D.; Yu, G. K., et al. The novel oral Hsp90 inhibitor NVP-HSP990 exhibits potent and broad-spectrum antitumor activities in vitro and in vivo. *Mol. Cancer Ther.* **2012**, *11*, 730–739.

[190] Novartis. A study of HSP990 administered by mouth in adult patients with advanced solid tumors. http://clinicaltrials.gov/ct2/show/NCT00879905?term=HSP990&rank=1.

[191] Fadden, P.; Huang, K. H.; Veal, J. M.; Steed, P. M.; Barabasz, A. F.; Foley, B., et al. Application of chemoproteomics to drug discovery: identification of a clinical candidate targeting Hsp90. *Chem. Biol.* **2010**, *17*, 686–694.

[192] Bryson, J.; Infante, J. R.; Ramanathan, R. K.; Jones, S. F.; Von Hoff, D. D.; Burris, H. A. A Phase 1 dose-escalation study of the safety and pharmacokinetics (PK) of the oral Hsp90 inhibitor SNX-5422. *J. Clin. Oncol.* **2008**, *26*, 14613 abstract.

[193] Rajan, A.; Kelly, R. J.; Trepel, J. B.; Kim, Y. S.; Alarcon, S. V.; Kummar, S., et al. A Phase 1 study of PF-04929113 (SNX-5422), an orally bioavailable heat shock protein 90 inhibitor in patients with refractory solid tumor malignancies and lymphomas. *Clin. Cancer Res.* **2011**, *17*, 6831–6839.

[194] Bussenius, J.; Blazey, C. M.; Aay, N.; Anand, N. K.; Arcalas, A.; Baik, T. G., et al. Discovery of XL888: A novel tropane-derived small molecule inhibitor of HSP90. *Bioorg. Med. Chem. Lett.* **2012**, *22*, 5396–5404.

[195] Exelixis. Safety study of pharmacokinetics of XL888 in adults with solid tumors. http://clinicaltrials.gov/ct2/show/NCT00796484?term=XL-888&rank=1.

[196] Gray, P. J., Jr.; Prince, T.; Cheng, J.; Stevenson, M. A.; Calderwood, S. K. Targeting the oncogene and kinome chaperone CDC37. *Nat. Rev. Cancer* **2008**, *8*, 491–495.

[197] Kino, T.; Hatanaka, H.; Hashimoto, M.; Nishiyama, M.; Goto, T.; Okuhara, M., et al. FK-506, a novel immunosuppressant isolated from a Streptomyces. I. Fermentation, isolation, and physico-chemical and biological characteristics. *J. Antibiot.* **1987**, *40*, 1249–1255.

[198] Schreiber, S. L. Chemistry and biology of the immunophilins and their immunosuppressive ligands. *Science* **1991**, *251*, 283–287.

[199] Snyder, S. H.; Sabatini, D. M.; Lai, M. M.; Steiner, J. P.; Hamilton, G. S.; Suzdak, P. D. Neural actions of immunophilin ligands. *Trends Pharmacol. Sci.* **1998**, *19*, 21–26.

[200] Kupina, N. C.; Detloff, M. R.; Dutta, S.; Hall, E. D. Neuroimmunophilin ligand V-10367 is neuroprotective after 24-hour delayed administration in a mouse model of diffuse traumatic brain injury. *J. Cerebral Blood Flow Metab.* **2002**, *22*, 1212–1221.

[201] Blackburn, E. A.; Walkinshaw, M. D. Targeting FKBP isoform with small molecule ligands. *Curr. Opin. Pharmacol.* **2011**, *11*, 365–371.

[202] Gold, B. G. FK506 and the role of the immunophilin FKBP-52 in nerve regeneration. *Drug Metab. Rev.* **1999**, *31*, 649–663.

[203] Gopalakrishnan, R.; Kozany, C.; Gaali, S.; Kress, C.; Hoogeland, B.; Bracher, A.; Hausch, F. Evaluation of synthetic FK506 analogues as ligands for the FK506-binding proteins 51 and 52. *J. Med. Chem.* **2012**, *55*, 4114–4122.

[204] Gopalakrishnan, R.; Kozany, C.; Wang, Y.; Schneider, S.; Hoogeland, B.; Bracher, A.; Hausch, F. Exploration of pipecolate sulfonamides as binders of the FK506-binding proteins 51 and 52. *J. Med. Chem.* **2012**, *55*, 4123–4131.

[205] Grad, I.; McKee, T. A.; Ludwig, S. M.; Hoyle, G. W.; Ruiz, P.; Wurst, W., et al. The Hsp90 cochaperone p23 is essential for perinatal survival. *Mol. Cell. Biol.* **2006**, *26*, 8976–8983.

[206] Chambraud, B.; Belabes, H.; Fontaine-Lenoir, V.; Fellous, A.; Baulieu, E. E. The immunophilin FKBP52 specifically binds to tubulin and prevents microtubule formation. *FASEB J.* **2007**, *21*, 2787–2797.

[207] Cau, W.; Konsolaki, M. FKBP immunophilins and Alzheimer's disease: a chaperoned affair. *J. Biosci.* **2011**, *36*, 493–498.

[208] Gandhi, L.; Harding, M. W.; Neubauer, M.; Langer, C. J.; Moore, M.; Ross, H. J., et al. A phase II study of the safety and efficacy of the multidrug resistance inhibitor VX-710 combined with doxorubicin and vincristine in patients with recurrent small cell lung cancer. *Cancer* **2007**, *109*, 924–932.

[209] Allison, A. C.; Cacabelos, R.; Lombardi, V. R. M.; Alvarez, X. A.; Vigo, C. Central nervous system effects of celastrol, a potent antioxidant and antiinflammatory agent. *CNS Drug Targ.* **2000**, *6*, 45–62.

[210] Sreeramulu, S., Gande, S. L., Goebel, M., Schwalbe, H., Eds. Molecular mechanism of inhibition of the human protein complex Hsp90–Cdc37, a kinome chaperone-cochaperone, by triterpene celastrol. *Angew. Chem. Int.* **2009**, *48*, 5853–5855.

[211] Zhang, T.; Li, Y.; Yu, Y.; Zou, P.; Jiang, Y.; Sun, D. Characterization of celastrol to inhibit Hsp90 and Cdc37 interaction. *J. Biol. Chem.* **2009**, *284*, 35381–35389.

[212] Paris, D.; Ganey, N. J.; Laporte, V.; Patel, N. S.; Beaulieu-Abdelahad, D.; Bachmeier, C., et al. Reduction of β-amyloid pathology by celastrol in a transgenic mouse model of Alzheimer's disease. *J. Neuroinflamm.* **2010**, *7*, 17.

[213] Cleren, C.; Calingasan, N. Y.; Chen, J.; Beal, M. F. Celastrol protects against MPTP and 3-nitropropionic acid neurotoxicity. *J. Neurochem.* **2005**, *94*, 995–1004.

[214] Kiaei, M.; Kipiani, K.; Petri, S.; Chen, J.; Calingasan, N. Y.; Beal, M. F. Celastrol blocks neuronal death and extends life in transgenic mouse model of amyotrophic lateral sclerosis. *Neurodeg. Dis.* **2005**, *2*, 246–254.

[215] Kannaiyan, R.; Shanmugam, M. K.; Sethi, G. Molecular targets of celastrol derived from Thunder of God Vine: potential role in the treatment of inflammatory disorders and cancer. *Cancer Lett.* **2011**, *303*, 9–20.

[216] Zhang, T.; Hamza, A.; Cao, X.; Wang, B.; Yu, S.; Zhan, C. -G.; Sun, D. A novel Hsp90 inhibitor to disrupt Hsp90/Cdc37 complex against pancreatic cancer cells. *Mol. Cancer Ther.* **2008**, *7*, 162–170.

[217] Cozza, G.; Bortolato, A.; Moro, S. How druggable is protein kinase CK2? *Med. Res. Rev.* **2010**, *30*, 419–462.

[218] Perez, D. I.; Gil, C.; Martinez, A. Protein kinases CK1 and CK2 as new targets for neurodegenerative diseases. *Med. Chem. Res.* **2011**, *31*, 924–954.

[219] Sarno, S.; Papinutto, E.; Franchin, C.; Bain, J.; Elliott, M.; Meggio, F., et al. ATP site-directed inhibitors of protein kinase CK2: an update. *Curr. Top. Med. Chem.* **2011**, *11*, 1340–1351.

[220] Zandomeni, R.; Zandomeni, M. C.; Shugar, D.; Weinmann, R. Casein kinase type II is involved in the inhibition by 5,6-dichloro-1-beta-D-ribofuranosylbenzimidazole of specific RNA polymerase II transcription. *J. Biol. Chem.* **1986**, *261*, 3414–3419.

[221] Pagano, M. A.; Andrzejewska, M.; Ruzzene, M.; Sarno, S.; Cesaro, L.; Bain, J., et al. Optimization of protein kinase CK2 inhibitors derived from 4,5,6,7-tetrabromobenzimidazole. *J. Med. Chem.* **2004**, *47*, 6239–6247.

[222] Pagano, M. A.; Meggio, F.; Ruzzene, M.; Andrzejewska, M.; Kazimierczuk, Z.; Pinna, L. A. 2-Dimethylamino-4,5,6,7-tetrabromo-1H-benzimidazole: a novel powerful and selective inhibitor of protein kinase CK2. *Biochem. Biophys. Res. Commun* **2004**, *321*, 1040–1044.

[223] Gianoncelli, A.; Cozza, G.; Orzeszko, A.; Meggio, F.; Kazimierczuk, Z.; Pinna, L. A. Tetraiodobenzimidazoles are potent inhibitors of protein kinase CK2. *Bioorg. Med. Chem.* **2009**, *17*, 7281–7289.

[224] Sarno, S.; Reddy, H.; Meggio, F.; Ruzzene, M.; Davies, S. P.; Donella-Deana, A. Selectivity of 4,5,6,7-tetrabromobenzotriazole, an ATP site-directed inhibitor of protein kinase CK2 ("casein kinase-2"). *FEBS Lett.* **2001**, *496*, 44–48.

[225] Pagano, M. A.; Bain, J.; Kazimierczuk, Z.; Sarno, S.; Ruzzene, M.; Di Maira, G., et al. The selectivity of inhibitors of protein kinase CK2: An update. *Biochem. J.* **2008**, *415*, 353–365.

[226] Wegiel, J.; Kaczmarski, W.; Barua, M.; Kuchna, I.; Nowicki, K.; Wang, K. -C., et al. Link between DYRK1A overexpression and several-fold enhancement of neurofibrillary degeneration with 3-repeat tau protein in Down syndrome. *J. Neuropathol. Exp. Neurol.* **2011**, *70*, 36–50.

[227] Moreno, H.; Yu, E.; Pigino, G.; Hernandez, A. I.; Kim, N.; Moreira, J. E., et al. Synaptic transmission block by presynaptic injection of oligomeric amyloid beta. *Proc. Natl. Acad. Sci. U.S.A.* **2009**, *106*, 5901–5906.

[228] Singh Atwal, R.; Desmond, C.; Caron, N.; Maiuri, T.; Xia, J.; Sipione, S.; Truant, R. Kinase inhibitors modulate huntingtin cell localization and toxicity. *Nat. Chem. Biol.* **2010**, *7*, 453–460.

[229] Manni, S.; Brancalion, A.; Quotti Tubi, L.; Colpo, A.; Pavan, L.; Cabrelle, A., et al. Protein kinase CK2 protects multiple myeloma cells from ER stress-induced apoptosis and from the cytotoxic effect of HSP90 inhibition through regulation of the unfolded protein response. *Clin. Cancer Res.* **2012**, *18*, 1888–1900.

[230] Miyata, Y.; Nishida, E. CK2 controls multiple protein kinases by phosphorylating a kinase targeting molecular chaperone. *Cdc37. Mol. Cell. Biol.* **2004**, *24*, 4065–4074.

[231] Yim, H.; Lee, Y. H.; Lee, C. H.; Lee, S. K. Emodin, an anthraquinone derivative isolated from the rhizomes of Rheum palmatum, selectively inhibits the activity of casein kinase II as a competitive inhibitor. *Planta Med.* **1999**, *65*, 9–13.

[232] Meggio, F.; Pagano, M. A.; Moro, S.; Zagotto, G.; Ruzzene, M.; Sarno, S., et al. Inhibition of protein kinase CK2 by condensed polyphenolic derivatives. *An in vitro and in vivo study. Biochemistry* **2004**, *43*, 12931–12936.

[233] Cozza, G.; Mazzorana, M.; Papinutto, E.; Bain, J.; Elliott, M.; di Maira, G., et al. Quinalizarin as a potent, selective and cell-permeable inhibitor of protein kinase CK2. *Biochem. J.* **2009**, *421*, 387–395.

[234] Sarno, S.; Moro, S.; Meggio, F.; Zagotto, G.; Dal Ben, D.; Ghisellini, P., et al. Toward the rational design of protein kinase casein kinase-2 inhibitors. *Pharmacol. Ther.* **2002**, *93*, 159–168.

[235] Patel, D.; Shukla, S.; Gupta, S. Apigenin and cancer chemoprevention: progress, potential and promise. *Int. J. Oncol.* **2007**, *30*, 233–245.

[236] Shukla, S.; Gupta, S. Suppression of constitutive and tumor necrosis factor alpha-induced nuclear factor (NF)-kappaB activation and induction of apoptosis by apigenin in human prostate carcinoma PC-3 cells: correlation with down-regulation of NF-kappaB-responsive genes. *Clin. Cancer Res.* **2004**, *10*, 3169–3178.

[237] Way, T. D.; Kao, M. C.; Lin, J. K. Degradation of HER2/neu by apigenin induces apoptosis through cytochrome c release and caspase-3 activation in HER2/neu-overexpressing breast cancer cells. *FEBS Lett.* **2005**, *579*, 145–152.

[238] Cozza, G.; Bonvini, P.; Zorzi, E.; Poletto, G.; Pagano, M. A.; Sarno, S., et al. Identification of ellagic acid as potent inhibitor of protein kinase CK2: a successful example of a virtual screening application. *J. Med. Chem.* **2006**, *49*, 2363–2366.

[239] Seeram, N. P.; Henning, S. M.; Zhang, Y.; Suchard, M.; Li, Z.; Heber, D. Pomegranate juice ellagitannin metabolites are present in human plasma and some persist in urine for up to 48 hours. *J. Nutr.* **2006**, *136*, 2481–2485.

[240] Sandholt, I. S.; Olsen, B. B.; Guerra, B.; Issinger, O. -G. Resorufin: a lead for a new protein kinase CK2 inhibitor. *Anti-Cancer Drugs* **2009**, *20*, 238–248.

[241] Klopffleisch, K.; Issinger, O. -G.; Niefind, K. Low-density crystal packing of human protein kinase CK2 catalytic subunit in complex with resorufin or other ligands: a tool to study the unique hinge-region plasticity of the enzyme without packing bias. *Acta Cryst., Section D: Biolog. Cryst.* **2012**, 883–892 D68.

[242] Nie, Z.; Perretta, C.; Erickson, P.; Margosiak, S.; Almassy, R.; Lu, J., et al. Structure-based design, synthesis, and study of pyrazolo[1,5-a][1,3,5]triazine derivatives as potent inhibitors of protein kinase CK2. *Bioorg. Med. Chem. Lett* **2007**, *17*, 4191–4195.

[243] Bettayeb, K.; Sallam, H.; Ferandin, Y.; Popowycz, F.; Fournet, G.; Hassan, M., et al. N-&-N, a new class of cell death-inducing kinase inhibitors derived from the purine roscovitine. *Mol. Cancer Ther.* **2008**, *7*, 2713–2724.

[244] Nie, Z.; Perretta, C.; Erickson, P.; Margosiak, S.; Lu, J.; Averill, A., et al. Structure-based design and synthesis of novel macrocyclic pyrazolo[1,5-a] [1,3,5]triazine compounds as potent inhibitors of protein kinase CK2 and their anticancer activities. *Bioorg. Med. Chem. Lett.* **2008**, *18*, 619–623.

[245] Sarno, S.; De Moliner, E.; Ruzzene, M.; Pagano, M. A.; Battistutta, R.; Bain, J., et al. Biochemical and three-dimensional-structural study of the specific inhibition of protein kinase CK2 by [5-oxo-5,6-dihydroindolo-(1,2-a)quinazolin-7-yl]acetic acid (IQA). *Biochem. J.* **2003**, *374*, 639–646.

[246] Sarno, S.; Ruzzene, M.; Frascella, P.; Pagano, M. A.; Meggio, F.; Zambon, A., et al. Development and exploitation of CK2 inhibitors. *Mol. Cell. Biochem.* **2005**, *274*, 69–76.

[247] Prudent, R.; Moucadel, V.; Lopez-Ramos, M.; Aci, S.; Laudet, B.; Mouawad, L., et al. Expanding the chemical diversity of CK2 inhibitors. *Mol. Cell. Biochem.* **2008**, *316*, 71–85.

[248] Novac, O.; Guenier, A. S.; Pelletier, J. Inhibitors of protein synthesis identified by a high throughput multiplexed translation screen. *Nucleic Acids Res.* **2004**, *32*, 902–915.

[249] Suzuki, Y.; Cluzeau, J.; Hara, T.; Hirasawa, A.; Tsujimoto, G.; Oishi, S., et al. Structure–activity relationships of pyrazine-based CK2 inhibitors: synthesis and evaluation of 2,6-disubstituted pyrazines and 4,6-disubstituted pyrimidines. *Arch. Pharm.* **2008**, *341*, 554–561.

[250] Hou, Z.; Nakanishi, I.; Kinoshita, T.; Takei, Y.; Yasue, M.; Misu, R., et al. Structure-based design of novel potent protein kinase CK2 (CK2) inhibitors with phenyl-azole scaffolds. *J. Med. Chem.* **2012**, *55*, 2899–2903.

[251] Suzuki, Y.; Oishi, S.; Takei, Y.; Yasue, M.; Misu, R.; Naoe, S., et al. Design and synthesis of a novel class of CK2 inhibitors: application of copper- and gold-catalysed cascade reactions for fused nitrogen heterocycles. *Org. Biomol. Chem.* **2012**, *10*, 4907–4915.

[252] Pierre, F.; Chua, P. C.; O'Brien, S. E.; Siddiqui-Jain, A.; Bourbon, P.; Haddach, M., et al. Discovery and SAR of 5-(3-chlorophenylamino) benzo[c][2,6]naphthyridine-

8-carboxylic acid (CX-4945), the first clinical stage inhibitor of protein kinase CK2 for the treatment of cancer. *J. Med. Chem.* **2011**, *54*, 635–654.

[253] Kim, J.; Kim, S. H. Druggability of the CK2 inhibitor CX-4945 as an anticancer drug and beyond. *Arch. Pharm. Res.* **2012**, *35*, 1293–1296.

[254] Pierre, F.; Chua, P. C.; O'Brien, S. E.; Siddiqui-Jain, A.; Bourbon, P.; Haddach, M., et al. Pre-clinical characterization of CX-4945, a potent and selective small molecule inhibitor of CK2 for the treatment of cancer. *Mol. Cell Biochem.* **2011**, *356*, 37–43.

[255] Padgett, C. S.; Lim, J. K. C.; Marschke, R. F.; Northfelt, D. W.; Andreopoulou, E.; Von Hoff, D. D., et al. Clinical pharmacokinetics and pharmacodynamics of CX-4945, a novel inhibitor of protein kinase CK2: interim report from the phase 1 clinical trial. *22nd EORTC-NCI-AACR symposium on "molecular targets and cancer therapeutics", Berlin.* **2010**.

[256] Battistutta, R.; Cozza, G.; Pierre, F.; Papinutto, E.; Lolli, G.; Sarno, S., et al. Unprecedented selectivity and structural determinants of a new class of protein kinase CK2 inhibitors in clinical trials for the treatment of cancer. *Biochemistry* **2011**, *50*, 8478–8488.

[257] Laudet, B.; Moucadel, V.; Prudent, R.; Filhol, O.; Wong, Y. S.; Royer, D.; Cochet, C. Identification of chemical inhibitors of protein-kinase CK2 subunit interaction. *Mol. Cell. Biochem.* **2008**, *316*, 63–69.

3

Targeting Proteasomal Degradation of Soluble, Misfolded Proteins
Ubi Major…

3.1 UPS-MEDIATED DEGRADATION OF MISFOLDED PROTEINS

The elimination of misfolded protein copies is an appealing option to tackle neurodegenerative diseases (NDDs) and tauopathies. The *ubiquitin–proteasome system* (*UPS*) [1,2] is the most important cellular pathway to dispose of soluble cytosolic proteins.

UPS is responsible for most regulated proteolytic events, and depends on the 76-mer protein ubiquitin (UBQ) [3,4]. UBQ is a stable, compact protein with the six C-terminus amino acids (AAs) arranged as a flexible tail [4]. A first UBQ molecule is anchored to the ε-NH_2 group of a Lys residue in the substrate proteins through an isopeptide bond involving the C-terminus of UBQ. Mono-ubiquitination may target a specific Lys residue [5], or a domain [6] on the substrate protein. Multiple mono-ubiquitination on different Lys residues of the substrate protein (multi-mono-ubiquitination) is also observed [7]. Ubiquitinated proteins are usually tagged with polyUBQ chains. UBQ chain elongation implies the formation of an isopeptide bond between one of seven Lys residues (K6, K11, K27, K29, K33, K48, and K63), or between the Met1 residue of a substrate-anchored proximal UBQ molecule and the C-terminus of a free, distal UBQ protein. Different UBQ chain elongation enzymes bind to interaction surfaces on UBQ with diverse specificities, and promote UBQ elongation on a specific anchoring point.

Once a di-UBQ chain is formed, the two UBQ molecules assume an anchoring residue-dependent conformation. Five out of the eight UBQ connections are structurally characterized by X-ray crystallography and/ or nuclear magnetic resonance (NMR) [8]. Among them, K48-linked di-UBQ chains adopt a "closed," compact conformation, where the two UBQ proteins interact with each other [9]. K63-linked di-UBQ chains adopt an "open" and more flexible conformation, where the two UBQ molecules do not interact beyond their isopeptide bond connection [10]. Highly flexible open conformations of di-UBQ chains provide even more alternative binding modes with their protein partners.

K48-linked polyUBQ chains are the most abundant homotypic poly-UBQ species [11], acting as labels on proteins to be degraded through the UPS [12]. Proteasome inhibition causes a fast increase of K48-polyUBQ proteins [13], and mutation of Lys residues in yeast UBQ show that K48 is the only essential residue among them [14].

There are eight Lys/Met anchoring points on UBQ, UBQ chain lengths varying between 1 (mono-UBQ) and >10 polyUBQs on each anchoring point, and >700 enzymes involved in a multi-step process (including UBQ activation, conjugation, transfer to a protein substrate, and chain trimming) [8]. The combinations of UBQ codes easily match the experimental observation of thousands of UBQ-labeled protein substrates on multiple sites, and ensure an exquisitely specific UBQ/UPS-dependent regulation of the functions reconducible to ubiquitinated proteins. But how can the UBQ machinery select the anchoring point, the UBQ chain length and nature, and the specific substrate to be ubiquitinated or deubiquitinated in a dynamic cellular environment?

UBQ is activated by two *UBQ-activating (E1) enzymes* through the formation of a high energy Cys–UBQ thioester bond [15,16]. E1 enzymes are relieved of their UBQ cargo by ≈40 *UBQ-conjugating (E2) enzymes* through a *trans*-thiolation reaction [17]. E2 enzymes contain a highly conserved 150–200 AA UBQ-conjugating (UBC) catalytic fold that acts as a scaffold for E1 enzymes, *E3 UBQ ligases*, and activated UBQ [18]. More than 600 E3 ligases receive UBQ from E2 enzymes and transfer it to substrates through three main mechanisms [19]. *Really interesting new gene (RING)* E3 ligases directly transfer UBQ from E2–UBQ complexes to RING E3-bound protein substrates [20]. *Homologous to the E6AP carboxyl terminus (HECT)* E3 ligases first bind UBQ onto a Cys residue of the HECT domain and release E2 enzymes, then bind protein substrates and transfer UBQ to them [21]. *RING-in-between-RING (RBR)* E3 ligases [22] act through a RING/HECT hybrid mechanism. Mono- and polyUBQ chains can be disassembled by a ≈100-membered class of isopeptide-specific *deubiquitinating enzymes (DUBs)* [23] that are essential to ensure proper processing of ubiquitinated proteins. Finally, the *26S proteasome complex* is the protein degradation terminal for UBQ-tagged proteins in eukaryotes [24]. It is made by a 20S barrel-shaped catalytic *core particle (CP)* composed by 28 subunits, structurally arranged in four stacked seven membered rings [25], and by two

19S *regulatory particles* (*RPs*), composed each by 19 subunits (a 9 subunit lid, and a 10 subunit base structure [26]). The UBQ activation–conjugation–ligation-trimming cycle is extensively described in the biology-oriented companion book [27].

Any E1, E2, E3, and DUB enzyme may be considered a suitable target to restore the UPS activity in an impaired cellular environment. Proteasome activity may also be targeted for a direct effect on UPS. NDDs and tauopathies require the potentiation/restoring of cellular mechanisms leading to the elimination of misfolded/aggregated proteins. UPS inhibition may appear to lead in the opposite direction—decreasing UPS-mediated elimination of tau and other misfolded proteins. Certain enzymes in the UPS system, though, contribute to the elimination of proteins hindering the rescuing/refolding of misfolded proteins. Their inhibition, thus, should be beneficial.

In particular, the *carboxy-terminus of Hsp70-interacting protein* (*CHIP*) [28] is an E3 UBQ ligase, due to a U-box domain at its C-terminus [29], and a Hsp70/Hsp90 co-chaperone, due to its three tandem tetratricopeptide (TPR) domains at the N-terminus [30]. CHIP is a key player in cellular management of misfolded proteins. Its role as an Hsp70-dependent, tau-ubiquitinating enzyme has been known for a decade [31,32]. The ubiquitin-specific protease *USP14* is a UBQ-trimming, proteasome-bound DUB that frees and recycles UBQ before substrate protein degradation by the UPS [33]. The role of USP14 in physiological and pathological events of the CNS is well known [34,35]. The next two sections describe small molecule modulators acting on each selected target.

3.2 CHIP

The role of CHIP in cellular management of misfolded proteins is due to its three N-terminal TPR/chaperone-binding domain repeats, and to its C-terminal U-box/UBQ-binding domain. Modulation of the whole set of CHIP functions would likely impact on many physiological processes. Specificity may be targeted through chaperone-mediated CHIP-misfolded protein interactions (through the TPR domains), aiming to regulate the target/substrate protein clearance (i.e., to increase it in NDDs) [36]. Specificity may also be obtained by targeting CHIP–E2-conjugating enzyme interactions (through the U-box domain), aiming to regulate the clearance of misfolded proteins through the specific CHIP–E2 couple that ubiquitinates a particular target/substrate protein [37].

NDDs often depend on pro-aggregation misfolded proteins. CHIP may contribute to their refolding, through complexation with Hsp90. When either neurotoxic protein oligomers exceed the refolding capacity of neurons, or the PQC machinery is impaired, CHIP may promote the aggregation of protein oligomers into insoluble, less toxic aggregates to be cleared *via* autophagy (see autophagy and aggrephagy, Chapters 4 and 5 here and in the biology-oriented companion book [27]). Either Hsp90 inhibition or CHIP overexpression increase the amount of protein clearance-directing Hsp70–CHIP complexes. Hsp90 inhibitors are described in Chapter 2, while small molecules capable of promoting CHIP induction in cells are unknown.

Protein regulators of CHIP, capable of orienting the fate of misfolded proteins, are well known. Members of the Bcl-2-associated athanogene co-chaperone family (BAG) may mediate the docking of a CHIP–chaperone–target protein complex at the proteasome, facilitating proteasomal degradation (BAG-1 [38]). They may suppress CHIP-mediated ubiquitination and degradation of a target protein by abrogation of the CHIP–E2 interaction (BAG-2 [39]). They may promote autophagic clearance when complexed with Hsp70 and CHIP in aging and/or protein aggregate-rich tissues (BAG-3 [40]). Finally, they may inhibit CHIP-mediated ubiquitination and degradation of a target protein through unclarified molecular mechanisms (BAG-5) [41]. The Hsp70-binding protein 1 (HspBP1) co-chaperone causes a conformational change in CHIP–chaperone complexes and prevents the ubiquitination of Hsc70-bound client proteins [42]. The S100A2 and S100P proteins inhibit CHIP-mediated ubiquitination and proteasomal degradation in a Ca^{2+}-dependent manner [43]. Stimulation of BAG-1- and BAG-3-promoted effects, and inhibition/prevention of CHIP negative regulation by BAG-2 and BAG-5 co-chaperones would be desirable in NDDs. Unfortunately, limited structural information on CHIP-containing chaperone complexes is available [44] to drive rational drug design projects.

Examples of selective modulation of CHIP–target protein complexes in the presence of Hsp chaperones are scarce [30]. Small molecules acting on TPR-mediated interactions of CHIP with chaperones, co-chaperones, and misfolded proteins are uncommon, and their selectivity is at best questionable. Thioflavin S (**3.1a,b**, Figure 3.1, see also **2.45a,b**, Figure 2.9 and section 2.3.2) is a mixture of benzothiazolylammonium salts showing sub-μM potency in preventing the Hsc/Hsp70–BAG-1 interaction [45]. BAG-1–CHIP interactions could be modulated by **3.1a,b**, although the cellular activity of thioflavin S could be due to Hsc/Hsp70–BAG-1-unrelated mechanisms.

The sulfonamide pifithrin-μ (PES, **3.2**, see also **2.44**, Figure 2.8 and section 2.3.1) shows multiple actions on the Hsp70 machinery [46]. Complexes between Hsp70 and CHIP, BAG-1 and other J-domain proteins are affected by PES. Its multiple and potent anticancer effects may be caused in part by the triple bond chemical reactivity, but probably are not entirely amenable to the Hsp70 chaperone network [46]. Naturally occurring gambogic acid (GA, **3.3**) causes the UPS-dependent degradation of mutant p53 [47]. GA decreases the Hsp90–mutant p53 interaction and in parallel

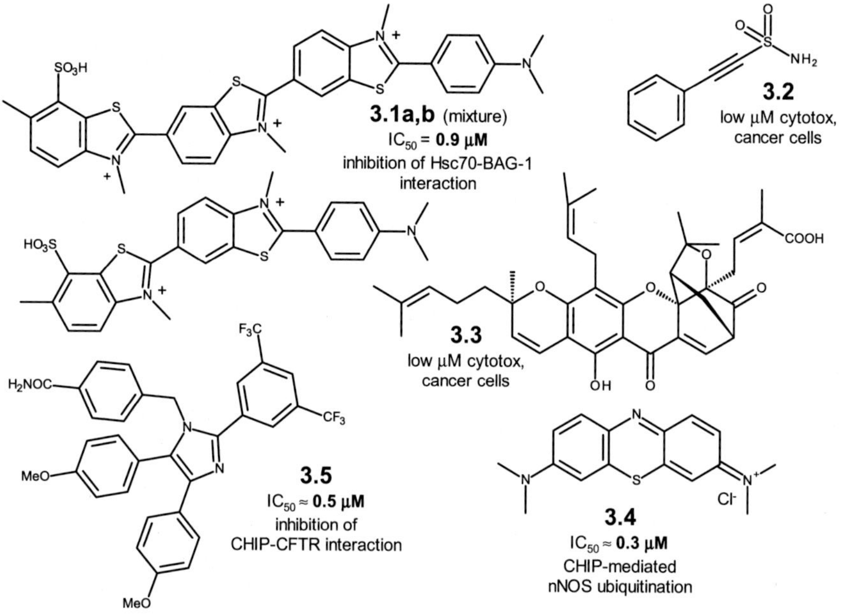

FIGURE 3.1 Small molecule modulators of chaperone-dependent CHIP activity: chemical structures, **3.1a–3.5**.

increases the levels of the ternary, UPS-oriented Hsp70–CHIP–mutant p53 complex through molecular interactions that are not yet elucidated. UPS-mediated degradation of polyUBQ mutant p53 *via* selective CHIP ubiquitination may be a GA-driven effect shared by other misfolded proteins [47]. The tricyclic phenothiazine methylene blue (MB, **3.4**), currently in clinical trials in AD patients as a tau aggregation inhibitor [48], negatively modulates polyQ protein degradation through Hsp70 binding and subsequent sub-μM inhibition of CHIP-mediated polyQ protein ubiquitination [49]. Its multi-targeted activity profile is discussed in detail in Chapter 6. Imidazole-based apoptozole (**3.5**, Figure 3.1) restores the defective cellular processing of ΔF508-CFTR (cystic fibrosis transmembrane conductance regulator), a mutant, misfolded protein involved in cystic fibrosis [50]. Its rescuing effect at sub-μM concentrations is likely due to the disruption of the tertiary Hsp70–CHIP–mutant protein complex, and to the prevention of CHIP-mediated ubiquitination and degradation of ΔF508-CFTR [50].

Small molecule modulators of chaperone-independent E3 ligase–target protein complexes are known [51]. Although they do not target CHIP-containing complexes, they indirectly prove the druggability of this protein–protein interaction (PPI) and must be mentioned. Two RING E3 ligases are targeted by candidates in clinical evaluation. The tetrasubstituted

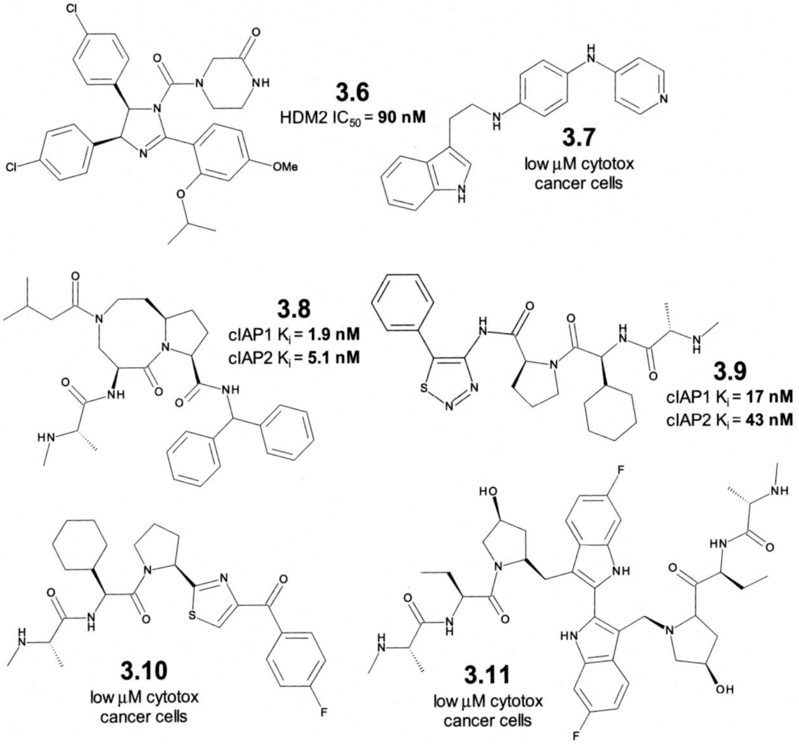

FIGURE 3.2 Small molecule modulators of chaperone-independent E3 ligase–target protein complexes in clinical trials: chemical structures, 3.6–3.11.

imidazolidine nutlin-3 (**3.6**, Figure 3.2) [52] and indole-based serdemetan (**3.7**) [53] are clinically tested PPI inhibitors/anticancer agents targeting the p53-human double minute 2 (HDM2) PPI [54]. HDM2 is a RING E3 ligase that modulates UPS-mediated degradation of p53, and compounds **3.6** and **3.7** inhibit HDM2-mediated ubiquitination of p53 [55].

Bicyclic (AT-406, **3.8**) [56], thiadiazole-, and thiazole-based monomeric (respectively GDC-0152 [57], **3.9** and LCL-161 [58], **3.10**) and dimeric antagonists of inhibitor of apoptosis proteins (IAPs) (TL32711, **3.11**, Figure 3.2) [59] are in clinical trials as anticancer agents. Their structure is inspired by the N-terminal AVPI sequence of the second mitochondria-derived activator of caspase/direct inhibitor of apoptosis-binding protein with low pI (Smac/DIABLO), an endogenous ligand of IAPs [60]. They act as PPI inhibitors/Smac mimetics/IAP antagonists, preventing the interaction between IAP proteins and caspases [61]. They also bind to RING domain-containing cellular IAPs (cIAPs), activate their E3 ligase activity, and induce their auto-ubiquitination and rapid proteasomal degradation [62].

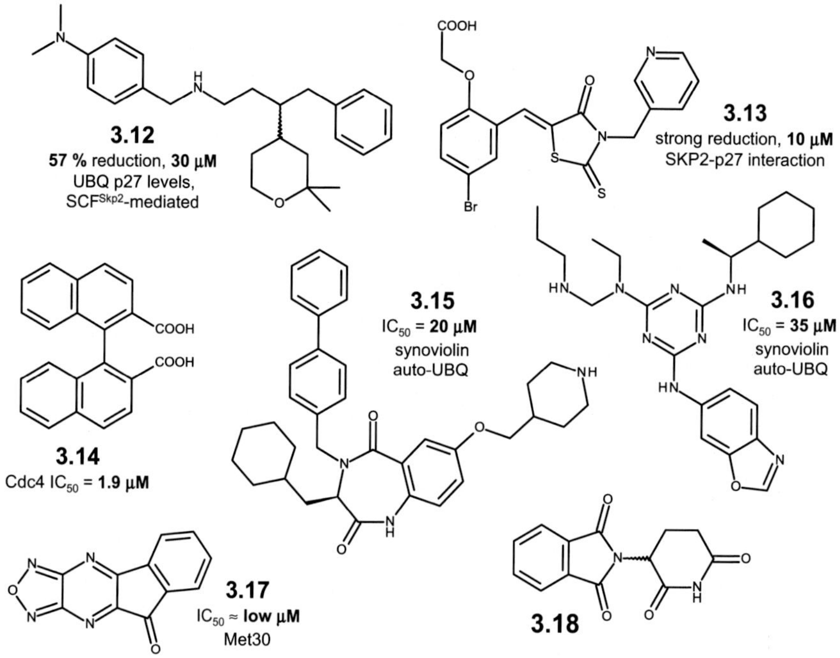

FIGURE 3.3   Small molecule modulators of chaperone-independent E3 ligase–target protein complexes in preclinical studies: chemical structures, **3.12–3.18**.

A few E3 ligase-containing complexes are targeted by small molecules at an earlier development stage. Such molecules are often discovered through high throughput screening (HTS) campaigns and assay formats that detect variations of E3 ligase activity [63]. The S-phase kinase-associated protein/Skp-cullin-F-box-containing (SCF) is the largest multiprotein RING E3 ligase family [64]. Diamino compound A (**3.12**, Figure 3.3) is identified through an HTS campaign targeted against inhibitors of the ubiquitination of p27^{Kip1} [65]. Enhanced UPS degradation of p27^{Kip1} is associated with poor prognosis in a variety of tumors, and SCF^{Skp2} is the E3 ligase that degrades p27^{Kip1} [66]. Compound A moderately increases p27^{Kip1} levels in cells through the exclusion of Skp2 from the SCF^{Skp2} E3 ligase complex, probably by inhibiting its binding with another complex member [65].

Inhibitors of SCF^{Skp2} ligase activity stem from structure-based virtual screening of a 315K compound data set against an Skp2–Csk1–p27 ternary complex [67]. An alkylidene thiazolidine (compound C1, **3.13**) reduces Skp2-mediated ubiquitination of p27 *in vitro* at low μM concentration, and increases p27 levels by decreasing its SCF^{Skp2}-mediated degradation in melanoma cells [67]. The diacid SCF-I2 (**3.14**) is identified as an inhibitor of SCF^{Cdc4} in an HTS campaign on a 50K member collection [68]. SCF-I2 binds to the WP40 domain of the F-box protein cell division control protein 4 (Cdc4), as shown by the X-ray structure of the Cdc4–Skp1–SCF-12 complex. The conformational change induced in Cdc4 by SCF-I2 allosterically inhibits substrate recognition and ubiquitination by SCF^{Cdc4}. Although WP40 domains are shared by F-box proteins in several SCF RING E3 ligases, SCF-I2 appears to be selective against SCF^{Cdc4}. SCF-I2 is not active in cellular assays, as its two carboxylates prevent cellular permeability [68]. LS-101 and LS-102 (respectively benzodiazepindione-based **3.15** and triazine-based **3.16**) are identified from an ≈4M compound collection in an HTS campaign [69] targeted against the auto-ubiquitination of synoviolin, a RING E3 ligase highly expressed in synoviocytes of patients suffering of rheumatoid arthritis [70]. Both compounds show moderate μM potency against synoviolin auto-ubiquitination. LS-102 shows complete specificity *vs.* three other RING E3 ligases, while LS-101 is non-selective. Both compounds show *in vivo* activity in collagene-induced arthritis models [69]. Tetracyclic SMER3 (**3.17**) is identified in a phenotype-based HTS looking for small molecule enhancers of the therapeutic effects of rapamycin (see also autophagy, Chapter 4 here and in the biology-oriented companion book [27]) [71]. SMER3 inhibits the RING E3 ligase SCF^{Met30}. It binds to the F-box motif in the Met30 protein, and it prevents its interaction with the SKC core protein Skp1. It is selective, as it is completely inactive against SCF^{Cdc4} [71]. Finally, the bis-imide thalidomide (**3.18**, Figure 3.3) binds to cereblon (CRBN), a component of a RING E3 ligase complex, and inhibits its interaction with damaged DNA binding protein 1 (DDBP1) and Cullin 4 (Cul4) [72]. It inhibits auto-ubiquitination in cells and *in vivo*, both in zebrafish and chicken models. Teratogenicity of thalidomide is at least partially due to CRBN binding and E3 ligase inhibition [72].

Interactions between CHIP and E2 conjugating enzymes are extensively studied. The X-ray structures of murine CHIP complexed with the E2 enzymes Ubc13 [73], and of zebrafish CHIP and E2D1/UbcH5 [74] are available. The conformational dynamics of the human CHIP–Ubc13 and CHIP–UbcH5 complexes, studied by amide hydrogen exchange mass spectrometry (HX-MS), highlight their differences and suggest that CHIP–E2 complexes in protein ubiquitination and chaperone interaction can be selectively modulated [75]. A systematic study [76] identifies a subset of E2 enzymes that bind CHIP through a common Ser-Pro-Ala motif, and promote target protein ubiquitination *via* activation of E2–UBQ conjugates. Ubiquitination of target proteins depends on the E2 conjugating enzyme in terms of point of attachment (K48, K63, others) and UBQ chain length [76]. For example, CHIP–UbcH5 preferentially catalyzes the mono-ubiquitination of target proteins through any Lys residue, while

CHIP–Ubc13 seems to be a polyUBQ-introducing complex with K63 specificity [77]. Finally, the tertiary E3:E2~UBQ complexes containing either breast cancer type 1 (BRCA1)/BRCA1–associated RING domain protein (BARD) or E4B/UFD2a as E3 ligases, and UbcH5c as E2 conjugating enzymes are studied by NMR [78]. The study provides valuable information on the role of RING/U-box E3 ligases such as CHIP in facilitating UBQ transfer and in promoting allosteric activation of E2~UBQ complexes [78]. The wealth of structural information is not yet translated into small molecules as regulators/modulators of CHIP:E2 binary, or CHIP:E2~UBQ tertiary complexes, and of their functions.

3.3 USP14

USP14 is a member of the largest USP/ubiquitin-specific protease subfamily, which contains ≈60 characterized family members [79,80]. Several USP enzymes are validated targets against various diseases [81], and CNS diseases in particular [34,35]. USP DUB inhibitors with varying degrees of *inter*-class selectivity are known [81–83]. UBQ analogs capable of specifically and irreversibly inactivating the thiol protease function of USPs include UBQ aldehyde [84] and UBQ vinyl sulfone [85]. They are useful probes to identify and characterize cysteine protease DUBs, but their peptidic nature and aspecificity hinder their use either as such, or as structural models for drug discovery efforts [81].

Aspecific USP inhibitors include electrophilic dienones, resulting from a computational pharmacophoric search on the National Cancer Institute (NCI) chemical database [86]. Curcumin (**3.19**, Figure 3.4), shikoccin (**3.20**), and Δ12-PGJ2 (**3.21**) are cytotoxic compounds whose cellular activity is at least partially due to DUB inhibition [86]. Δ12-PGJ2 is the most potent representative among electrophilic prostaglandins [87], whose activity against UCH DUBs is also reported [88]. Curcumin is covered in detail for its anti-aggregating properties on amyloidogenic peptides in Chapter 6 here and in the biology-oriented companion book [27].

The dienone NSC 632839 (**3.22**) shows similar, aspecific DUB inhibition [89]. Bis-isothiocyanate PR-619 (**3.23**) is an aspecific DUB inhibitor isolated from a pan-DUB-targeted HTS campaign [90]. It causes protein aggregation in neuronal cells and stabilizes microtubules (MTs), possibly with some effects on tau [91].

The tricyclic dinitrile HBX-41,108 (**3.24a**) results from structural optimization of hits from an HTS campaign targeted against the USP family member USP7 [92]. It stabilizes p53 and induces p53-dependent apoptosis in cancer cells through inhibition of the p53-deubiquitinating enzyme USP7 [92]. It inhibits at least five other USPs, and an UCH

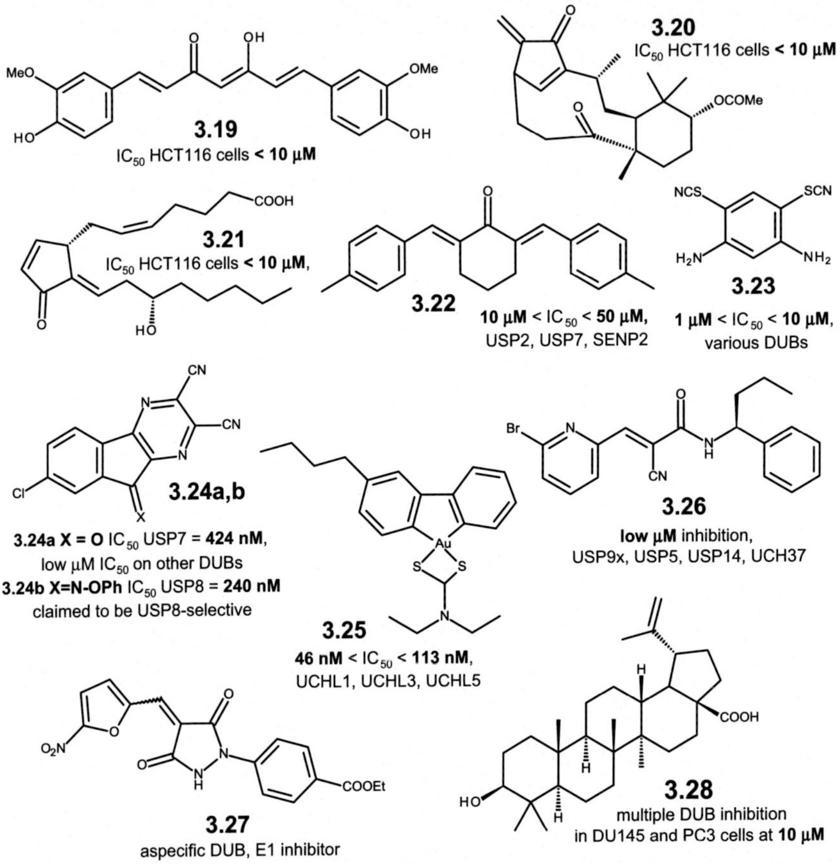

FIGURE 3.4   Small molecule inhibitors of USP14 and other deubiquitinases: chemical structures, **3.19–3.28**.

DUB family member [93]. Compound **3.24b** is reported as a selective USP8 inhibitor [82,94] but its structural similarity with **3.24a** induces to suspect a limited selectivity against other DUBs. Gold complexes such as **3.25** potently inhibit DUBs and are endowed with cytotoxic activity [95].

The cyanoamide WP-1130 (**3.26**) is a member of a synthetic tyrphostin-like library [96]. It is active in a cell-based screen targeted towards the Janus-activated kinase (JAK)/signal transducer and activator of transcription (STAT) pathway [97]. WP-1130 is a partially selective, cell permeable USP inhibitor, active against USP5, USP9x, USP14, UCH-L1, and UCH37 [97]. It shows pro-apototic effects through up-regulation of p53 and down-regulation of myeloid cell leukemia sequence 1 (MCL-1) levels, and promotes aggresome formation in cancer cells [97]. The cellular effects of

WP-1130 are due to inhibition of the unknown DUB responsible for JAK-2 deubiquitination [98]. It also enhances bacterial killing *via* localization of inducible nitric oxide synthase (iNOS) to the macrophage phagosome [99], and shows antiviral activity through USP14-mediated induction of the unfolded protein response (UPR) [100].

The alkyliden-pyrazolidindione PYR-41 (**3.27**), originally reported as a selective and irreversible E1 UBQ-activating enzyme [101], inhibits several DUBs, and even unrelated Cys-containing enzymes, through displacement of their nitro function by Cys residues [102]. Betulinic acid (**3.28**, Figure 3.4) shows multiple DUB-inhibiting activities and cytotoxicity on proliferating cancer cells, while it does not have similar effects on normal cells [103]. This may be due to a general increase of DUB levels in proliferating *vs.* non-proliferating cells, or to partially selective inhibition by betulinic acid of a subset of DUBs that are highly overexpressed/much more active in cancer cells [103].

A chalcone-based library contains cytotoxic, UPS-inhibiting representatives [104]. Some of its members, such as RA-9 (**3.29**, Figure 3.5), are partially selective DUB inhibitors, inhibiting >50% overall DUB activity in cervical cancer HeLa cells at 10 μM [105]. UPS2, UPS5, UPS8, UCH-L1, and UCH-L3 are among the DUBs inhibited by RA-9, while USP7 is not affected. RA-9 and its analogs induce polyUBQ accumulation, deplete the free UBQ pool, and promote apoptosis in cancer cells, while being non-toxic to normal cells [105].

The anti-psychotic, phthalimide-based pimozide (**3.30**) is a selective, allosteric, low μM inhibitor of the DUB enzyme USP1–USP1 associated factor 1 (UAF1) complex [106]. Pimozide weakly inhibits USP7, and is selective against a wide set of cysteine proteases. It restores cisplatin sensitivity to cisplatin-resistant small lung cancer cells [106]. A phenotype screen aimed towards autophagy inhibitors identifies the specific and potent autophagy inhibitor-1 (spautin-1, **3.31**) [107]. Spautin-1 is a selective nM inhibitor of USP10 and USP13. USP10 and USP13 are the DUBs acting on beclin-1, a key component of the autophagy-regulating kinase vacuolar protein sorting 34 (Vps34) complex (see also autophagy, Chapter 4 here and in the biology-oriented companion book [27]). Conversely, beclin-1 and the Vsp34 complex regulate the levels of USP10 and USP13. Spautin-1-caused autophagy inhibition increases p53 levels and may represent a novel anticancer approach [107].

Thiophene-based P5091 (**3.32**) is a USP7–USP47-selective DUB inhibitor with low μM potency identified in a pan-DUB-targeted HTS campaign [90]. P5091 increases ubiquitinated HDM2, which is then degraded and leads to p53-mediated cytotoxicity in cancer cells [108]. P5091 restores sensitivity to bortezomib-resistant cancer cells, and shows synergistic effects when combined with proteasome and histone deacetylase (HDAC) inhibitors [108]. A structurally related analog (**3.33**) is more potent in cellular

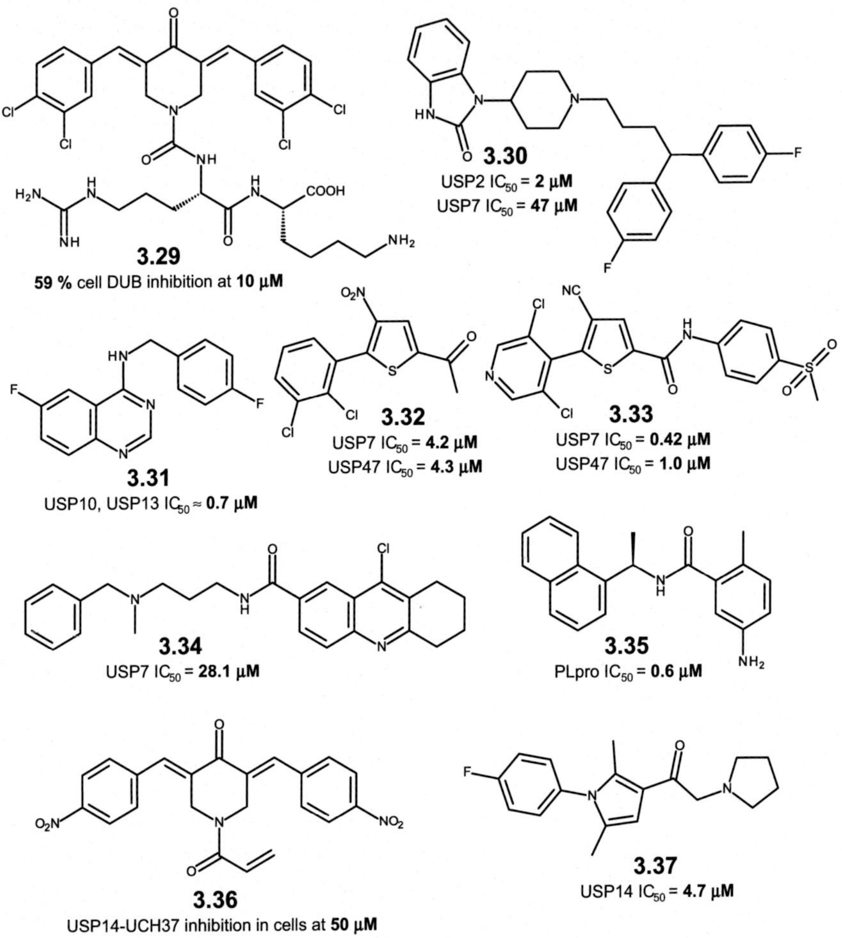

FIGURE 3.5 Small molecule inhibitors of USP14 and other deubiquitinases: chemical structures, **3.29–3.37**.

assays, causing an increase in p53 levels and an induction of p21 [109]. The aminotetrahydroacridine HBX-19,818 (**3.34**) is a cell permeable, selective, moderately potent USP7 inhibitor identified through an HTS campaign [93]. HDM2-p53 regulation and subsequent cytotoxicity are observed in human colon carcinoma (HCT116) cancer cells [93].

The naphthylamide GRL0617 (**3.35**) is an nM, non-covalent inhibitor of the papain-like protease (PLpro) from the severe acute respiratory syndrome (SARS)-causing coronavirus [110]. PLpro acts as a DUB, and the antiviral properties of GRL0617 (the result of structural optimization on an HTS hit) stem from inhibition of the deubiquitinase activity of PLpro.

GRL0617 does not inhibit human DUBs, and its specificity is explained by the X-ray structure of the PLpro–GRL0617 complex [110].

Two USP14-targeted small molecule inhibitors are known. The electrophilic dienone NSC687852/b-AP15 (**3.36**) is a pro-apoptotic compound identified in an HTS campaign on an NCI chemical collection [111]. It induces caspase-dependent apoptosis, and increases the levels of poly-UBQ proteins through DUB inhibition [112]. b-AP15 selectively inhibits with moderate potency two structurally unrelated, proteasomal DUBs, USP14 and UCH37, probably because of their common association with the proteasome [113]. It shows *in vivo* potency in animal models of leukemia, colon, lung, and breast carcinoma [113,114]. Pyrrole-based IU1 (**3.37**, Figure 3.5) is a selective, low μM inhibitor of USP14. It enhances *in vitro* UBQ–transactive response/TAR DNA binding protein 43 (TDP-43), UBQ–ataxin-3, and UBQ–tau levels, and increases their UPS-mediated proteolysis in murine embryonic fibroblasts (MEFs) [115]. It shows antiviral properties against Dengue virus, most likely through UPS enhancement [116]. The tau- and ataxin-3-regulating role of USP14 (and consequently the putative therapeutic effect of IU1) is questionable [117], but a compensatory increase of USP14 activity is observed in elderly cells [118].

Additional basic studies and potent, selective compounds are needed to elucidate the potential of "clean" and "mixed" USP14 inhibitors against neurodegeneration, as even small selectivity profile changes may induce major alterations in cellular effects. For example, USP14–UCH37-targeted bAP15 regulates caspase-1 activation and interleukin (IL)-1 release in an inflammation model, while IU1, a "clean" USP14 inhibitor, is inactive in the same model [119]. The existence of an additional, unknown DUB target for bAP15 cannot be ruled out, but the importance of cell-permeable DUB inhibitors with finely tuned poly-DUB pharmacology is evident. The available information regarding the structure of DUBs [79], and in particular the crystal structure of the 45 kDa catalytic domain of USP14 in isolation and complexed with UBQ aldehyde [120], should assist the rational design and synthesis of USP14 inhibitors with varying selectivity profiles.

3.4 RECAP

This chapter deals with small molecule modulators of neuropathological alterations caused by protein misfolding and aggregation in general, and by tau and/or tau-connected events in particular. A potential therapeutic mechanism was examined in detail in the biology-oriented companion book [27], and two targets were arbitrarily chosen. Thirty-eight scaffolds shown in Figures 3.1 to 3.5, acting on selected targets, are described in detail in this chapter, and are briefly summarized in Table 3.1.

TABLE 3.1   Compounds **3.1–3.37**: Chemical Class, Target, Developing Organization, Development Status

Number	Chemical cpd./class	Target	Organization	Dev. status
3.1a,b	Thioflavin S	Hsp70–BAG-1	Cancer Research, UK	DD
3.2	Pifithrin-m, PES	Hsp70, plus others	University of Pennsylvania	LO
3.3	Gambogic acid, GA	Hsp70–, Hsp90–CHIP regulation	Jangsu University, China	TM
3.4	Methylene blue, MB	Hsp70–CHIP regulation	TauRX Therapeutics	Ph III
3.5	Apoptozole	Hsp70, ATPase inhib.	Yonsei University, South Korea	LO
3.6	Nutlin-3	HDM2–p53	Roche	Ph I
3.7	Serdemetan	HDM2–p53	Johnson & Johnson	Ph I
3.8	AT-406	IAPs	Ascenta	Ph I
3.9	GDC-0152	IAPs	Genentech	Ph I
3.10	LCL-161	IAPs	Novartis	Ph II
3.11	TL32711	IAPs	Tetralogic	Ph II
3.12	Diamines, compound A	Skp2	University of North Carolina	DD
3.13	Alkylidene thiazolidines, compound C1	Skp2	NY University	DD
3.14	Diacids, SKP-I2	Cdc4	Mount Sinai Hosp., Toronto, Canada	DD
3.15	Benzodiazepindiones, LS-101	Synoviolin	Tokyo Medical Univ.	LO
3.16	Triazines, LS-102	RING E3 ligases	Tokyo Medical Univ.	LO
3.17	Tetracycles, SMER3	Met30	UCLA	LO
3.18	Thalidomide	Cereblon	Tokyo Institute of Technology	Ph III
3.19	Curcumin	Pan-DUB inhibition	University of Utah	TM
3.20	Shikoccin	Pan-DUB inhibition	University of Utah	DD
3.21	Δ12-PGJ2	Pan-DUB inhibition	Karolinska Institute	DD
3.22	Dienones, NSC 632839	Pan-DUB inhibition	Progenra	DD

TABLE 3.1   Compounds 3.1–3.37: Chemical Class, Target, Developing Organization, Development Status (cont.)

Number	Chemical cpd./class	Target	Organization	Dev. status
3.23	Bis-isothiocyanate, PR-419	Pan-DUB inhibition	Oldenburg University, Germany	DD
3.24a,b	Tricyclic dinitriles, HBX-41,108 (3.24a)	USP DUBs	Hybrigenics	LO
3.25	Gold complexes	Pan-DUB inhibition	University of Hong Kong	DD
3.26	Tyrposthin-like WP-1130	USP5, USP9x, USP14, UCH-L1, UCH37	University of Michigan	LO
3.27	Alkyliden-pyrazolidindiones, PYR41	Cys DUBs	University of Michigan	DD
3.28	Betulinic acid	Pan-DUB inhibition	University of Miami	PE
3.29	Chalcones, RA-9	UPS2, UPS5, UPS8, UCH-L1, UCH-L3	University of Minnesota	DD
3.30	Phthalimide based, pimozide	UAF1, USP7	University of Delaware	DD
3.31	Spautin-1	USP10, USP13	Chinese Academy of Sciences	DD
3.32	Thiophene based, P5091	USP7, USP47	Harvard Med. School	LO
3.33	Thiophene based	USP7, USP47	Progenra	LO
3.34	Aminotetrahydroacridines, HBX-19,818	USP7	Hybrigenics	Ph I
3.35	Naphthylamides, GRL0617	PLpro	University of Illinois	LO
3.36	Electrophilic dienones, NSC687852/b-AP15	USP14	Karolinska Institute	LO
3.37	Pyrrole based, IU1	USP14	Harvard Med. School	LO

Not progressed, NP; early discovery, DD; lead optimization, LO; preclinical evaluation, PE; clinical Phase I-II-III, Ph I–Ph III; marketed, MKTD; traditional medicine, TM. Please note that the most advanced status for NDD-targeted experiments is listed: for example, candidates in clinical trials for non-CNS indications with early in vitro characterization against proteinopathies/tauopathies are classified as DD.

The chemical core of each scaffold/compound is structurally defined; its molecular target is mentioned; the developing laboratory (either public or private) is listed; and the development status—according to publicly available information—is finally provided.

References

[1] Hershko, A.; Ciechanover, A. The ubiquitin system. *Annu. Rev. Biochem.* **1998**, *67*, 425–479.

[2] Weissman, A. M.; Shabek, N.; Ciechanover, A. The predator becomes the prey: regulating the ubiquitin system by ubiquitylation and degradation. *Nat. Rev. Mol. Cell Biol.* **2011**, *12*, 605–620.

[3] Goldstein, G.; Scheid, M.; Hammerling, U.; Boyse, E. A.; Schlesinger, D. H.; Niall, H. D. Isolation of a polypeptide that has lymphocyte-differentiating properties and is probably represented universally in living cells. *Proc. Natl. Acad. Sci. U.S.A.* **1975**, *72*, 11–15.

[4] Vijay-Kumar, S.; Bugg, C. E.; Cook, W. J. Structure of ubiquitin refined at 1.8 A resolution. *J. Mol. Biol.* **1987**, *194*, 531–544.

[5] Hoege, C.; Pfander, B.; Moldovan, G. L.; Pyrowolakis, G.; Jentsch, S. RAD6-dependent DNA repair is linked to modification of PCNA by ubiquitin and SUMO. *Nature* **2002**, *419*, 135–141.

[6] Carter, S.; Bischof, O.; Dejean, A.; Vousden, K. H. C-terminal modifications regulate MDM2 dissociation and nuclear export of p53. *Nat. Cell Biol.* **2007**, *9*, 428–435.

[7] Haglund, K.; Sigismund, S.; Polo, S.; Szymkiewicz, I.; Di Fiore, P. P.; Dikic, I. Multiple monoubiquitination of RTKs is sufficient for their endocytosis and degradation. *Nat. Cell Biol.* **2003**, *5*, 461–466.

[8] Komander, D.; Rape, R. The ubiquitin code. *Annu. Rev. Biochem.* **2012**, *81*, 203–229.

[9] Cook, W. J.; Jeffrey, L. C.; Carson, M.; Chen, Z.; Pickart, C. M. Structure of a diubiquitin conjugate and a model for interaction with ubiquitin conjugating enzyme (E2). *J. Biol. Chem.* **1992**, *267*, 16467–16471.

[10] Varadan, R.; Assfalg, M.; Haririnia, A.; Raasi, S.; Pickart, C.; Fushman, D. Solution conformation of Lys63-linked di-ubiquitin chain provides clues to functional diversity of polyubiquitin signaling. *J. Biol. Chem.* **2004**, *279*, 7055–7063.

[11] Kaiser, S. E.; Riley, B. E.; Shaler, T. A.; Trevino, R. S.; Becker, C. H.; Schulman, H.; Kopito, R. R. Protein standard absolute quantification (PSAQ) method for the measurement of cellular ubiquitin pools. *Nat. Methods* **2011**, *8*, 691–696.

[12] Finley, D. Recognition and processing of ubiquitin-protein conjugates by the proteasome. *Annu. Rev. Biochem.* **2009**, *78*, 477–513.

[13] Kim, W.; Bennett, E. J.; Huttlin, E. L.; Guo, A.; Li, J.; Possemato, A., et al. Systematic and quantitative assessment of the ubiquitin-modified proteome. *Mol. Cell* **2011**, *44*, 325–340.

[14] Finley, D. Inhibition of proteolysis and cell cycle progression in a multiubiquitination-deficient yeast mutant. *Mol. Cell. Biol.* **1994**, *14*, 5501–5509.

[15] Kulkarni, M.; Smith, H. E. E1 ubiquitin-activating enzyme UBA-1 plays multiple roles throughout *C. elegans* development. *PLoS Genet.* **2008**, *4*, e1000131.

[16] Gavin, J. M.; Chen, J. J.; Liao, H.; Rollins, N.; Yang, X.; Xu, Q., et al. Mechanistic studies on activation of ubiquitin and di-ubiquitin-like protein, FAT10, by ubiquitin-like modifier activating enzyme 6, Uba6. *J. Biol. Chem.* **2012**, *287*, 15512–15522.

[17] Bedford, L.; Lowe, J.; Dick, L. R.; Mayer, R. J.; Brownell, J. E. Ubiquitin-like protein conjugation and the ubiquitin-proteasome system as drug targets. *Nat. Rev. Drug Discov.* **2011**, *10*, 29–46.

[18] Burroughs, A. M.; Jaffee, M.; Iyer, L. M.; Aravind, L. Anatomy of the E2 ligase fold: implications for enzymology and evolution of ubiquitin/Ub-like protein conjugation. *J. Struct. Biol.* **2008**, *162*, 205–218.

[19] Li, W.; Bengtson, M. H.; Ulbrich, A.; Matsuda, A.; Reddy, V. A.; Orth, A., et al. Genome-wide and functional annotation of human E3 ubiquitin ligases identifies MULAN, a mitochondrial E3 that regulates the organelle's dynamics and signaling. *PLoS ONE* **2008**, *3*, e1487.

[20] Budhidarmo, R.; Nakatani, Y.; Day, C. L. RINGs hold the key to ubiquitin transfer. *Trends Biochem. Sci.* **2012**, *37*, 58–65.

[21] Rotin, D.; Kumar, S. Physiological functions of the HECT family of ubiquitin ligases. *Nat. Rev. Mol. Cell. Biol.* **2009**, *10*, 398–409.

[22] Wenzel, D. M.; Klevit, R. E. Following Ariadne's thread: a new perspective on RBR ubiquitin ligases. *BMC Biol.* **2012**, *10*, 24.

[23] Fraile, J. M.; Quesada, V.; Rodrıguez, D.; Freije, J. M. P.; Lopez-Otın, C. Deubiquitinases in cancer: new functions and therapeutic options. *Oncogene* **2012**, *31*, 2373–2388.

[24] Gallastegui, N.; Groll, M. The 26S proteasome: assembly and function of a destructive machine. *Trends Biochem. Sci.* **2010**, *35*, 634–642.

[25] Groll, M.; Dtizel, L.; Lowe, J.; Stock, D.; Bochtler, M.; Wolf, D. H.; Huber, R. Structure of the 20S proteasome from yeast at 2.4 A resolution. *Nature* **1997**, *386*, 463–471.

[26] Glickman, M. H.; Rubin, D. M.; Coux, O.; Wefes, I.; Pfeifer, G.; Cjeka, Z., et al. A subcomplex of the proteasome regulatory particle required for ubiquitin-conjugate degradation and related to the COP9-signalosome and eIF3. *Cell* **1998**, *94*, 615–623.

[27] Seneci, P. *Protein misfolding and neurodegenerative diseases: Focus on disease-modifying targets.* Academic Press, **2014**, 278 pages.

[28] Ballinger, C. A.; Connell, P.; Wu, Y.; Hu, Z.; Thompson, L. J.; Yin, L. Y.; Patterson, C. Identification of CHIP, a novel tetratricopeptide repeat-containing protein that interacts with heat shock proteins and negatively regulates chaperone functions. *Mol. Cell. Biol.* **1999**, *19*, 4535–4545.

[29] Murata, S.; Minami, Y.; Minami, M.; Chiba, T.; Tanaka, K. CHIP is a chaperone-dependent E3 ligase that ubiquitylates unfolded protein. *EMBO Rep.* **2001**, *2*, 1133–1138.

[30] Connell, P.; Ballinger, C. A.; Jiang, J.; Wu, Y.; Thompson, L. J.; Hohfeld, J.; Patterson, C. The co-chaperone CHIP regulates protein triage decisions mediated by heat-shock proteins. *Nat. Cell Biol.* **2001**, *3*, 93–96.

[31] Petrucelli, L.; Dickson, D.; Kehoe, K.; Taylor, J.; Snyder, H.; Grover, A., et al. CHIP and Hsp70 regulate tau ubiquitination, degradation and aggregation. *Human Mol. Genet.* **2004**, *13*, 703–714.

[32] Hatakeyama, S.; Matsumoto, M.; Kamura, T.; Murayama, M.; Chui, D. -H.; Planel, E., et al. U-box protein carboxyl terminus of Hsc70-interacting protein (CHIP) mediates polyUbiquitylation preferentially on four-repeat Tau and is involved in neurodegeneration of tauopathy. *J. Neurochem.* **2004**, *91*, 299–307.

[33] Liu, C. -W.; Jacobson, A. D. Functions of the 19S complex in proteasomal degradation. *Trends Biochem. Sci.* **2013**, *38*, 103–110.

[34] Todi, S. V.; Paulson, H. L. Balancing act: deubiquitinating enzymes in the nervous system. *Tr. Neurosci.* **2011**, *34*, 370–382.

[35] Kowalski, J. R.; Juo, P. The role of deubiquitinating enzymes in synaptic function and nervous system diseases. *Neural Plastic.* **2012**, *2012*, 892749.

[36] Cook, C.; Petrucelli, L. Tau triage decisions mediated by the chaperone network. *J. Alzheimer's Dis.* **2012**, *30*, 1–7.

[37] Deshaies, R. J.; Joazeiro, C. A. P. Ring domain E3 ubiquitin ligases. *Annu. Rev. Biochem.* **2009**, *78*, 399–434.

[38] Alberti, S.; Demand, J.; Esser, C.; Emmerich, N.; Schild, H.; Hohfeld, J. Ubiquitylation of BAG-1 suggests a novel regulatory mechanism during the sorting of chaperone substrates to the proteasome. *J. Biol. Chem.* **2002**, *277*, 45920–45927.

[39] Arndt, V.; Daniel, C.; Nastainczyk, W.; Alberti, S.; Höhfeld, J. BAG-2 acts as an inhibitor of the chaperone-associated ubiquitin ligase CHIP. *Mol. Biol. Cell* **2005**, *16*, 5891–5900.

[40] Gamerdinger, M.; Carra, S.; Behl, C. Emerging roles of molecular chaperones and co-chaperones in selective autophagy: focus on BAG proteins. *J. Mol. Med.* **2011**, *89*, 1175–1182.

[41] Kalia, L. V.; Kalia, S. K.; Chau, H.; Lozano, A. M.; Hyman, B. T.; McLean, P. J. Ubiquitinylation of α-synuclein by carboxyl terminus Hsp70-interacting protein (CHIP) is regulated by Bcl-2-associated athanogene 5 (BAG5). *PLoS ONE* **2011**, *6*, e14695.

[42] Alberti, S.; Böhse, K.; Arndt, V.; Schmitz, A.; Höhfeld, J. The cochaperone HspBP1 inhibits the CHIP ubiquitin ligase and stimulates the maturation of the cystic fibrosis transmembrane conductance regulator. *Mol. Biol. Cell* **2004**, *15*, 4003–4010.

[43] Shimamoto, S.; Kubota, Y.; Yamaguchi, F.; Tokumitsu, H.; Kobayashi, R. $Ca^{2+}/S100$ proteins act as upstream regulators of the chaperone-associated ubiquitin ligase CHIP (C Terminus of Hsc70-interacting Protein). *J. Biol. Chem.* **2013**, *288*, 7158–7168.

[44] Cichero, E.; Basile, A.; Turco, M. C.; Mazzei, M.; Fossa, P. Scouting new molecular targets for CFTR therapy: the HSC70/BAG-1 complex. A computational study. *Med. Chem. Res.* **2012**, *21*, 4430–4436.

[45] Sharp, A.; Crabb, S. J.; Johnson, P. W. M.; Hague, A.; Cutress, R.; Townsend, P. A., et al. Thioflavin S (NSC71948) interferes with Bcl-2-associated athanogene (BAG-1)-mediated protein-protein interactions. *J. Pharmacol. Exp. Ther.* **2009**, *331*, 680–689.

[46] Leu, J. I.; Pimkina, J.; Frank, A.; Murphy, M. E.; George, D. L. A small molecule inhibitor of inducible heat shock protein 70. *Mol. Cell* **2009**, *36*, 15–27.

[47] Wang, J.; Zhao, Q.; Qi, Q.; Gu, H. Y.; Rong, J. J.; Mu, R., et al. Gambogic acid-induced degradation of mutant p53 is mediated by proteasome and related to CHIP. *J. Cell. Biochem.* **2011**, *112*, 509–519.

[48] Wischik, C. TauRX Therapeutics, Sept 10th, 2012. Press release announcing the initiation of a global Phase 3 clinical trial in a type of Frontotemporal Dementia (FTD) also known as Pick's Disease.

[49] Wang, A. M.; Morishima, Y.; Clapp, K. M.; Peng, H. -M.; Pratt, W. B.; Gestwicki, J. E., et al. Inhibition of Hsp70 by methylene blue affects signaling protein function and ubiquitination and modulates polyglutamine protein degradation. *J. Biol. Chem.* **2010**, *285*, 15714–15723.

[50] Cho, H. J.; Gee, H. Y.; Baek, K. -H.; Ko, S. -K.; Park, J. -M.; Lee, H., et al. A small molecule that binds to an ATPase domain of Hsc70 promotes membrane trafficking of mutant cystic fibrosis transmembrane conductance regulator. *J. Am. Chem. Soc.* **2011**, *133*, 20267–20276.

[51] Wang, L.; Liu, Y. T.; Hao, R.; Chen, L.; Chang, Z.; Wang, H. R., et al. Molecular mechanism of the negative regulation of Smad1/5 protein by carboxyl terminus of Hsc70-interacting protein (CHIP). *J. Biol. Chem.* **2011**, *286*, 15883–15894.

[52] Tovar, C.; Rosinski, J.; Filipovic, Z.; Higgins, B.; Kolinsky, K.; Hilton, H., et al. Small-molecule MDM2 antagonists reveal aberrant p53 signaling in cancer: Implications for therapy. *Proc. Natl. Acad. Sci. U.S.A.* **2006**, *103*, 1888–1893.

[53] Tabernero, J.; Dirix, L.; Schoffski, P.; Cervantes, A.; Lopez-Martin, J. A.; Capdevila, J., et al. A Phase I first-in-human pharmacokinetic and pharmacodynamic study of serdemetan in patients with advanced solid tumors. *Clin. Cancer Res.* **2011**, *17*, 6313–6321.

[54] Millard, M.; Pathania, D.; Grande, F.; Xu, S.; Neamati, N. Small molecule inhibitors of p53-MDM2 interaction: the 2006-2010 update. *Curr. Pharm. Des.* **2011**, *17*, 536–559.

[55] Haupt, Y.; Maya, R.; Kazaz, A.; Oren, M. Mdm2 promotes the rapid degradation of p53. *Nature* **1997**, *387*, 296–299.

[56] Cai, Q.; Sun, H.; Peng, Y.; Lu, J.; Nikolovska-Coleska, Z.; McEachern, D., et al. A potent and orally active antagonist (SM-406/AT-406) of multiple Inhibitor of Apoptosis Proteins (IAPs) in clinical development for cancer treatment. *J. Med. Chem.* **2011**, *54*, 2714–2726.

[57] Flygare, J. A.; Beresini, M.; Budha, N.; Chan, H.; Chan, I. T.; Cheeti, S., et al. Discovery of a potent small-molecule antagonist of inhibitor of apoptosis (IAP) proteins

and clinical candidate for the treatment of cancer (GDC-0152). *J. Med. Chem.* **2012**, *55*, 4101–4113.

[58] Weisberg, E.; Ray, A.; Barrett, R.; Nelson, E.; Christie, A. L.; Porter, D., et al. Smac mimetics: implications for enhancement of targeted therapies in leukemia. *Leukemia* **2011**, *24*, 2100–2109.

[59] Condon, S.M.; Deng, Y.; LaPorte, M.G.; Rippin, S.R. Smac mimetic. US 2011/0003877 A1.

[60] Wu, G.; Chai, J.; Suber, T. L.; Wu, J. -W.; Du, C.; Wang, X.; Shi, Y. Structural basis of IAP recognition by Smac/DIABLO. *Nature* **2000**, *408*, 1008–1012.

[61] Sun, H.; Nikolovska-Coleska, Z.; Lu, J.; Qiu, S.; Yang, C. -Y.; Gao, W., et al. Design, synthesis, and evaluation of a potent, cell-permeable, conformationally constrained second mitochondria derived activator of caspase (Smac) mimetic. *J. Med. Chem.* **2006**, *49*, 7916–7920.

[62] Varfolomeev, E.; Blankenship, J. W.; Wayson, S. M.; Fedorova, A. V.; Kayagaki, N.; Garg, P., et al. IAP antagonists induce autoubiquitination of c-IAPs, NF-κB activation, and TNFα-dependent apoptosis. *Cell* **2007**, *131*, 669–681.

[63] Goldenberg, S. J.; Marblestone, J. G.; Mattern, M. R.; Nicholson, B. Strategies for the identification of ubiquitin ligase inhibitors. *Biochem. Soc. Trans.* **2010**, *38*, 132–136.

[64] Nakayama, K. I.; Nakayama, K. Regulation of the cell cycle by SCF-type ubiquitin ligases. *Semin. Cell. Dev. Biol.* **2005**, *16*, 323–333.

[65] Chen, Q.; Xie, W.; Kuhn, D. J.; Voorhees, P. M.; Lopez-Girona, A.; Mendy, D., et al. Targeting the p27 E3 ligase SCFSkp2 results in p27- and Skp2-mediated cell-cycle arrest and activation of autophagy. *Blood* **2008**, *111*, 4690–4699.

[66] Filipits, M.; Pohl, G.; Stranzl, T.; Kaufmann, H.; Ackermann, J.; Gisslinger, H., et al. Low p27Kip1 expression is an independent adverse prognostic factor in patients with multiple myeloma. *Clin. Cancer Res.* **2003**, *9*, 820–826.

[67] Wu, L.; Grigoryan, A. V.; Li, Y.; Hao, B.; Pagano, M.; Cardozo, T. J. Specific small molecule inhibitors of Skp2-mediated p27 degradation. *Chem. Biol.* **2012**, *19*, 1515–1524.

[68] Orlicky, S.; Tang, X.; Neduva, V.; Elowe, N.; Brown, E. D.; Sicheri, F.; Tyers, M. An allosteric inhibitor of substrate recognition by the SCF(Cdc4) ubiquitin ligase. *Nat. Biotechnol.* **2010**, *28*, 733–737.

[69] Yagishita, N.; Aratani, S.; Leach, C.; Amano, T.; Yamano, Y.; Nakatani, K., et al. RING-finger type E3 ubiquitin ligase inhibitors as novel candidates for the treatment of rheumatoid arthritis. *Int. J. Mol. Med.* **2012**, *30*, 1281–1286.

[70] Amano, T.; Yamasaki, S.; Yagishita, N.; Tsuchimochi, K.; Shin, H.; Kawahara, K. -I., et al. Synoviolin/Hrd1, an E3 ubiquitin ligase, as a novel pathogenic factor for arthropathy. *Genes Dev.* **2003**, *17*, 2436–2449.

[71] Aghajan, M.; Jonai, N.; Flick, K.; Fu, F.; Luo, M.; Cai, X., et al. Chemical genetics screen for enhancers of rapamycin identifies a specific inhibitor of an SCF family E3 ubiquitin ligase. *Nat. Biotech.* **2010**, *28*, 738–742.

[72] Ito, T.; Ando, H.; Suzuki, T.; Ogura, T.; Hotta, K.; Imamura, Y., et al. Identification of a primary target of thalidomide teratogenicity. *Science* **2010**, *327*, 1345–1350.

[73] Zhang, M.; Windheim, M.; Roe, S. M.; Peggie, M.; Cohen, P.; Prodromou, C.; Pearl, L. H. Chaperoned ubiquitylation-crystal structures of the CHIP U box E3 ubiquitin ligase and a CHIP-Ubc13-Uev1a complex. *Mol. Cell* **2005**, *20*, 525–538.

[74] Xu, Z.; Kohli, E.; Devlin, K. I.; Bold, M.; Nix, J. C.; Misra, S. Interactions between the quality control ubiquitin ligase CHIP and ubiquitin conjugating enzymes. *BMC Struct. Biol.* **2008**, *8*, 26.

[75] Graf, C.; Stankiewicz, M.; Nikolay, R.; Mayer, M. P. Insights into the conformational dynamics of the E3 ubiquitin ligase CHIP in complex with chaperones and E2 enzymes. *Biochemistry* **2010**, *49*, 2121–2129.

[76] Soss, S. E.; Yue, Y.; Dhe-Paganon, S.; Chazin, W. J. E2 conjugating enzyme selectivity and requirements for function of the E3 ubiquitin ligase CHIP. *J. Biol. Chem.* **2011**, *286*, 21277–21286.

[77] Windheim, M.; Peggie, M.; Cohen, P. Two different classes of E2 ubiquitin-conjugating enzymes are required for the monoubiquitination of proteins and elongation by polyubiquitin chains with a specific topology. *Biochem. J.* **2008**, *409*, 723–729.

[78] Pruneda, J. N.; Littlefield, P. J.; Soss, S. E.; Nordquist, K. A.; Chazin, W. J.; Brzovic, P. S.; Klevit, R. E. Structure of an E3:E2~Ub complex reveals an allosteric mechanism shared among RING/U-box ligases. *Mol. Cell* **2012**, *47*, 933–942.

[79] Komander, D.; Clague, M. J.; Urbe, S. Breaking the chains: structure and function of the deubiquitinases. *Nat. Rev. Mol. Cell Biol.* **2009**, *10*, 550–563.

[80] Faesen, A. C.; Luna-Vargas, M. P. A.; Sixma, T. K. The role of UBL domains in ubiquitin-specific proteases. *Biochem. Soc. Trans.* **2012**, *40*, 539–545.

[81] Daviet, L.; Colland, F. Targeting ubiquitin specific proteases for drug discovery. *Biochimie* **2008**, *90*, 270–283.

[82] Colland, F. The therapeutic potential of deubiquitinating enzyme inhibitors. *Biochem. Soc. Trans.* **2010**, *38*, 137–143.

[83] Mattern, M. R.; Wu, J.; Nicholson, B. Ubiquitin-based anticancer therapy: Carpet bombing with proteasome inhibitors vs surgical strikes with E1, E2, E3, or DUB inhibitors. *Biochim. Biophys. Acta* **1823**, *2012*, 2014–2021.

[84] Marchenko, N. D.; Wolff, S.; Erster, S.; Becker, K.; Moll, U. M. Monoubiquitylation promotes mitochondrial p53 translocation. *EMBO J.* **2007**, *26*, 923–934.

[85] Ovaa, H.; Kessler, B. M.; Rolen, U.; Galardy, P. J.; Ploegh, H. L.; Masucci, M. G. Activity-based ubiquitin-specific protease (USP) profiling of virus-infected and malignant human cells. *Proc. Natl. Acad. Sci. U.S.A.* **2004**, *101*, 2253–2258.

[86] Mullally, J. E.; Fitzpatrick, F. A. Pharmacophore model for novel inhibitors of ubiquitin isopeptidases that induce p53-independent cell death. *Mol. Pharmacol.* **2002**, *62*, 351–358.

[87] Mullally, J. E.; Moos, P. J.; Edes, K.; Fitzpatrick, F. A. Cyclopentenone prostaglandins of the J series inhibit the ubiquitin isopeptidase activity of the proteasome pathway. *J. Biol. Chem.* **2001**, *276*, 30366–30373.

[88] Li, Z.; Melandri, F.; Berdo, I.; Jansen, M.; Hunter, L.; Wright, S., et al. Delta12-Prostaglandin J2 inhibits the ubiquitin hydrolase UCH-L1 and elicits ubiquitin-protein aggregation without proteasome inhibition. *Biochem. Biophys. Res. Commun.* **2004**, *319*, 1171–1180.

[89] Nicholson, B.; Leach, C. A.; Goldenberg, S. J.; Francis, D. M.; Kodrasov, M. P.; Tian, X., et al. Characterization of ubiquitin and ubiquitin-like-protein isopeptidase activities. *Prot. Sci.* **2008**, *17*, 1035–1043.

[90] Altun, M.; Kramer, H. B.; Willems, L. I.; McDermott, J. L.; Leach, C. A.; Goldenberg, S. J., et al. Activity-based chemical proteomics accelerates inhibitor development for deubiquitylating enzymes. *Chem. Biol.* **2011**, *18*, 1401–1412.

[91] Seiberlich, V.; Goldbaum, O.; Zhukareva, V.; Richter-Landsberg, C. The small molecule inhibitor PR-619 of deubiquitinating enzymes affects the microtubule network and causes protein aggregate formation in neural cells: Implications for neurodegenerative diseases. *Biochim. Biophys. Acta* **1823**, *2012*, 2057–2068.

[92] Colland, F.; Formstecher, E.; Jacq, X.; Reverdy, C.; Planquette, C.; Conrath, S., et al. Small molecule inhibitor of USP7/HAUSP ubiquitin protease stabilizes and activates p53 in cells. *Mol. Cancer Ther.* **2009**, *8*, 2286–2295.

[93] Reverdy, C.; Conrath, S.; Lopez, R.; Planquette, C.; Atmanene, C.; Collura, V., et al. Discovery of specific inhibitors of human USP7/HAUSP deubiquitinating enzyme. *Chem. Biol.* **2012**, *19*, 467–477.

[94] Colombo, M.; Vallese, S.; Peretto, I.; Jacq, X.; Rain, J. -C.; Colland, F.; Guedat, P. Synthesis and biological evaluation of 9-oxo-9H-indeno[1,2-b]pyrazine-2,3-dicarbonitrile analogues as potential inhibitors of deubiquitinating enzymes. *ChemMedChem* **2010**, *5*, 552–558.

[95] Zhang, J. -J.; Ng, K. -M.; Lok, C. -N.; Wai-Yin Sun, R.; Che, C. M. Deubiquitinases as potential anti-cancer targets for gold(III) complexes. *Chem. Commun.* **2013**, *49*, 5153–5155.

[96] Peng, Z.; Pal, A.; Han, D.; Wang, S.; Maxwell, D.; Levitzki, A., et al. Tyrphostin-like compounds with ubiquitin modulatory activity as possible therapeutic agents for multiple myeloma. *Bioorg. Med. Chem.* **2011**, *19*, 7194–7204.

[97] Kapuria, V.; Peterson, L. F.; Fang, D.; Bornmann, W. G.; Talpaz, M.; Donato, N. J. Deubiquitinase inhibition by small-molecule WP1130 triggers aggresome formation and tumor cell apoptosis. *Cancer Res.* **2010**, *70*, 9265–9276.

[98] Kapuria, V.; Levitzki, A.; Bornmann, W. G.; Maxwell, D.; Priebe, W.; Sorenson, R. J., et al. A novel small molecule deubiquitinase inhibitor blocks Jak2 signaling through Jak2 ubiquitination. *Cell. Signalling* **2011**, *23*, 2076–2085.

[99] Burkholder, K. M.; Perry, J. W.; Wobus, C. E.; Donato, N. J.; Showalter, H. D.; Kapuria, V.; O'Riordan, M. X. D. A small molecule deubiquitinase inhibitor increases localization of inducible nitric oxide synthase to the macrophage phagosome and enhances bacterial killing. *Infect. Immun.* **2011**, *79*, 4850–4857.

[100] Perry, J. W.; Ahmed, M.; Chang, K. -O.; Donato, N. J.; Showalter, H. D.; Wobus, C. E. Antiviral activity of a small molecule deubiquitinase inhibitor occurs via induction of the unfolded protein response. *PLoS Pathogens* **2012**, *8*, e1002783.

[101] Yang, Y.; Kitagaki, J.; Dai, R. -M.; Tsai, Y. C.; Lorick, K. L.; Ludwig, R. L., et al. Inhibitors of ubiquitin-activating enzyme (E1), a new class of potential cancer therapeutics. *Cancer Res.* **2007**, *67*, 9472–9481.

[102] Kapuria, V.; Peterson, L. F.; Showalter, H. D. H.; Kirchhoff, P. D.; Talpaz, M.; Donato, N. J. Protein cross-linking as a novel mechanism of action of a ubiquitin-activating enzyme inhibitor with anti-tumor activity. *Biochem. Pharmacol.* **2011**, *82*, 341–349.

[103] Reiner, T.; Parrondo, R.; de las Pozas, A.; Palenzuela, D.; Perez-Stable, C. Betulinic acid selectively increases protein degradation and enhances prostate cancer-specific apoptosis: Possible role for inhibition of deubiquitinase activity. *PLoS ONE* **2013**, *8*, e56234.

[104] Bazzaro, M.; Anchoori, R. K.; Mudiam, M. K. R.; Issaenko, O.; Kumar, S.; Karanam, B., et al. α,β-Unsaturated carbonyl system of chalcone-based derivatives is responsible for broad inhibition of proteasomal activity and preferential killing of human papilloma virus (HPV) positive cervical cancer cells. *J. Med. Chem.* **2011**, *54*, 449–456.

[105] Issaenko, O. A.; Amerik, A. Y. Chalcone-based small-molecule inhibitors attenuate malignant phenotype via targeting deubiquitinating enzymes. *Cell Cycle* **2012**, *11*, 1804–1817.

[106] Chen, J.; Dexheimer, T. S.; Ai, Y.; Liang, Q.; Villamil, M. A.; Inglese, J., et al. Selective and cell-active inhibitors of the USP1/UAF1 deubiquitinase complex reverse cisplatin resistance in non-small cell lung cancer cells. *Chem. Biol.* **2011**, *18*, 1390–1400.

[107] Liu, J.; Xia, H.; Kim, M.; Xu, L.; Li, Y.; Zhang, L., et al. Beclin1 controls the levels of p53 by regulating the deubiquitination activity of USP10 and USP13. *Cell* **2011**, *147*, 223–234.

[108] Chauhan, D.; Tian, Z.; Nicholson, B.; Suresh Kumar, K. G.; Zhou, B.; Carrasco, R., et al. A small molecule inhibitor of ubiquitin-specific protease-7 induces apoptosis in multiple myeloma cells and overcomes bortezomib resistance. *Cancer Cell* **2012**, *22*, 345–358.

[109] Weinstock, J.; Wu, J.; Cao, P.; Kingsbury, W. D.; McDermott, J. L.; Kodrasov, M. P., et al. Selective dual inhibitors of the cancer-related deubiquitylating proteases USP7 and USP47. *ACS Med. Chem. Lett.* **2012**, *3*, 789–792.

[110] Ratia, K.; Pegan, S.; Takayama, J.; Sleeman, K.; Coughlin, M.; Baliji, S., et al. A noncovalent class of papain-like protease/deubiquitinase inhibitors blocks SARS virus replication. *Proc. Natl. Acad. Sci. U.S.A.* **2008**, *105*, 16119–16124.

[111] Erdal, H.; Berndtsson, M.; Castro, J.; Brunk, U.; Shoshan, M. C.; Linder, S. Induction of lysosomal membrane permeabilization by compounds that activate p53-independent apoptosis. *Proc. Natl. Acad. Sci. U.S.A.* **2005**, *102*, 192–197.

[112] Berndtsson, M.; Beaujouin, M.; Rickardson, L.; Mandic Havelka, A.; Larsson, R.; Westman, J., et al. Induction of the lysosomal apoptosis pathway by inhibitors of the ubiquitin-proteasome system. *Int. J. Cancer* **2009**, *124*, 1463–1469.

[113] D'Arcy, P.; Brnjic, S.; Hägg Olofsson, M.; Fryknäs, M.; Lindsten, K.; De Cesare, M., et al. Inhibition of proteasome deubiquitinating activity as a new cancer therapy. *Nat. Med.* **2011**, *17*, 1636–1640.

[114] D'Arcy, P.; Linder, S. Proteasome deubiquitinases as novel targets for cancer therapy. *Int. J. Biochem. Cell Biol.* **2012**, *44*, 1729–1738.

[115] Lee, B. -H.; Lee, M. J.; Park, S.; Oh, D. -C.; Elsasser, S.; Chen, P. C., et al. Enhancement of proteasome activity by a small-molecule inhibitor of USP14. *Nature* **2010**, *467*, 179–184.

[116] Nag, D. K.; Finley, D. A small-molecule inhibitor of deubiquitinating enzyme USP14 inhibits Dengue virus replication. *Virus Res.* **2012**, *165*, 103–106.

[117] Jin, Y. N.; Chen, P. -C.; Watson, J. A.; Walters, B. J.; Phillips, S. E.; Green, K., et al. USP14 deficiency increases tau phosphorylation without altering tau degradation or causing tau-dependent deficits. *PLoS ONE* **2012**, *7*, e47884.

[118] Ponnappan, S.; Palmieri, M.; Sullivan, D. H.; Ponnappan, U. Compensatory increase in USP14 activity accompanies impaired proteasomal proteolysis during aging. *Mechan. Ageing Dev.* **2013**, *134*, 53–59.

[119] Lopez-Castejon, G.; Luheshi, N. M.; Compan, V.; High, S.; Whitehead, R. C.; Flitsch, S., et al. Deubiquitinases regulate the activity of caspase-1 and interleukin-1β secretion via assembly of the inflammasome. *J. Biol. Chem.* **2013**, *288*, 2721–2733.

[120] Hu, M.; Li, P.; Song, L.; Jeffrey, P. D.; Chernova, T. A.; Wilkinson, K. D., et al. Structure and mechanisms of the proteasome-associated deubiquitinating enzyme USP14. *EMBO J.* **2005**, *24*, 3747–3756.

4

Targeting Unselective Autophagy of Cellular Aggregates
A Shotgun Approach

4.1 MACROAUTOPHAGY MEDIATED DEGRADATION OF PROTEIN AGGREGATES

Autophagy is the self-degradation of cellular components spanning between proteins and whole organelles. Its crucial role has been fully appreciated for almost a decade [1,2].

Non-selective autophagy deals, *inter alia*, with long-lived proteins and insoluble/ubiquitin–proteasome system (UPS)-resistant aggregates [3]. *Macroautophagy* (*MA* from now on [4]) is the most common non-selective autophagic process. MA is a cellular process induced by stress stimuli [5].

Five MA steps—and the putative targets implied—can be identified. Initiation of autophagy, nucleation to form phagophores/isolation membranes, expansion to form autophagosomes (APs), and finally AP maturation and fusion to yield fully degradative autolysosomes and complete MA [6–8]. *Initiation* starts at phagophore assembly sites (PASs [9]), and entails the assembly of vesicles and cellular material into pre-APs. It is achieved through the action of the UNC-51-like kinase (ULK) complex [10], whose translocation creates cytoplasmic PASs. *Nucleation* depends on the PAS-bound class III phosphatidylinositol-3-kinase (PI3K) complex [11] to produce a phagophore/isolation membrane. The PI3K complex phosphorylates phosphatidylinositol (PI) to yield phosphatidylinositol-3-phosphate (PI3P), a promoter of membrane nucleation [12]. *Expansion* entails transport and incorporation of membrane material into the phagophore [13]. The *trans*-membrane protein carrier mammalian autophagy-related gene 9 (mAtg9) protein promotes ULK complex-dependent cycling

Chemical Modulators of Protein Misfolding and Neurodegenerative Disease. http://dx.doi.org/10.1016/B978-0-12-801944-3.00004-7

of mAtg9-containing vescicles between the *trans*-Golgi network and the endosomes (ESs) [14]. *Elongation* happens through two ubiquitin (UBQ)-like (UBL) conjugating systems [15]. Atg5–Atg12–Atg16 multimeric complexes selectively associate to the pre-autophagosomal membrane, with a preference for PI3P-containing membranes [16]. Their role in elongation is due to the Atg12 UBL-like protein, and to two UBL domains on Atg5, that recruit and activate components needed for phagophore expansion and closure [17]. Among them, the complex centered around Atg8-like UBL proteins, such as the light chain 3 (LC3) protein [18]. The complex conjugates Atg8-like UBL proteins with the membrane phospholipid phosphatidylethanolamine (PE), forming an LC3-II conjugate [19]. LC3-II is an essential element in phagophore expansion and AP formation. *Maturation* initiates the removal of LC3-II from the outer surface of mature APs by the protease Atg4 [20]. Then, fully formed APs enter a multi-step fusion process with one or more endosomal vescicles to form an amphisome [21]. The final fusion step merges amphisomes with dense lysosomes (LSs), with the formation of fully degradative autolysosomes [22].

A thorough description of non-selective autophagy can be accessed in the biology-oriented companion book [23]. Any enzyme, receptor or protein complex involved in non-selective autophagy may be considered a suitable target to modulate autophagic activity in an impaired cellular environment. The vast majority of them have not been targeted yet in drug discovery projects, possibly because more information is needed before rationally targeting a single component of an extremely complex biochemical pathway.

The induction of autophagy by inhibition of the *mammalian target of rapamycin/mTOR complex 1* (*mTORC1*) is a validated molecular target in autophagy. mTORC1 small molecule inhibitors are described in detail here, together with *small molecule enhancers of rapamycin* (*SMERs*) identified through phenotypic high-throughput screening (HTS) campaigns.

4.2 mTORC1

The mTOR story starts with the naturally occurring macrolide rapamycin (**4.1a**, Figure 4.1) [24]. The identification of the molecular target of rapamycin/mTOR in yeast [25] and humans [26], and the characterization of the molecular interactions between rapamycin and mTOR, are the gateway to the rational design of mTOR inhibitors with varying degrees of selectivity. A number of semi-synthetic rapamycin analogs, or rapalogs (**4.1b–f**), are currently marketed in several disease areas.

Rapamycin interacts with the immunophilin 12 kDa FK506 binding protein (FKBP12) in mammalian cells. The FKBP12–rapamycin complex then binds the FKBP–rapamycin-binding (FRB) domain in the mTOR

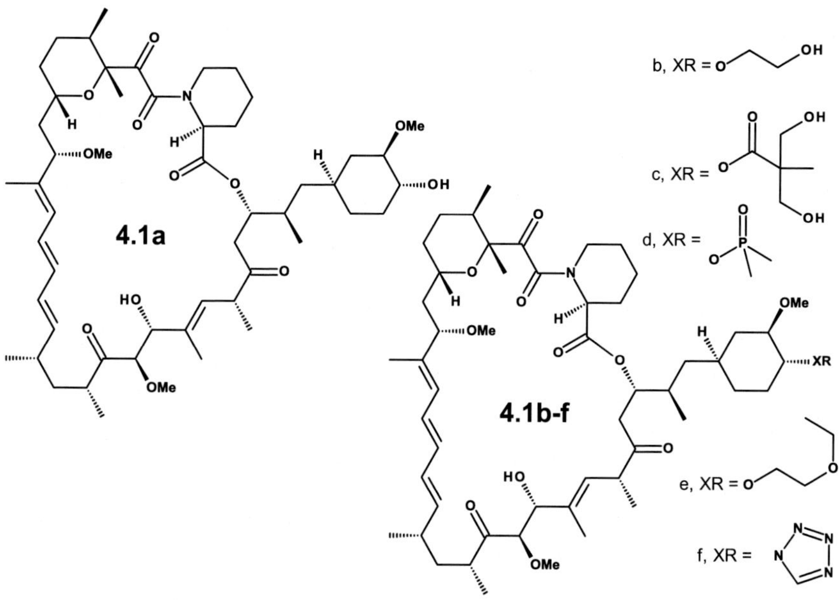

FIGURE 4.1 Rapamycin and rapalogs: chemical structures, **4.1a–f**.

kinase [27]. Thus, rapamycin is an allosteric mTOR inhibitor. The FKBP12–rapamycin complex binds to and causes the dissociation of the mTORC1 dimeric complex through a stepwise process [28]. A first binding event weakens the mTORC1 dimer complex and sterically hinders the binding/phosphorylation of bulkier mTORC1 substrates, such as S6 kinase 1 (S6K1). Either a second binding event or a first event-dependent, slower dissociation of the mTORC1 complex leads to the inhibition of binding/phosphorylation of the eukaryotic initiation factor 4E-binding protein 1 (EIF4E-BP1). Thus, rapamycin differentially inhibits the kinase activity of mTORC1 on various substrates [28].

The FKBP12–rapamycin complex does not bind to the mTORC2 complex, due to the inaccessibility of its FRB domain when mTORC2 is assembled [29]. Long-term exposure to rapamycin causes cell type-dependent inhibition of mTORC2 kinase activity, possibly because the FKBP12–rapamycin complex binds to free mTOR kinase and makes it unavailable for the formation of mTORC2. Some cells retain a limited concentration of mTORC2 complex and carry out mTORC2-dependent phosphorylation of Akt/protein kinase B even when exposed to rapamycin for days [29].

Rapamycin (sirolimus, Rapamune™, **4.1a**) has been marketed since 1999 to prevent rejection after kidney transplantation [30]. Its chronic use is allowed by its low kidney toxicity, when compared to calcineurin inhibitors [31]. Its use as a coating agent for coronary stents against restenosis has lasted several years [32]. *In vivo* testing of rapamycin shows therapeutic effects against cancer [33], obesity [34], cerebral ischemia [35], food allergy [36], muscular dystrophy [37], autism [38], and kidney diseases [39]. Anti-aging/neuroprotectant, autophagy induction-dependent effects of rapamycin and rapalogs are observed *in vivo* [40]. Rapamycin is neuroprotective *in vitro* and *in vivo* in acute 1-methyl-4-phenyl-1,2,3,6-tetrahydropyridine (MPTP) mice models of *Parkinson's disease (PD)*, and is more effective than pan-TOR inhibitors—see below [41]. Rapamycin restores the autophagic flux, attenuates dopaminergic cell death and terminal loss in the *substantia nigra* of MPTP mice [42]. Rapamycin attenuates levodopa-induced diskynesia in 6-OH dopamine (6-OHDA)-treated mice, through reduction of mTOR hyperactivity in D1 receptor-containing neurons [43]. It protects mitochondria from oxidative stress and apoptosis in a rat model of PD [44]. Autophagy induction caused by rapamycin contributes to the degradation of overexpressed, neurotoxic, PD-associated α-synuclein in a transgenic (TG) mouse model [45]. Inhibition of mTOR by rapamycin induces autophagy and reduces the toxicity of polyQ expansions in fly and mouse models of *Huntington disease (HD)* [46], and in other polyQ and polyA disease models [47,48]. Rapamycin inhibits the aggregation of mutant huntingtin fragments in a cellular system through inhibition of protein synthesis, as a result of mTOR-dependent and -independent molecular mechanisms [49]. Rapamycin prevents neurotoxicity and cognition impairments in a TG mouse model carrying a transactive response/TAR DNA-binding protein 43 (TDP-43) transgene related to *frontotemporal lobar dementia (FTLD)* and *amyotrophic lateral sclerosis (ALS)* [50]. It activates autophagy and reverses the pathological phenotype in models of *Hutchinson–Gilford progeria syndrome (HGPS)* [51]. Namely, increased autophagy reduces accumulation of progerin, an alternate splicing form of lamin A/C. Rapamycin treatment clears insoluble toxic aggregates and prevents their effects (reduction of HGPS-associated nuclear blebbing, growth inhibition, epigenetic dysregulation, and genomic instability) [52]. Conversely, rapamycin promotes motor neuron degeneration in another ALS mouse model and does not induce autophagy-driven elimination of mutant superoxide dismutase (SOD1) [53].

Rapamycin shows significant effects—mostly neuroprotective, seldom neurotoxic—on age-dependent cognitive degeneration (*Alzheimer's disease/ AD* in particular) in animal models [54,55]. Chronic feeding with encapsulated rapamycin delays cognitive decline associated with aging in old C57BL/6 mice, has anxiolytic and antidepressant activity, and stimulates monoamine pathways in brain through autophagy induction [56]. Two-month-old C57BL/6 mice chronically treated (16 months) with encapsulated rapamycin show significantly better spatial learning and memory compared to age-matched mice on control diet [57]. Rapamycin does not improve cognition when given to 15-month-old C57BL/6 mice with pre-existing, age-dependent learning and memory deficits. Rapamycin-mediated decrease in interleukin (IL-1β) levels, and increase in N-methyl-D-aspartate (NMDA) signaling contribute to the observed learning and memory improvements [57]. Antioxidant [44,58] and anti-inflammatory [59] properties contribute to neuroprotective effects of rapamycin in AD models. mTORC1, and consequently rapamycin, regulate long-term synaptic plasticity in the hippocampus [60]. Contradictory, acute/neurotoxic, and chronic/neuroprotectant effects on synaptic plasticity are observed following treatment with rapamycin of mutant KM670/671NL amyloid precursor protein (APP)-carrying TG Tg2576 mice recapitulating AD [61]. This apparent dichotomy mirrors acute, detrimental effects of rapamycin in mice observed during early stages of treatment—up to 6 weeks [62]. Prolonged rapamycin treatment—around 20 weeks or more—leads to beneficial metabolic alterations, contributing to life extension [62].

Rapamycin, through inhibition of mTORC1/activation of autophagy, has a beneficial influence on amyloid β (Aβ)-driven pathologies. Physiological Aβ production and aggregation is taken care of by autophagy [63], which is impaired in AD [64–66]. Thus, Aβ accumulates and aggregates in AD brains without being removed [67], and causes mTOR hyperactivation and signaling through a positive feedback loop [68]. Rapamycin treatment decreases $A\beta_{42}$ levels and rescues cognitive functions in a mutant V717I APP-carrying TG mouse model [69]. It acts through autophagy induction, as autophagy is not induced and endogenous Aβ levels are not reduced in wild type (WT) mice [69]. Rapamycin decreases $A\beta_{42}$ levels (without changing soluble, non-toxic $A\beta_{40}$ levels) and rescues cognitive functions also in a triple TG (APP–tau–presenilin 1 (PS1) mutations) mouse model through mTOR inhibition and autophagy induction [70]. Autophagy up-regulation by rapamycin in $A\beta_{42}$-treated cell lines is mediated by beclin-1 [71]. Conversely, rapamycin increases the sensitivity of neural cells to Aβ neurotoxicity [72]. It enhances amyloidogenic processing of APP/increases Aβ levels in the brains of Tg2576 TG mice, due to the decreased activation of the Aβ-processing α-secretase disintegrin and metalloproteinase domain 10 (ADAM10) [73]. As rapamycin treatment is acute (2 weeks), the observed detrimental effects could be reversed by a prolonged rapamycin treatment.

Rapamycin acts on tau *in vitro* and *in vivo*, through mTORC1-dependent and -independent effects [54,55,74]. It acts *in vitro* as expected from an mTORC1 inhibitor on tau phosphorylation [75–77]. Rapamycin decreases tau levels (total and phosphorylated) in a cellular assay [78], possibly acting on translation and autophagy. It reduces tau neurotoxicity in R406W mutant tau-carrying *Drosophila* flies [47]. In addition to cognitive recovery and beneficial effects against Aβ-caused pathologies in AD, rapamycin reduces soluble and hyperphosphorylated (HP)-tau levels in the hippocampal region of triple APP–tau–PS1 TG mice [70]. Chronic rapamycin treatment (6 months) of 2-month-old, mutant P301L tau-carrying TG mice rescues tau-dependent motor deficits (open field activity test, rotarod) [79]. Rapamycin treatment reduces tau levels, particularly in the hippocampus, cortex and brain stem. Phosphorylation of tau is particularly reduced on Thr181, Ser202/Thr205, and Ser232/Thr235 residues [79]. Rapamycin reduces p-Thr389 levels on p70 S6K1 and p-Ser65 on EIF4E-BP1, through reduction of mTORC1 activity in the same brain areas. p-Ser9 levels on glycogen synthase kinase 3β (GSK-3β) are also reduced, linking mTORC1 inhibition- and GSK-3β inhibition-mediated reduction of HP-tau [79]. The same mice, submitted to a 5-month long term preventive (3-week-old, asymptomatic mice) and to a 1.5-month short term therapeutic (3-month-old, diseased mice) treatment with rapamycin, show similar reduction ($\geq$50%) of cortical tangles [80]. Rapamycin causes a brain area-dependent reduction of sarkosyl insoluble tau levels, and a marked increase of autophagy markers [80]. Rapamycin, when given at a low dosage between 1.5 and 3.5 months of age, reduces total tau levels and p-Ser181 tau levels in OXYS rats, and increases total S6K1 and p-S6K1 levels [81].

Semi-synthetic rapalogs (**4.1b–f**, Figure 4.1) are C43-substituted rapamycin analogs with an improved pharmacokinetic (PK) profile coupled with rapamycin-like pharmacological activity. Everolimus (RAD-001, Afinitor™, **4.1b**) is an orally bioavailable rapamycin hydroxyethyl ether [82] marketed as a treatment for allograft rejection, tuberous sclerosis, advanced kidney cancer, and pancreatic neuroendocrine tumors [83]. Everolimus testing in neuronal [84] and mouse models of HD [85] fails to show efficacy in inducing autophagy and in reducing mutant huntingtin levels. Everolimus reduces the levels of brain p-S6K1, and shows induction of autophagy and a decrease in soluble mutant huntingtin in quadriceps muscle [85]. Limited, positive behavioral effects observed in polyQ-carrying R6/2 TG mice are likely to be due to an activity of everolimus in muscle [85]. Pan-mTOR inhibitors binding to the ATP binding site of mTOR (see below) induce neuronal autophagy, at least in part due to an mTORC2-dependent mechanism [84]. Temsirolimus (CCI-779, Torisel™, **4.1c**) is an orally bioavailable, water soluble rapamycin ester [86] marketed as a treatment for mantle cell lymphoma and advanced kidney cancer [87]. Temsirolimus reduces neurotoxicity in a Ross–Borchelt mouse HD model expressing mutant huntingtin [46]. It decreases the number and the size of mutant huntingtin aggregates, due to increased autophagy and reduced mTORC1 signaling [46]. Temsirolimus improves motor performance in a TG mouse model of spinocerebellar ataxia (2 months' treatment starting at 5 weeks of age) [88]. It reduces the number of mutant ataxin-3 aggregates in the brain, and the cytoplasmic levels of soluble protein—an indication of increased cytoplasmic autophagy [88]. Ridaforolimus (AP23573, deforolimus, **4.1d**) is an orally available phosphinate [89] developed for the maintenance treatment of patients with soft tissue sarcoma or primary malignant bone tumor.

Umirolimus (TRM-986, Biolimus A9, **4.1e**) [90] and zotarolimus (ABT-578, **4.1f**, Figure 4.1) [91] are lipophilic compounds with poor aqueous solubility, used as coatings for coronary eluting stents against restenosis.

Rapalogs inhibit mTORC1-mediated phosphorylation of S6K1 and EIF4E-BP1 with varying strength and kinetics, and show limited activity against mTORC2 only when chronically administered. Slow EIF4E-BP1 inhibition may cause only partial inhibition of protein translation, while efficient S6K1 inhibition suppresses the S6K1-insulin receptor signaling 1 (IRS1) feedback loop and causes the activation of the PI3K/Akt pathway, with potential tumor-promoting side effects [92]. The therapeutic relevance of mTORC complexes in cancer has stimulated the rational design, synthesis, and characterization of ATP-competitive mTOR inhibitors. They can be active against both mTORC complexes (*pan-mTOR inhibitors*), and against other kinases (*dual PI3K/mTOR inhibitors*). Both inhibitor classes prevent PI3K/Akt activation.

Dual PI3K/mTOR (DPmT) inhibitors stem from PI3K drug discovery projects, where closely related mTOR kinase is a likely off-target. Tricyclic pyrimidine/morpholine-containing PI-103 (**4.2**, Figure 4.2) is the first identified DPmT inhibitor [93]. Its structural optimization, aiming to improve its PK and to generate a quantitative structure–activity relationship (QSAR) around the PI-103 core, leads to PI-540 and PI-620 (**4.3a,b**). They are active in a glioblastoma xenograft model in athymic mice [94]. Closely related GDC-0980 (**4.4**) shows preclinical efficacy against a variety of cancers [95], and is currently undergoing clinical evaluation [96]. Imidazo[4,5-c]quinolones NVP-BEZ235 [97] and NVP-BGT226 [98] (respectively **4.5** and **4.6**) result from a structure-based design project. Both compounds are being clinically evaluated in patients with advanced solid tumors [99,100]. The quinoxaline XL765 (SAR245408, **4.7**, Figure 4.2) [101] shows preliminar clinical efficacy against several cancers [102].

The quinolone GSK2126458 (**4.8**, Figure 4.3) [103] is currently tested in clinical trials against lymphomas, and against solid tumors in combination with a mitogen-activated protein kinase kinase (MEK) inhibitor [27].

The triazolopyrimidine PKI-402 (**4.9**) [104] shows preclinical efficacy *in vivo* against breast and lung cancer, and glioblastoma multiforme [105]. Structurally related morpholinotriazine PKI-587 (PF-05212384, **4.10**) [106] is under clinical evaluation against neoplasms as a standalone treatment and in combination with irinotecan [27,107]. The morpholylbenzopyranone LY294002 (**4.11**) [108] is a moderately active DPmT inhibitor with poor PK properties. Its water soluble pro-drug/conjugate SF1126 (**4.12**, Figure 4.3) [109] is selectively delivered by the integrin-binding RGDS sequence to cancer cells, where LY294002 is released. SF1126 is a potent anticancer agent, currently in Phase I/II clinical trials against multiple myeloma [110].

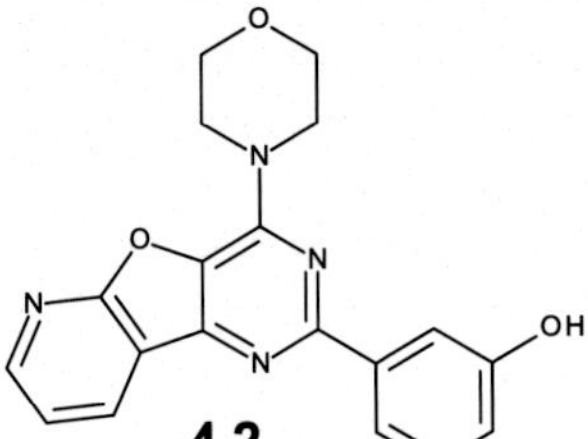

4.2

IC_{50} PI3K isoforms: **2-15 nM**
IC_{50} mTOR: **20-80 nM**

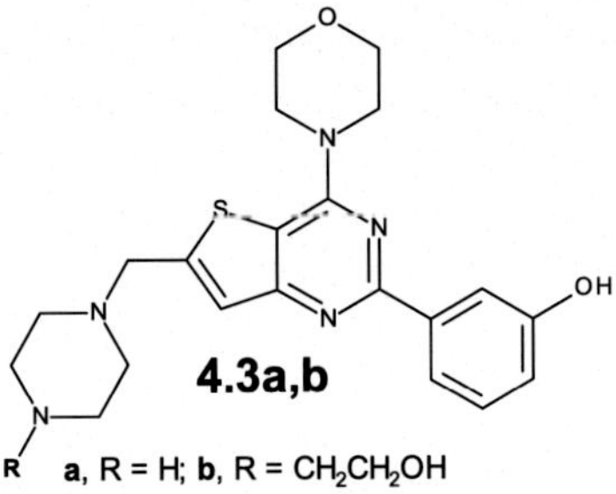

4.3a,b

a, R = H; **b**, R = CH_2CH_2OH

IC_{50} PI3K isoforms: **4-331 nM (a)**, **7-672 nM (b)**
IC_{50} mTOR: **61 nM (a)**, **231 nM (b)**

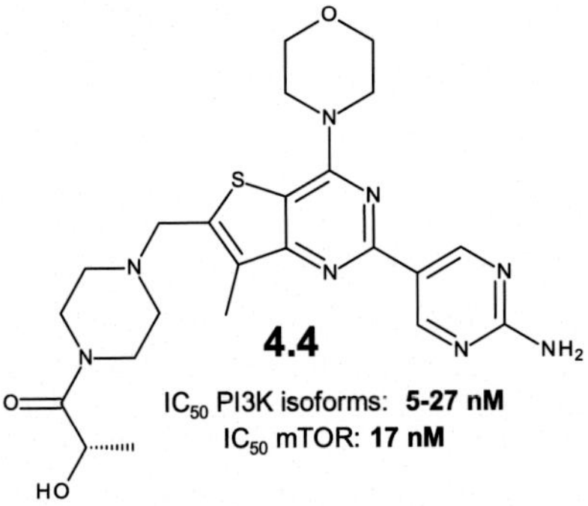

4.4

IC_{50} PI3K isoforms: **5-27 nM**
IC_{50} mTOR: **17 nM**

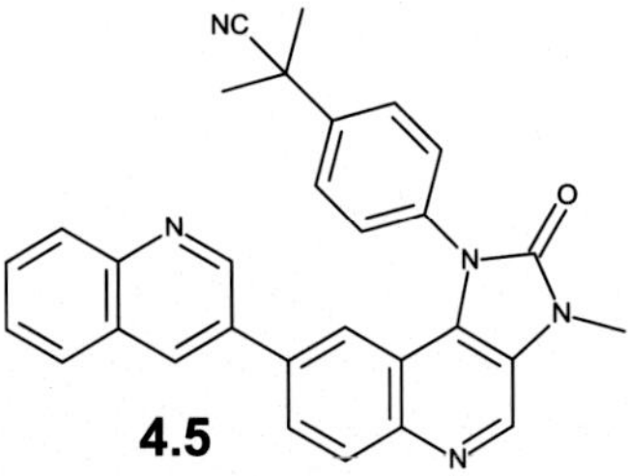

4.5

IC_{50} PI3K isoforms: **4-75 nM**
IC_{50} mTOR: **21 nM**

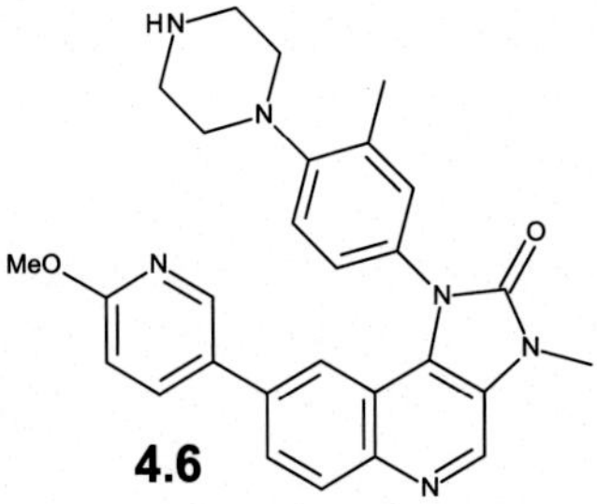

4.6

IC_{50} PI3K isoforms: ≈ **1nM**
IC_{50} mTOR: ≈ **1 nM**

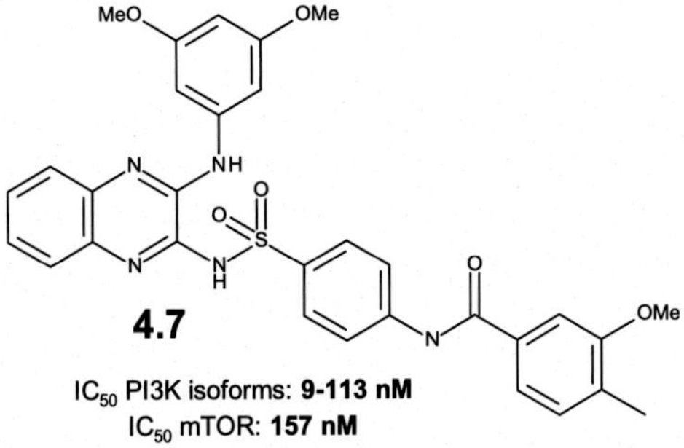

4.7

IC_{50} PI3K isoforms: **9-113 nM**
IC_{50} mTOR: **157 nM**

FIGURE 4.2 Dual PI3K/mTOR (DPmT) small molecule inhibitors: chemical structures, **4.2–4.7**.

DPmT inhibitors are popular in cancer therapy, as their efficacy is significant and their side effects profile is acceptable. Their use as chronic treatments in age-related NDDs would require an adverse event (AE)-free profile, which may be difficult to achieve as DPmT inhibitors show PI3K-dependent immunosuppressant activity [111]. Selective pan-mTOR inhibitors, with marginal PI3K inhibition, are more promising with regard to AEs, and are targeted by several research groups.

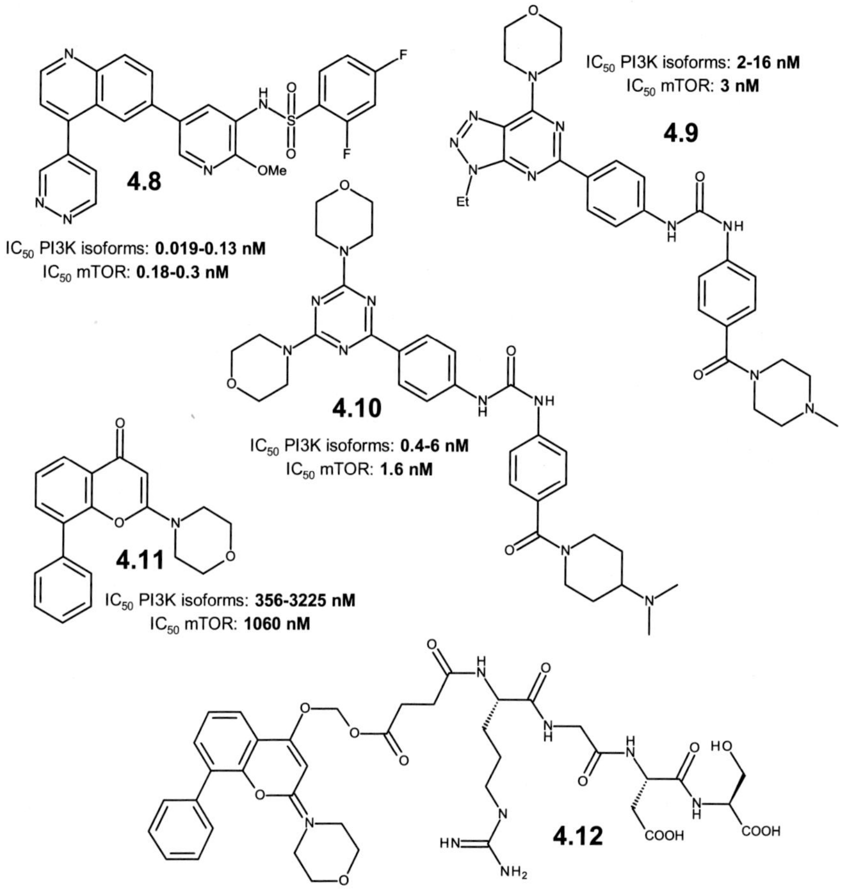

FIGURE 4.3   Dual PI3K/mTOR (DPmT) small molecule inhibitors: chemical structures, **4.8–4.12**.

Pyrazolo[3,4-d]pyrimidine PP242 (**4.13**, Figure 4.4) is the first reported pan-mTOR inhibitor devoid of activity against PIK3/Akt, identified from a library of aspecific kinase inhibitors [112]. It is potent *in vivo* in a mouse model of leukemia, inducing its regression, while it does not show any undesired immunosuppressant activity [111]. It does not inhibit PI3K/Akt, but shows nM potency against Janus-activated kinase (JAK) kinases and the rearranged during transfection (RET) receptor [113]. Closely related MLN0128 (INK128, **4.14**) [114] is a more potent and selective analog. It is clinically tested against lymphoma and multiple myeloma as a standalone treatment and in combination with paclitaxel and trastuzumab [115]. Pyrazolo[3,4-d]pyrimidines WAY600 (**4.15**) [116] and WYE-132 (WYE125132, **4.16**) [117] result from the structural optimization of the DPmT inhibitor

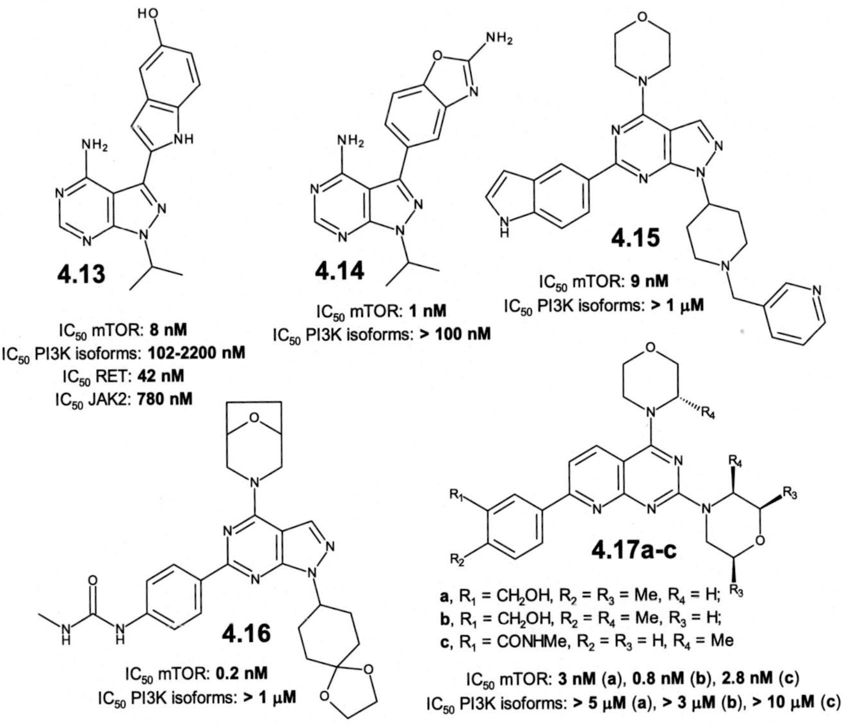

FIGURE 4.4   Small molecule pan-mTOR inhibitors: chemical structures, **4.13–4.17c**.

PI-103. WYE125132 is orally active in a panel of cancer mouse models, either alone or in combination with bevacizumab [117]. Closely related pyrido[2,3-d]pyrimidines Ku0063794 (**4.17a**) [118], AZD8055 (**4.17b**) [119], and AZD2014 (**4.17c**, Figure 4.4) [120] share a common bicyclic chemotype [121]. Ku0063794 suppresses mTORC2-mediated Thr308 phosphorylation of Akt in WT cells, but not in cells where mTORC2 is inactivated [118], suggesting that alternative pathways may ensure undesirable Akt activation when mTORC2 is inactivated. AZD8055 results from an HTS campaign aimed towards inhibitors of p70 S6K1 phosphorylation by mTORC1 [119]. It is clinically tested against breast cancer and lymphoma [122]. The clinical candidate AZD2014 is targeted against advanced solid tumors [123].

The imidazo[1,5-f][1,2,4]triazine OSI-027 (**4.18**, Figure 4.5) [124] is a promising treatment against chronic myelogenous leukemia (CML) patients carrying the Gleevac™-resistant T315I BCR-ABL mutation. Its clinical evaluation against advanced solid tumors and lymphoma is ongoing [125]. The benzochromenone palomid 529 (**4.19**) [126] is a blood–brain barrier (BBB)-permeable pan-mTOR inhibitor under clinical evaluation against age-related macular degeneration [127]. The benzonaphthyridinones torin1 [128] and torin2 [129] (respectively **4.20** and **4.21**, Figure 4.5)

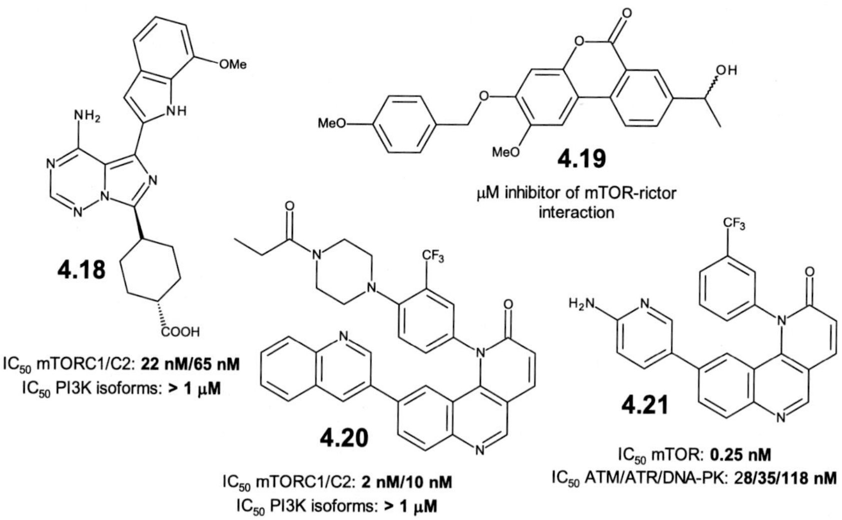

FIGURE 4.5 Small molecule pan-mTOR inhibitors: chemical structures, **4.18–4.21**.

are potent pan-mTOR inhibitors. The former compound results from an mTOR kinase inhibition-targeted HTS campaign on a heterocyclic library [130]. Torin1 has a slow mTOR dissociation constant and a suboptimal PK profile, but shows preclinical efficacy in several cancer models [130]. Orally bioavailable torin2 inhibits ataxia telangiectasia-mutated (ATM) and ATM and RAD3-related (ATR) kinases, has a better PK profile, and shows *in vivo* efficacy in combination with MEK inhibitors against lung tumors [131]. It shows potent antimalarial activity, blocking the dynamic trafficking of the *Plasmodium* exported protein 1 (EXP1) and up-regulated in sporozoites 4 (UIS4) proteins. It causes efficient parasite eradication from hepatocytes [132].

Torin1 is developed in an academic setting, so that its preclinical characterization is fully reported. Torin1 impairs cancer cell growth and proliferation to a far greater degree than rapamycin. This is due to the suppression of rapamycin-resistant functions of mTORC1 that are necessary for translation and suppression of autophagy, rather than to mTORC2 inhibition [130]. Torin1 activates the lysosomal function through complete inhibition of mTORC1, while allosteric-incomplete mTORC1 inhibition by temsirolimus does not produce the same result [133]. Torin1 and rapamycin stimulate autophagy (increased AP formation and AP–LS fusion) in a diabetic mouse model, protecting the mice and preventing β-cell apoptosis in an autophagy-dependent manner [134]. Torin1 and temsirolimus alleviate mechanical hypersensitivity in mouse models of inflammatory and neuropathic pain, inhibiting mTORC1 in sensory axons and in the spinal dorsal horn [135]. Rapamycin protects neurons in cellular and animal toxin-induced models of PD, due to blocked translation of the regulated in development and DNA damage responses 1 (REDD1) protein [41]. REDD1 is induced in affected neurons of PD patients and causes neuron death by dephosphorylation of Akt [41]. Rapamycin does not dephosphorylate Akt at the Thr308 residue and does not prevent its survival functions. Torin1, conversely, inhibits mTORC2, causes Thr308 dephosphorylation, and does not protect neurons in cellular and animal toxin-induced models of PD [41].

Small molecules with multiple pharmacological effects may directly or indirectly regulate mTOR activity. mTOR regulation contributes to their overall efficacy in clinical settings. The xanthine alkaloid caffeine (**4.22**, Figure 4.6) is a weak mTOR inhibitor that induces autophagy *via* inhibition of mTOR signaling [136], increases the lifespan of yeasts [137], and may have mTOR/autophagy-related beneficial effects in humans. The phospholipid phosphatidic acid (**4.23**) binds to the FRB domain of mTOR, competing with the FKBP12–rapamycin complex, and activates both mTORC complexes [138]. The X-ray structure of phosphatidic acid complexed with mTOR FRB could allow the design of structurally similar

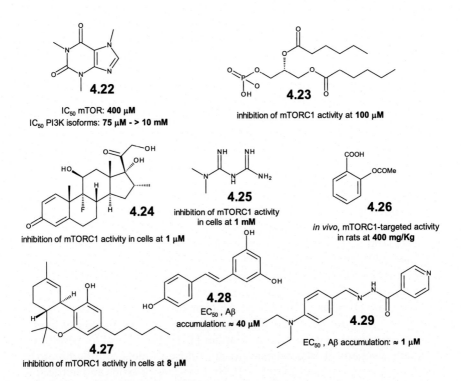

4.22

IC$_{50}$ mTOR: **400 μM**
IC$_{50}$ PI3K isoforms: **75 μM - > 10 mM**

4.23

inhibition of mTORC1 activity at **100 μM**

4.24

inhibition of mTORC1 activity in cells at **1 μM**

4.25

inhibition of mTORC1 activity
in cells at **1 mM**

4.26

in vivo, mTORC1-targeted activity
in rats at **400 mg/Kg**

4.27

inhibition of mTORC1 activity in cells at **8 μM**

4.28

EC$_{50}$, Aβ
accumulation: ≈ **40 μM**

4.29

EC$_{50}$, Aβ accumulation: ≈ **1 μM**

FIGURE 4.6   Small molecule regulators of mTOR activity: chemical structures, **4.22–4.29**.

mTOR inhibitors [139], which could possess a different mTOR regulation profile than rapalogs and ATP-competitive mTOR inhibitors [140].

The glucocorticoid dexamethasone (**4.24**) increases/up-regulates REDD1, inhibits mTORC1 in a tuberous sclerosis 2 (TSC2)-dependent manner, and activates autophagy [141]. The biguanide metformin (**4.25**) inhibits phosphorylation of mTORC1 substrates S6K1 and EIF4E-BP1 and decreases translation [142]. Inhibition of the mTORC1 activator AMP-mediated protein kinase (AMPK) [143], of the Ras-related GTP binding (Rag) GTPases [144] that also activate mTOR, and up-regulation of REDD1 [145] by metformin lead to mTORC1 inhibition. Metformin induces autophagy [146] and shows a different mTOR regulation profile from rapalogs and ATP-competitive mTOR inhibitors [147].

Aspirin (**4.26**) shows AMPK-dependent and -independent effects on mTOR signaling *in vivo*, leading to reduction of S6K1 phosphorylation and stimulation of autophagy [148]. Its beneficial effects on longevity in rodents [149], and on mortality in humans [150], could be mTOR signalling-related. The cannabinoid Δ^9-tetrahydrocannabinol (Δ^9-THC, **4.27**) shows autophagy-enhancing effects attributable to AMPK and mTORC1 [151]. The former implies calmodulin-activated kinase kinase β (CaMKKβ)-promoted phosphorylation and activation of AMPK. The latter proceeds through stimulation of the endoplasmic reticulum (ER) stress pathway, up-regulation of pseudokinase homolog tribbles 3 (TRB3) and Akt-dependent mTORC1 inhibition [151].

Polyphenols are known autophagy inducers [152]. The stilbenetriol resveratrol (**4.28**) modulates several pathways connected with autophagy. Resveratrol increases the intracellular concentration of Ca^{2+}, causing CaMKKβ-promoted phosphorylation and activation of AMPK. The resulting mTORC1 inhibition leads to stimulation of autophagy [153]. A small resveratrol-inspired bisaryl hydrazine library includes **4.29** (Figure 4.6), that inhibits Aβ accumulation in a cell model at low μM concentration through the same autophagy-promoting Ca^{2+}/CaMKKβ/AMPK/mTORC1 pathway [154]. Surprisingly, resveratrol shows *in vivo* mTORC2- and autophagy-dependent cardioprotective effects, as low doses of resveratrol induce rictor expression, increase the levels of mTORC2, and stimulate autophagy in an ischemia-reperfusion rat model [155].

Dynamic complexation of mTOR with proteins composing the mTORC1 and the mTORC2 complex determine the concentration and the activity of mTOR complexes at a given time. Direct intervention on mTOR, either with allosteric/rapalogs or with ATP-competitive/pan-mTOR inhibitors, provides information on the role of mTORC1 in physiological and pathological conditions, and further elucidates mTORC2 functions [156]. The identification and characterization of mTORC2-specific modulators would be possible, if we could selectively target the complex-specific protein–protein interactions (PPIs) that assemble mTORC2.

Targeted mTORC2 disruption may lead to unknown therapeutic outcomes, and even selective PPI-targeted mTORC1 disruption may allow selective interference with some among mTOR functions [27]. Unfortunately, the rational design of mTOR-specific PPI inhibitors is prevented by the lack of structural information about their "hot spot" binding regions [157]. Phenotypic screens [158] on drug-like chemical collections could identify chemical tools to discover new molecular targets influencing mTOR signaling and autophagy stimulation, and may provide access to new drug-like chemotypes—see section 4.3.

The use of rapalogs and pan-mTOR inhibitors in NDDs is disease- and disease progression-dependent, as specific functions of mTORC1 and mTORC2 (the last ones still to be elucidated) play different roles in each aging-related disease at different disease stages. We can safely assume that mTORC1 is a key target in neurodegeneration, that its modulation (together with mTORC2 and PI3K, when needed) has great therapeutic potential against chronic central nervous system (CNS) diseases such as tauopathies, and that mTOR-driven autophagy stimulation is a major therapeutic avenue. Key issues remain to be solved—among others, the risk of immunosuppression [159] and dysregulation of insulin signaling [160] associated to long-term use of rapalogs; the toxic effects of chronic treatment with pan-mTOR inhibitors [92]; the physiological role of mTOR in the brain [161]; the consequences of chronic activation of autophagy in the brain [162]; and the nature and the magnitude of beneficial effects of mTOR modulators on quality of life and lifespan in humans [40].

4.3 SMALL MOLECULE ENHANCERS OF RAPAMYCIN (SMERs)

Autophagy is a complex network of interconnected pathways, whose full elucidation is far from being completed. Its therapeutic relevance in many disease states is now accepted, and autophagy-regulating drugs are being evaluated as treatments for acute diseases—oncology in particular [163]. Validated and safe molecular targets are scarce, though, as even the answer to several key questions to fully understand the effects of mTORC1 regulators *in vivo* is still awaited. Switching to chronic diseases and neurodegeneration may increase the value of putative autophagy-dependent therapies. Unfortunately, the selection of a molecular target with appropriate validation and compliance (no AEs on patients), and the rational design of small molecule effectors targeted against it, is an extremely complex challenge.

Rapamycin is a useful example. Much of what we know today about mTORC1 and mTORC2, and about their regulation of autophagy and other essential cellular processes, is either directly or indirectly due to rapamycin [24]. Natural products are instrumental to discover and validate therapeutically relevant targets (i.e., taxol and microtubule polymerization in oncology [164]), indications (i.e., rapamycin itself and immunosuppression/organ transplant [165]), and even previously unimaginable therapeutic areas (i.e., penicillin and antibacterials [166]). Chemical biology [167] and phenotype-based screening [168] define similar, man-driven processes: i.e., a cellular assay that determines the influence of each member of a suitable/drug-like chemical collection on a complex phenotype, such as induction of autophagy. Once a positive/hit is found, it is used to identify its molecular target in the cell, and to elucidate the influence of such target on the appearance and development of the phenotype. The identification and validation of a novel molecular target and of a drug-like target effector to be structurally optimized in a drug discovery project are the ultimate goals.

Phenotype-based screening campaigns targeting small molecule enhancers (SMERs) or inhibitors (SMIRs) of rapamycin are known [169]. The LC3-II conjugate [170] is a known autophagy marker. LC3-II can be coupled with green fluorescent proteins (GFPs) or GFPs and red fluorescent proteins (RFPs) to give fluorescence microscopy-detectable fusion proteins [171]. They can be quantitated as a whole cytosol pool of fusion proteins, providing information on autophagy induction (increase of FP-LC3-II conjugates) or inhibition (decrease of FP-LC3-II conjugates) [172]. Detailed observations may determine phenotypic differences, such as the number of APs and/or autolysosomes. GFP-LC3-II constructs cannot discriminate between inhibitors of late stages of autophagy and stimulators of autophagy, as both cause an accumulation of APs and an increase of GFP-LC3-II levels [169,173]. Secondary assays to determine the true nature of screening hits are needed [174]. RFP-GFP-LC3-II constructs show both GFP and RFP signals in APs and only RFP signals in autolysosomes [175]. RFP-GFP-LC3-II-based assays (and other sensitive/specific assay formats taking advantage of easily quantifiable LC3-II constructs) are validated in a low-throughput format [176,177]. Their adaptation to the screening of medium-size chemical collections is ongoing [178,179].

Alternative screening formats include the search for enhancers or inhibitors of the cytostatic effects of rapamycin in yeast (primary HTS assay), followed by human autophagy-specific secondary assays [180]. A recently validated high content screen (HCS) format takes advantage of the inverse correlation between autophagy and cellular levels of lipid droplets (LDs, lower levels correspond to autophagy induction), and discriminates between late stage autophagy inhibition and autophagy induction [181]. Screening methods based on the degradation of aggregate-prone proteins (identifying putative hits against NDDs) are also reported [182]. Their throughput is generally lower, and they are rather used as secondary-confirmation assays.

The use of phenotype-based assay formats in ≈10 screening campaigns is described in this section. The properties of identified hits (including their confirmation as autophagy inducers rather than late stage autophagy inhibitors, and their potency against protein aggregation), and the identification of novel targets/pathways influencing autophagy, are also described in detail here.

A collection of 3854 chemicals, including drugs and clinically tested candidates, contains four compounds (**4.30–4.33**, Figure 4.7) that increase

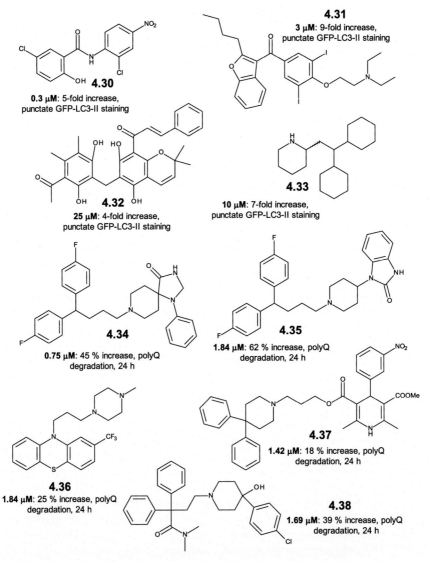

FIGURE 4.7 Small molecule enhancers of rapamycin (SMERs): chemical structures of mTORC1-dependent (**4.30–4.33**) and mTORC1-independent (**4.34–4.38**) SMERs.

GFP-LC3-II fluorescence (>three-fold $vs.$ basal levels) without major toxicity in human breast cancer MCF-7 cells [183]. A panel of secondary screens confirms their stimulation of autophagy, and their mTORC1-dependent, mTORC2-independent mechanism of action. They differ from rapamycin, as the latter irreversibly inhibits mTORC1 signaling, while the four drugs are reversible inhibitors. Rapamycin rapidly stimulates autophagy in cells (a few minutes), while compounds **4.30–4.33** require hours of incubation to achieve full mTORC1 inhibition. Rapamycin has stronger effects on cell viability and proliferation [183]. The four drugs show increased staining/punctate GFP-LC3-II fluorescence from high μM (amiodarone, antianginal–antiarrhythmic, **4.31**) to sub-μM concentrations (niclosamide, anthelmintic, **4.30**). Amiodarone is slowly reversible and shows autophagy-unrelated side effects, while the other drugs are selective and rapidly reversible [183]. Niclosamide rapidly achieves partial mTORC1 inhibition, while the other drugs require several hours' incubation to fully inhibit mTORC1 signaling. Rottlerin (multiple kinase inhibitor, **4.32**) shows a TSC2-dependent mTORC1 inhibition, failing to inhibit mTORC1 in TSC2$^{-/-}$ cells, although it stimulates autophagy in the same cells (increased GFP-LC3-II fluorescence). Niclosamide, amiodarone, and perhexiline (antianginal, **4.33**) act through TSC2-independent mechanisms, possibly acting through upstream regulation of PI3K/mTORC1 signaling [183].

Rottlerin induces autophagy in pancreatic cancer stem cells in a PI3K/Akt/mTORC1-dependent manner, and prolonged rottlerin treatment (>24 hr) leads to apoptotic cell death [184]. Other studies imply a protein kinase C (PKC)-δ/transglutaminase 2 (TG2)-dependent mechanism [185]. Early autophagy induction and late apoptosis caused by rottlerin, thus, are due to multiple cellular effects [186]. Niclosamide induces cytoplasmic acidification, leading to mTORC1 indirect inhibition, through its protonophoric/proton gradient regulator properties [187]. Interestingly, niclosamide (but not rottlerin or perhexiline) promotes the disposal of UBQ-tagged protein aggregates in neuronal cells when the UPS degradation system is impaired, possibly through LS redistribution [188]. This putative link between UPS (see Chapter 3 here and in the biology-oriented companion book [23]) and autophagy should be further studied for therapeutic purposes, taking advantage of a clinically compliant chemical tool.

Screening of 480 bioactive compounds in human glioblastoma H4 cells leads to 72 positives/increasing GFP-LC3-II fluorescence (>50% increase $vs.$ basal levels) [172]. A set of secondary screens (toxicity, autophagy markers, mTOR activation, degradation of polyQ aggregates) confirms six hits (including amiodarone). They induce autophagy in an mTOR-independent manner, promote polyQ disposal (increased % of degraded polyQ at 24 hr, see Figure 4.7) and do not cause major toxicity [172]. Diphenylbutylpiperidine (DPBP)-based fluspirilene and pimozide (respectively **4.34** and **4.35**, Figure 4.7) are antipsychotic drugs, as is the phenothiazine-based trifluoperazine (**4.36**). The latter also inhibits the Ca^{2+}-binding protein calmodulin. Antiarrhythmic Ca^{2+} channel blockers include the dihydropyridine niguldipine (**4.37**) and amiodarone. Loperamide (**4.38**, Figure 4.7) is a piperidine agonist of μ-opioid receptors with Ca^{2+} channel-blocking activity [172]. The authors propose a role for intracellular Ca^{2+} regulation in the activation of autophagy.

Fluspirilene induces autophagy in H4 cells at low μM concentrations, with a peak around 4 hr [189]. Fluspirilene (and compounds **4.35–4.38**) causes an increase of the levels of core autophagy protein Atg5, and consequently of Atg12–Atg5 conjugates, by complete inhibition of Ca^{2+} channel/Ca^{2+} influx in cells at 10 μM. The effects of fluspirilene and structural analogs on autophagy and on Ca^{2+} influx are similar [188]. Fluspirilene-induced intracellular Ca^{2+} depletion impairs the activity of calpains, a family of Ca^{2+}-dependent ubiquitous non-lysosomal cysteine proteases. Calpain-1 controls autophagy in physiological/non-starving conditions through regulation of Atg12–Atg5 conjugate levels [189]. Trifluoperazine reduces the invasiveness of breast cancer MBA-MD-231 cells through induction of autophagy and autophagy-dependent degradation of hypoxia factor 1α (HIF-1α) [190].

Several mTOR-independent autophagy inducers, including compounds **4.34–4.38**, share a similar pharmacophore [191]. The DPBP moiety (**4.34**, **4.35**), closely related diphenylpropylpiperidines (**4.38**), and 4,4-diphenylpiperidines (**4.37**) are moderately active "open" autophagy stimulators, while phenothiazine-N-propylamines (**4.36**, **4.39a–d**, Figure 4.8) and phenoxazine-N-butylamine (10-NCP, **4.40**) are "closed/cyclic" autophagy stimulators.

Computational studies show that two hydrophobic/aromatic groups and a positively charged group are shared by compounds **4.34–4.40**. The model fits at best with two aromatic groups in a tricyclic ring system (higher potency, **4.36**, **4.39a–d**, **4.40**), while "open" compounds must twist their structure to fit the model (lower potency, **4.34**, **4.35**, **4.37**, **4.38**) [191]. 10-NCP protects against mutant huntingtin (Htt) toxicity in a primary culture model of HD, decreases the levels of diffuse and aggregated forms of mutant Htt and does not show toxicity [191]. A library of "open" and "close" compounds shows that substitution on the aromatic rings may increase the potency of compounds as autophagy inducers (e.g., compare penfluridol, **4.41a**, with Ph-substituted compounds **4.41b–e**) [192,193].

HTS of a collection made by >50 K compounds in a yeast assay identifies SMERs and SMIRs [180]. Hit reconfirmation for SMERs in autophagy-dependent, protein aggregation/disposal assays on mutant synuclein and huntingtin in yeast and mammalian cells leads to the identification of three hits (SMER10/aminopyrimidinone, SMER18/vinylogous amide, and SMER28/bromoquinazoline, respectively **4.42–4.44**, Figure 4.8) [180]. They stimulate autophagy (>LC3-II levels) in COS-7 and HeLa cells, and

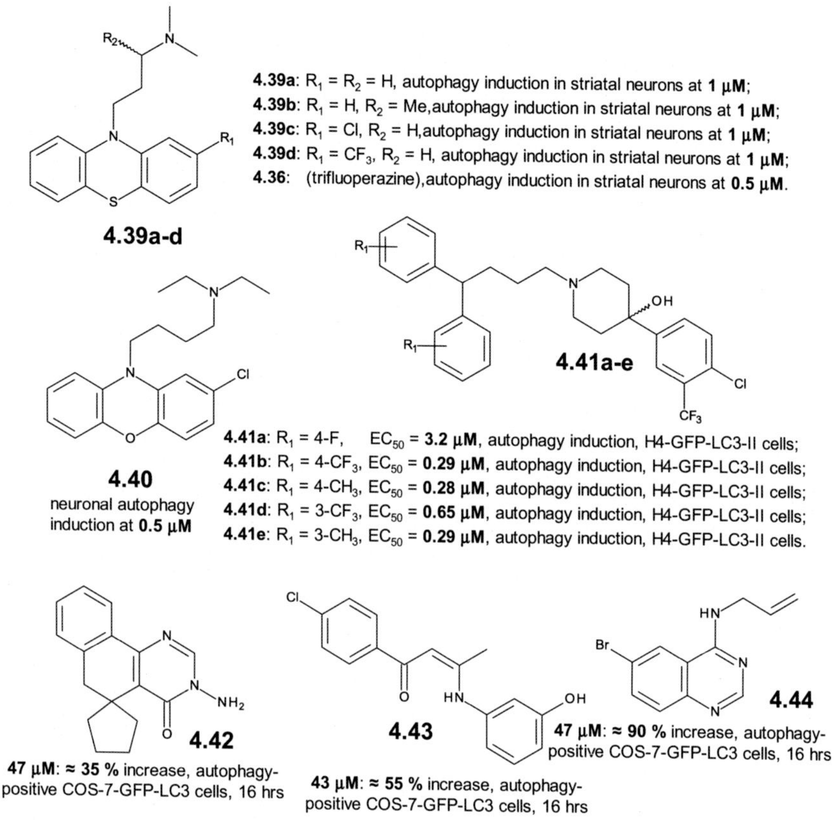

4.39a-d

4.39a: $R_1 = R_2 = H$, autophagy induction in striatal neurons at **1 μM**;
4.39b: $R_1 = H$, $R_2 = Me$, autophagy induction in striatal neurons at **1 μM**;
4.39c: $R_1 = Cl$, $R_2 = H$, autophagy induction in striatal neurons at **1 μM**;
4.39d: $R_1 = CF_3$, $R_2 = H$, autophagy induction in striatal neurons at **1 μM**;
4.36: (trifluoperazine), autophagy induction in striatal neurons at **0.5 μM**.

4.41a-e

4.40
neuronal autophagy
induction at **0.5 μM**

4.41a: $R_1 = 4$-F, $EC_{50} = 3.2$ μM, autophagy induction, H4-GFP-LC3-II cells;
4.41b: $R_1 = 4$-CF_3, $EC_{50} = 0.29$ μM, autophagy induction, H4-GFP-LC3-II cells;
4.41c: $R_1 = 4$-CH_3, $EC_{50} = 0.28$ μM, autophagy induction, H4-GFP-LC3-II cells;
4.41d: $R_1 = 3$-CF_3, $EC_{50} = 0.65$ μM, autophagy induction, H4-GFP-LC3-II cells;
4.41e: $R_1 = 3$-CH_3, $EC_{50} = 0.29$ μM, autophagy induction, H4-GFP-LC3-II cells.

4.42

47 μM: ≈ **35 %** increase, autophagy-
positive COS-7-GFP-LC3 cells, 16 hrs

4.43

43 μM: ≈ **55 %** increase, autophagy-
positive COS-7-GFP-LC3 cells, 16 hrs

4.44

47 μM: ≈ **90 %** increase, autophagy-
positive COS-7-GFP-LC3 cells, 16 hrs

FIGURE 4.8 mTOR-independent small molecule enhancers of rapamycin (SMERs): chemical structures, **4.39a–4.44**.

increase AP formation. They act *in vivo* in an mTOR-independent manner to prevent neurodegeneration in a *Drosophila* model. A limited structural exploration on their scaffolds identifies compounds with equal or even better activity [180].

SMER28 reduces the accumulation of Aβ fibrils in SK-N-SH human neuroblastoma cells [194]. It reduces $A\beta_{40}$ secretion and the amount of C-terminal fragments of the precursor APP peptide (APP-CTF) in APP-CTF-overexpressing murine embryonic fibroblasts (MEFs), inducing autophagy in a γ-secretase-independent, Atg5-dependent manner [195].

A medium-throughput screening campaign on 253 pharmacologically active compounds for their promotion of autophagy-driven clearance of A30P (primary screen) and A53T mutant α-synuclein (secondary/reconfirmation assay) in PC-12 cells [182] identifies loperamide, amiodarone, two loperamide-like L-type Ca^{2+} channel activity blockers (the phenylalkylamine verapamil **4.45** and the dihydropyridine nimodipine **4.46**,

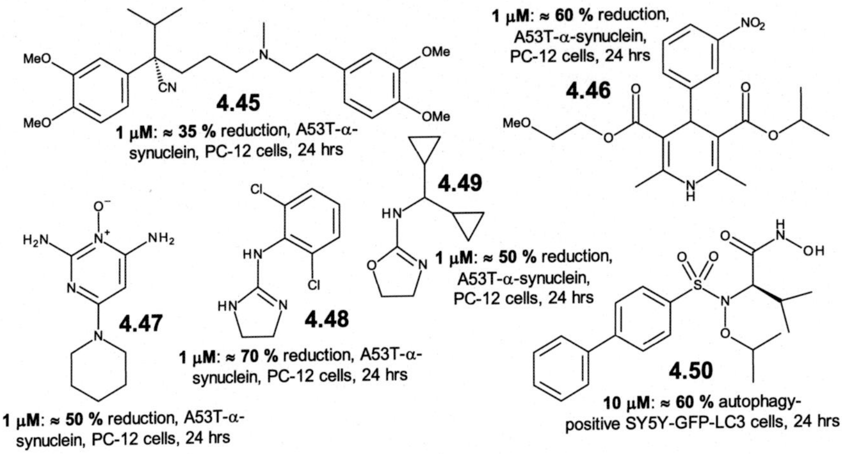

FIGURE 4.9 mTOR-independent small molecule enhancers of rapamycin (SMERs): chemical structures, **4.45–4.50**.

Figure 4.9), the pyrimidine-N-oxide, ATP-sensitive K⁺ channel agonist minoxidil (**4.47**), and the cyclic guanidine mimic clonidine, a ligand of α_2-adrenergic and type I imidazoline receptors (**4.48**) [182], as confirmed hits.

Clonidine allows the identification of a novel, mTOR-independent pathway leading to autophagy induction/protein aggregation disposal. Clonidine induces autophagy as an imidazoline 1 (I1R) receptor agonist, as does the specific I1R agonist rilmenidine (**4.49**) [182]. They reduce cyclic AMP (cAMP) levels and negatively regulate the exchange protein directly activated by cAMP (Epac)-ras-related protein 2B (Rap2B)-phospholipase C ε (PLCε) cascade. cAMP activates Epac, which activates Rap2B, which activates PLCε. PLCε activity produces inositol triphosphate (IP3), which causes an increase of cytosolic Ca²⁺ levels [182]. Thus, I1R-agonistic, L-type Ca²⁺ channel activity blockers and minoxidil, which also acts on L-type Ca2⁺ channel activity, inhibit calpains. Calpain inhibition induces autophagy and reduces htt aggregation in an mTOR-independent manner in cells and in *Drosophila* [182]. Rilmenidine shows therapeutic efficacy and tolerability in a mouse HD model [196].

In addition to fluspirilene-mediated effects of calpain1 inhibition on Atg5 [87], another calpain1 substrate contributes to the induction of autophagy. The α subunit of heterotrimeric G proteins ($G_{s\alpha}$) is activated by calpain cleavage and, when inhibited by small molecules or by siRNA, reduces htt aggregation and increases GFP-LC3-II levels [73]. A GFP-LC3-II-based HTS campaign on 2400 drugs and clinically tested compounds identifies the hydroxamic acid ARP101 (**4.50**, Figure 4.9), a selective matrix metalloproteinase 2 (MMP-2) inhibitor, as an autophagy inducer in MCF-7,

HTC116, and SH-SY5Y cells [197]. ARP101 shows an Atg5-dependent effect on autophagy, while unspecific MMP inhibitors do not have similar potency and cell line-independent influence on autophagy [197].

The antipsychotic phenylindole sertindole (**4.51**, Figure 4.10) is an Atg5-dependent autophagy inducer identified in an HTS campaign on 1120 drug-like compounds in SH-SY5Y cells [198]. Its effect on autophagy

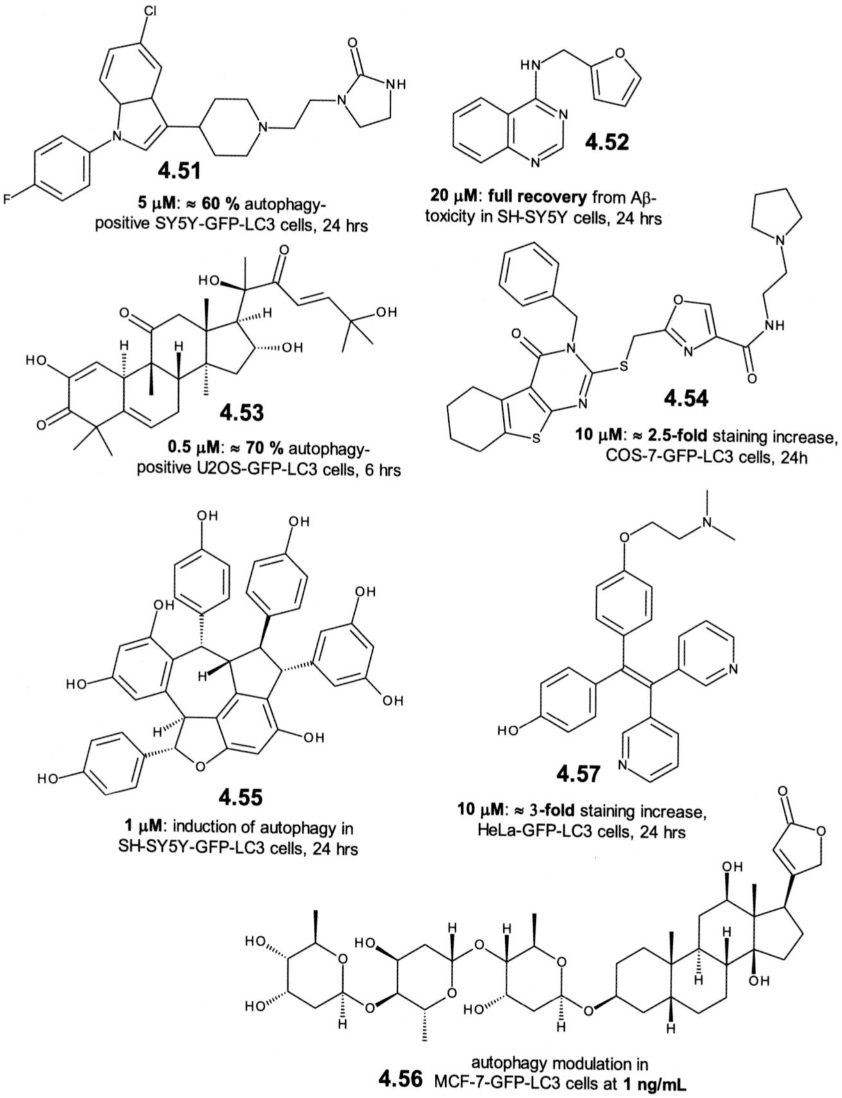

4.51

5 µM: ≈ 60 % autophagy-positive SY5Y-GFP-LC3 cells, 24 hrs

4.52

20 µM: full recovery from Aβ-toxicity in SH-SY5Y cells, 24 hrs

4.53

0.5 µM: ≈ 70 % autophagy-positive U2OS-GFP-LC3 cells, 6 hrs

4.54

10 µM: ≈ 2.5-fold staining increase, COS-7-GFP-LC3 cells, 24h

4.55

1 µM: induction of autophagy in SH-SY5Y-GFP-LC3 cells, 24 hrs

4.57

10 µM: ≈ 3-fold staining increase, HeLa-GFP-LC3 cells, 24 hrs

autophagy modulation in
4.56 MCF-7-GFP-LC3 cells at **1 ng/mL**

FIGURE 4.10 mTOR-independent small molecule enhancers of rapamycin (SMERs): chemical structures, **4.51–4.57**.

is linked to its antagonistic activity on dopamine D_2 receptors. D_2 receptors inhibit the production of reactive oxygen species (ROS), while the effect on autophagy of sertindole requires production of ROS [198].

A smaller, 120-member collection of heterocycles tested on a similar GFP-LC3-II-based assay format in SH-SY5Y cells identifies the quinazoline GTM-1 (4.52) as an mTOR-independent autophagy inducer [199]. GTM-1 prevents the aggregation of Aβ oligomers in an Aβ precursor-overexpressing cell line, and fully protects SH-SY5Y cells from exogenous Aβ toxicity through autophagy induction at μM concentrations. GTM-1 reduces both soluble Aβ levels and Aβ deposition without being toxic when administered for 2 months to triple APP-tau-PS1 TG mice [199]. The structural similarity between quinazoline-containing SMER28 and GTM-1, and their *in vitro* and *in vivo* activity on Aβ pathologies, should stimulate further studies on similar compounds.

A screening campaign on GFP-LC3-II-expressing osteosarcoma U2OS cells provides the signal transducer and activator of transcription 3 (STAT3) inhibitor JSI-124 (cucurbitacin I, 4.53) as an autophagy inducer [200] out of a 1400-membered compound collection [201]. Autophagy induction is confirmed in a secondary *in vitro* and *in vivo* profiling panel for JSI-124 and two other STAT3 inhibitors. STAT3 inhibitors prevent the interaction between STAT3 and protein kinase R (PKR, EIF2AK2). When unbound, PKR phosphorylates the translational regulator eIF2α and activates autophagy [200]. Further studies determine that autophagy inducers, such as palmitate and other fatty acids, affect the newly validated, pro-autophagy STAT3–PKR–eIF2α pathway [202]. In particular, fatty acids disrupt the STAT3–PKR interaction and increase the levels of phosphorylated eIF2α and c-jun-N-terminal kinase 1 (JNK-1) [202].

HTS on a 2000-membered collection of monkey kidney CV-1 in origin SV40-carrying (COS-7) cells, expressing the GFP-LC3-II construct, and hit confirmation on human fibrosarcoma HT 1080 cells, identifies the tetrahydrobenzothienopyrimidine autophagonizer (4.54) [203]. A smaller screening (100 natural products from plants) on GFP-LC3-II-expressing HEK293 cells produces amurensin G (4.55) [204]. Amurensin G reduces neurotoxicity (accumulation of ubiquitinated proteins, including α-synuclein) through autophagy activation in a cellular model [204].

1120 Drug-like compounds are tested on an MCF-7 cell line expressing multiple autophagy-measuring constructs (GFP-LC3-II/primary screen; dual RFP-GFP-LC3-II and GFP-Rab7/secondary screen), using flow cytometry to evaluate their effect on autophagic and endolysosomal flux [178]. Out of 38 confirmed autophagy inducers, eight cardiac glycosides (in particular digoxin, 4.56) show potency and specificity. Cardiac glycosides are known Na^+-K^+-ATPase inhibitors, leading to increased intracellular calcium levels that may be responsible for their effects on autophagy [178,205]. In an independent study on non-small cell lung carcinoma

(NSCLC) cells digoxin induces autophagy, although in an AMPK/mTOR- and extracellular-regulated signal kinase 1 (ERK1)/ERK2-dependent manner [206]. A putative multi-targeted action on autophagy should grant further efforts to elucidate the effects of digoxin and other cardiac glycosides.

Finally, the LD-based HCS of a 1400-member polyheterocyclic library on HeLa cells identifies P29A03 (**4.57**, Figure 4.10) as an mTOR-independent autophagy inducer that increases the expression levels of beclin and the formation of phagophores in cells [181].

4.4 RECAP

This chapter deals with small molecule modulators of neuropathological alterations related to autophagy-related protein misfolding and aggregation in general, and to their effects on tau and/or tau-connected events in particular. A single molecular target and a chemical biology-driven approach are arbitrarily chosen, and discussed in detail in the companion biology book [23]. Seventy-two scaffolds acting on selected mechanisms are reported in Figures 4.1 to 4.10, and are briefly summarized in Table 4.1. The chemical core of each scaffold/compound is structurally defined; its molecular target is mentioned; the developing laboratory (either public or private) is listed; and the development status—according to publicly available information—is finally provided.

TABLE 4.1   Compounds **4.1–4.57**: Chemical Class, Target, Developing Organization, Development Status

Number	Chemical cpd./class	Target	Organization	Dev. status
4.1a	Rapamycin-sirolimus-rapamune™	mTOR inhibition	Pfizer	MKTD
4.1b	Everolimus-RAD-001-afinitor™	mTOR inhibition	Novartis	MKTD
4.1c	Temsirolimus-CCI-779-Torisel™	mTOR inhibition	Wyeth	MKTD
4.1d	Ridaforolimus-AP23573-deforolimus	mTOR inhibition	Merck/ARIAD	Ph III
4.1e	Umirolimus-TRM-986-Biolimus A9	mTOR inhibition	Biosensors Intl.	MKTD
4.1f	Zotarolimus-ABT-578	mTOR inhibition	Medtronics	MKTD
4.2	Tricyclic pyrimidine-morpholines, PI-103	PI3K/mTOR dual inhibitors	Piramed Pharma	LO

(Continued)

TABLE 4.1   Compounds **4.1–4.57:** Chemical Class, Target, Developing Organization, Development Status (*cont.*)

Number	Chemical cpd./class	Target	Organization	Dev. status
4.3a,b	Tricyclic pyrimidine-morpholines PI-540 (a), PI-620 (b)	PI3K/mTOR dual inhibitors	Piramed Pharma	PE
4.4	Tricyclic pyrimidine-morpholine GDC-0980	PI3K/mTOR dual inhibitors	Genentech	Ph II
4.5	Imidazo[4,5-c]quinolones, NVP-BEZ235	PI3K/mTOR dual inhibitors	Novartis	Ph II
4.6	Imidazo[4,5-c]quinolones, NVP-BGT226	PI3K/mTOR dual inhibitors	Novartis	PhI/II
4.7	Quinoxalines, XL765-SAR245408	PI3K/mTOR dual inhibitors	Sanofi	Ph I/II
4.8	Quinolones, GSK2126458	PI3K/mTOR dual inhibitors	GSK	Ph I
4.9	Triazolopyrimidines, PKI-402	PI3K/mTOR dual inhibitors	Wyeth	PE
4.10	Morpholinotriazines, PKI-587/PF-05212384	PI3K/mTOR dual inhibitors	Pfizer	Ph II
4.11, 4.12	Morpholylbenzopyra-nones, LY294002 (4.11), pro-drug (SF1129, 4.12)	PI3K/mTOR dual inhibitors	Eli Lilly	Ph I
4.13	Pyrazolo[3,4-d]pyrimidines, PP242	Pan-mTOR inhibition	University of San Francisco	PE
4.14	Pyrazolo[3,4-d]pyrimidines, MLN0128-INK128	Pan-mTOR inhibition	Millennium	Ph I/II
4.15	Pyrazolo[3,4-d]pyrimidines, WAY600	Pan-mTOR inhibition	Wyeth	LO
4.16	Pyrazolo[3,4-d]pyrimidines, WYE-132-WYE125132	Pan-mTOR inhibition	Wyeth	PE
4.17a	pyrido[2,3-d]pyrimidines, Ku0063794	Pan-mTOR inhibition	AstraZeneca	LO
4.17b	pyrido[2,3-d]pyrimidines, AZD8055	Pan-mTOR inhibition	AstraZeneca	Ph I
4.17c	pyrido[2,3-d]pyrimidines, AZD2014	Pan-mTOR inhibition	AstraZeneca	Ph II
4.18	Imidazo[1,5-f][1,2,4]tri-azines, OSI-027	Pan-mTOR inhibition	Astellas	Ph I

TABLE 4.1 Compounds **4.1–4.57**: Chemical Class, Target, Developing Organization, Development Status *(cont.)*

Number	Chemical cpd./class	Target	Organization	Dev. status
4.19	Benzochromenones, palomid 529	Pan-mTOR inhibition	Paloma Pharmaceuticals	Ph I
4.20	Benzonaphthiridinones, torin1	Pan-mTOR inhibition	Dana Farber Cancer Inst., Boston, USA	LO
4.21	Benzonaphthiridinones, torin2	Pan-mTOR inhibition	Dana Farber Cancer Inst., Boston, USA	PE
4.22	Caffeine	mTORC1 regulation	Juntendo University, Tokyo	FS
4.23	Phosphatidic acid	mTORC1 regulation	University of Leicester, UK	FS
4.24	Dexamethasone	mTORC1 regulation	Case Western Reserve Univ., Cleveland, USA	MKTD
4.25	Metformin	mTORC1 regulation	Several	MKTD
4.26	Aspirin	mTORC1 regulation	Several	MKTD
4.27	Δ^9-tetrahydrocannabinol	mTORC1 regulation	Alcala Univ., Madrid, Spain	MKTD
4.28, 4.29	Resveratrol (4.28), resveratrol analogs (4.29)	mTORC1 regulation	Feinstein Institute, New York, USA	LO
4.30	Niclosamide	mTOR-dependent SMER	University of British Columbia, Canada	DD
4.31	Amiodarone	mTOR-dependent SMER	University of British Columbia, Canada	DD
4.32	Rottlerin	mTOR-dependent SMER	University of British Columbia, Canada	DD
4.33	Perhexiline	mTOR-dependent SMER	University of British Columbia, Canada	DD
4.34	Fluspirilene	mTOR-independent SMER	Chinese Academy of Sciences, Shanghai, China	DD

(Continued)

TABLE 4.1   Compounds **4.1–4.57:** Chemical Class, Target, Developing Organization, Development Status (*cont.*)

Number	Chemical cpd./class	Target	Organization	Dev. status
4.35	Pimozide	mTOR-independent SMER	Chinese Academy of Sciences, Shanghai, China	DD
4.36	Trifluoperazine	mTOR-independent SMER	Chinese Academy of Sciences, Shanghai, China	DD
4.37	Niguldipine	mTOR-independent SMER	Chinese Academy of Sciences, Shanghai, China	DD
4.38	Loperamide	mTOR-independent SMER	Chinese Academy of Sciences, Shanghai, China	DD
4.39a–d	Phenothiazine-N-propylamines	mTOR-independent SMER	University of San Francisco, USA	DD
4.40	Phenoxazine-N-butylamines, 10-NCP	mTOR-independent SMER	University of San Francisco, USA	DD
4.41a	Penfluridol	mTOR-independent SMER	Chinese Academy of Sciences, Shanghai, China	DD
4.41b-e	Diphenylpropylpiperidines	mTOR-independent SMERs	Chinese Academy of Sciences, Shanghai, China	DD
4.42	Aminopyrimidinones, SMER10	mTOR-independent SMER	University of Cambridge, UK	DD
4.43	Vinylogous amides, SMER18	mTOR-independent SMER	University of Cambridge, UK	DD
4.44	Bromoquinazolines, SMER28	mTOR-independent SMER	University of Cambridge, UK	DD
4.45	Verapamil	mTOR-independent SMER	University of Cambridge, UK	DD
4.46	Nimodipine	mTOR-independent SMER	University of Cambridge, UK	DD

TABLE 4.1 Compounds **4.1–4.57**: Chemical Class, Target, Developing Organization, Development Status *(cont.)*

Number	Chemical cpd./class	Target	Organization	Dev. status
4.47	Minoxidil	mTOR-independent SMER	University of Cambridge, UK	DD
4.48	Clonidine	mTOR-independent SMER	University of Cambridge, UK	DD
4.49	Rilmenidine	mTOR-independent SMER	University of Cambridge, UK	DD
4.50	Hydroxamic acids, ARP101	mTOR-independent SMER	Asan Medical Center, Seoul, Korea	DD
4.51	Sertindole	mTOR-independent SMER	Kyung Hee University, Korea	DD
4.52	Quinazolines, GTM-1	mTOR-independent SMER	Second Military Medical University, Shanghai, China	DD
4.53	JSI-124-cucurbitacin I	mTOR-independent SMER	INSERM Villejuif, France	DD
4.54	Tetrahydrobenzothieno-pyrimidines, autophagonizer	mTOR-independent SMERs	Yonsei University, Korea	DD
4.55	Amurensin G	mTOR-independent SMER	Yonsei University, Korea	DD
4.56	Digoxin	mTOR-independent SMER	University of Heidelberg	DD
4.57	P29A03	mTOR-independent SMER	Seoul National University, Korea	DD

Not progressed, NP; early discovery, DD; lead optimization, LO; preclinical evaluation, PE; clinical Phase I-II-III, Ph I–Ph III; marketed, MKTD; food supplement, FS.

References

[1] Codogno, P.; Mehrpour, M.; Proikas-Cezanne, T. Canonical and non-canonical autophagy: variations on a common theme of self-eating? *Nat. Rev. Mol. Cell Biol.* **2012**, *13*, 7–12.

[2] Korolchuk, V. I.; Rubinsztein, D. C. On signals controlling autophagy: It's time to eat yourself healthy. *Biochemist* **2012**, *34*, 8–13.

[3] Mizushima, N.; Levine, B.; Cuervo, A. M.; Klionsky, D. J. Autophagy fights disease through cellular self-digestion. *Nature* **2008**, *451*, 1069–1075.

[4] He, C.; Klionsky, D. J. Regulation mechanisms and signaling pathways of autophagy. *Annu. Rev. Genet.* **2009**, *43*, 67–93.

[5] Kroemer, G.; Marino, G.; Levine, B. Autophagy and the integrated stress response. *Mol. Cell* **2010**, *40*, 280–293.

[6] Ravikumar, B.; Sarkar, S.; Davies, J. E.; Futter, M.; Garcia-Arencibia, M.; Green-Thompson, Z. W., et al. Regulation of mammalian autophagy in physiology and pathophysiology. *Physiol. Rev.* **2010**, *90*, 1383–1435.

[7] Puyal, J.; Ginet, V.; Grishchuk, Y.; Truttmann, A. C.; Clarke, P. G. H. Neuronal autophagy as a mediator of life and death: contrasting roles in chronic neurodegenerative and acute neural disorders. *The Neuroscientist* **2012**, *18*, 224–236.

[8] Choi, A. M. K.; Ryter, S. W.; Levine, B. Autophagy in human health and disease. *N. Engl. J. Med.* **2013**, *368*, 651–662.

[9] Suzuki, K.; Ohsumi, Y. Current knowledge of the pre-autophagosomal structure (Pas). *FEBS Lett* **2010**, *584*, 1280–1286.

[10] Mizushima, N. The role of the Atg1/ULK1 complex in autophagy regulation. *Curr. Opin. Cell Biol.* **2010**, *22*, 132–139.

[11] Backer, J. M. The regulation and function of Class III PI3Ks: novel roles for Vps34. *Biochem. J.* **2008**, *410*, 1–17.

[12] He, C.; Levine, B. The Beclin 1 interactome. *Curr. Opin. Cell Biol.* **2010**, *22*, 140–149.

[13] Orsi, A.; Razi, M.; Dooley, H. C.; Robinson, D.; Weston, A. E.; Collinson, L. M.; Tooze, S. A. Dynamic and transient interactions of Atg9 with autophagosomes, but not membrane integration, are required for autophagy. *Mol. Biol. Cell* **2012**, *23*, 1860–1873.

[14] Young, A. R. J.; Chan, E. Y. W.; Hu, X. W.; Köchl, R.; Crawshaw, S. G.; High, S., et al. Starvation and ULK1-dependent cycling of mammalian Atg9 between the TGN and endosomes. *J. Cell Sci.* **2006**, *119*, 3888–3900.

[15] Esclatine, A.; Chaumorcel, M.; Codogno, P. Macroautophagy signaling and regulation. *Curr. Top. Microbiol. Immunol.* **2009**, *335*, 33–70.

[16] Kihara, A.; Noda, T.; Ishihara, N.; Ohsumi, Y. Two distinct Vps34 phosphatidylinositol 3-kinase complexes function in autophagy and carboxypeptidase Y sorting in *Saccharomyces cerevisiae*. *J. Cell Biol.* **2001**, *152*, 519–530.

[17] Rogov, V.; Doetsch, V.; Johansen, T.; Kirkin, V. Interactions between autophagy receptors and ubiquitin-like proteins form the molecular basis for selective autophagy. *Mol. Cell* **2014**, *53*, 167–178.

[18] Shpilka, T.; Weidberg, H.; Pietrokovski, S.; Elazar, Z. Atg8: an autophagy-related ubiquitin-like protein family. *Genome Biol* **2011**, *12*, 226.

[19] Ichimura, Y.; Kirisako, T.; Takao, T.; Satomi, Y.; Shimonishi, Y.; Ishihara, N., et al. A ubiquitin like system mediates protein lipidation. *Nature* **2000**, *408*, 488–492.

[20] Tanida, I.; Sou, Y. S.; Ezaki, J.; Minematsu-Ikeguchi, N.; Ueno, T.; Kominami, E. HsAtg4B/HsApg4B/autophagin-1 cleaves the carboxyl termini of three human Atg8 homologues and delipidates microtubule-associated protein light chain 3- and GABAA receptor associated protein-phospholipid conjugates. *J. Biol. Chem.* **2004**, *279*, 36268–36276.

[21] Longatti, A.; Tooze, S. A. Vesicular trafficking and autophagosome formation. *Cell Death Differ* **2009**, *16*, 956–965.

[22] Fengsrud, M.; Lunde Sneve, M.; Øverbye, A.; Seglen, P. O. Structural aspects of mammalian autophagy. In *Autophagy*; Klionsky, D. J., Ed.; Landes Bioscience: Georgetown, TX, 2009; pp. 11–25.

[23] Seneci, P. *Protein misfolding and neurodegenerative diseases: Focus on disease-modifying targets.* Academic Press, **2014**, 278 pages.

[24] Vezina, C.; Kudelski, A.; Sehgal, S. N. Rapamycin (AY–22,989), a new antifungal antibiotic. I. Taxonomy of the producing streptomycete and isolation of the active principle. *J. Antibiot.* **1975**, *28*, 721–726.

[25] Heitman, J.; Movva, N. R.; Hall, M. N. Targets for cell cycle arrest by the immunosuppressant rapamycin in yeast. *Science* **1991**, *253*, 905–909.

[26] Sabatini, D. M.; Erdjument-Bromage, H.; Lui, M.; Tempst, P.; Snyder, S. H. RAFT1: a mammalian protein that binds to FKBP12 in a rapamycin-dependent fashion and is homologous to yeast TORs. *Cell* **1994**, *78*, 35–43.

[27] Benjamin, D.; Colombi, M.; Moroni, C.; Hall, M. N. Rapamycin passes the torch: a new generation of mTOR inhibitors. *Nature Rev. Drug Discov.* **2011**, *10*, 868–880.

[28] Yip, C. K.; Murata, K.; Walz, T.; Sabatini, D. M.; Kang, S. A. Structure of the human mTOR complex I and its implications for rapamycin inhibition. *Mol. Cell* **2010**, *38*, 768–774.

[29] Sarbassov, D. D.; Ali, S. M.; Sengupta, S.; Sheen, J. -H.; Hsu, P. P.; Bagley, A. F., et al. Prolonged rapamycin treatment inhibits mTORC2 assembly and Akt/PKB. *Mol. Cell* **2006**, *22*, 159–168.

[30] http://www.rapamune.com/Index.aspx.

[31] http://en.wikipedia.org/wiki/Sirolimus.

[32] http://www.reuters.com/article/2011/06/15/us-johnsonandjohnson-idUSTRE75E2PK20110615.

[33] Efeyan, A.; Sabatini, D. M. mTOR and cancer: many loops in one pathway. *Curr. Opin. Cell Biol.* **2010**, *22*, 169–176.

[34] Spilman, P.; Podlutskaya, N.; Hart, M. J.; Debnath, J.; Gorostiza, O.; Bredesen, D., et al. Inhibition of mTOR by rapamycin abolishes cognitive deficits and reduces amyloid-beta levels in a mouse model of Alzheimer's disease. *PLoS One* **2010**, *5*, e9979.

[35] Yin, L.; Ye, S.; Chen, Z.; Zeng, Y. Rapamycin preconditioning attenuates transient focal cerebral ischemia/reperfusion injury in mice. *Int. J. Neurosci.* **2012**, *122*, 748–756.

[36] Yamaki, K.; Yoshino, S. Preventive and therapeutic effects of rapamycin, a mammalian target of rapamycin inhibitor, on food allergy in mice. *Allergy* **2012**, *67*, 1259–1270.

[37] Ramos, F. J.; Chen, S. C.; Garelick, M. G.; Dai, D. -F.; Liao, C. -Y.; Schreiber, K. H., et al. Rapamycin reverses elevated mTORC1 signaling in lamin A/C-deficient mice, rescues cardiac and skeletal muscle function, and extends survival. *Sci. Transl. Med.* **2012**, *4*, 144ra103.

[38] Ehninger, D.; Han, S.; Shilyansky, C.; Zhou, Y.; Li, W.; Kwiatkowski, D. J.; Ramesh, V.; Silva, A. J. Reversal of learning deficits in a Tsc2 +/− mouse model of tuberous sclerosis. *Nat. Med.* **2008**, *14*, 843–848.

[39] Lieberthal, W.; Levine, J. S. Mammalian target of rapamycin and the kidney. II. Pathophysiology and therapeutic implications. *Am. J. Physiol. Renal Physiol.* **2012**, *303*, F180–F191.

[40] Lamming, D. W.; Ye, L.; Sabatini, D. M.; Baur, J. A. Rapalogs and mTOR inhibitors as anti-aging therapeutics. *J. Clin. Invest.* **2013**, *123*, 980–989.

[41] Malagelada, C.; Jin, Z. H.; Jackson-Lewis, V.; Przedborski, S.; Greene, L. A. Rapamycin protects against neuron death in in vitro and in vivo models of Parkinson's disease. *J. Neurosci.* **2010**, *30*, 1166–1175.

[42] Dehay, B.; Bové, J.; Rodríguez-Muela, N.; Perier, C.; Recasens, A.; Boya, P.; Vila, M. Pathogenic lysosomal depletion in Parkinson's disease. *J. Neurosci.* **2010**, *30*, 12535–12544.

[43] Santini, E.; Heiman, M.; Greengard, P.; Valjent, E.; Fisone, G. Inhibition of mTOR signaling in Parkinson's disease prevents L–DOPA–induced dyskinesia. *Sci. Signal.* **2009**, *2*, ra36.

[44] Jiang, J.; Jiang, J.; Zuo, Y.; Gu, Z. Rapamycin protects the mitochondria against oxidative stress and apoptosis in a rat model of Parkinson's disease. *Int. J. Mol. Med.* **2013**, *31*, 825–832.

[45] Crews, L.; Spencer, B.; Desplats, P.; Patrick, C.; Paulino, A.; Rockenstein, E., et al. Selective molecular alterations in the autophagy pathway in patients with Lewy body disease and in models of α-synucleinopathy. *PLoS ONE* **2010**, *5*, e9313.

[46] Ravikumar, B.; Vacher, C.; Berger, Z.; Davies, J. E.; Luo, S.; Oroz, L. G., et al. Inhibition of mTOR induces autophagy and reduces toxicity of polyglutamine expansions in fly and mouse models of Huntington disease. *Nat. Genet.* **2004**, *36*, 585–595.

[47] Sarkar, S.; Ravikumar, B.; Floto, R. A.; Rubinsztein, D. C. Rapamycin and mTOR-independent autophagy inducers ameliorate toxicity of polyglutamine-expanded huntingtin and related proteinopathies. *Cell Death Differ* **2009**, *16*, 46–56.

[48] Berger, Z.; Ravikumar, B.; Menzies, F. M.; Garcia Oroz, L.; Underwood, B. R.; Pangalos, M. N., et al. Rapamycin alleviates toxicity of different aggregate-prone proteins. *Hum. Mol. Genet.* **2006**, *15*, 433–442.

[49] Wyttenbach, A.; Hands, S.; King, M. A.; Lipkow, K.; Tolkovsky, A. M. Amelioration of protein misfolding disease by rapamycin: translation or autophagy? *Autophagy* **2008**, *4*, 542–545.

[50] Wang, I. -F.; Guo, B. -S.; Liu, Y. -C.; Wu, C. -C.; Yang, C. -H.; Tsai, K. -J.; Chen, C. -K. J. Autophagy activators rescue and alleviate pathogenesis of a mouse model with proteinopathies of the TAR DNA-binding protein 43. *Proc. Natl. Acad. Sci. U.S.A.* **2012**, *109*, 15024–15029.

[51] Graziotto, J. J.; Cao, K.; Collins, F. S.; Krainc, D. Rapamycin activates autophagy in Hutchinson-Gilford progeria syndrome: implications for normal aging and age-dependent neurodegenerative disorders. *Autophagy* **2012**, *8*, 147–151.

[52] Mendelsohn, A. R.; Larrick, J. W. Rapamycin as an antiaging therapeutic? Targeting mammalian target of rapamycin to treat Hutchinson-Gilford progeria and neurodegenerative diseases. *Rejuvenation Res* **2011**, *14*, 437–441.

[53] Zhang, X.; Li, L.; Chen, S.; Yang, D.; Wang, Y.; Zhang, X., et al. Rapamycin treatment augments motor neuron degeneration in SOD1G93A mouse model of amyotrophic lateral sclerosis. *Autophagy* **2011**, *7*, 412–425.

[54] Cai, Z.; Yan, L. -J. Rapamycin, autophagy, and Alzheimer's disease. *J. Biochem. Pharmacol. Res.* **2013**, *1*, 84–90.

[55] Santos, R. X.; Correia, S. C.; Cardoso, S.; Carvalho, C.; Santos, M. S.; Moreira, P. I. Effects of rapamycin and TOR on aging and memory: implications for Alzheimer's disease. *J. Neurochem.* **2011**, *117*, 927–936.

[56] Halloran, J.; Hussong, S. A.; Burbank, R.; Podlutskaya, N.; Fischer, K. E.; Sloane, L. B., et al. Chronic inhibition of mammalian target of rapamycin by rapamycin modulates cognitive and non-cognitive components of behavior throughout lifespan in mice. *Neuroscience* **2012**, *223*, 102–113.

[57] Majumder, S.; Caccamo, A.; Medina, D. X.; Benavides, A. D.; Javors, M. A.; Kraig, E., et al. Lifelong rapamycin administration ameliorates age-dependent cognitive deficits by reducing IL-1beta and enhancing NMDA signaling. *Aging Cell* **2012**, *11*, 326–335.

[58] Marobbio, C. M.; Pisano, I.; Porcelli, V.; Lasorsa, F. M.; Palmieri, L. Rapamycin reduces oxidative stress in frataxin-deficient yeast cells. *Mitochondrion* **2012**, *12*, 156–161.

[59] Chen, H. C.; Fong, T. H.; Hsu, P. W.; Chiu, W. T. Multifaceted effects of rapamycin on functional recovery after spinal cord injury in rats through autophagy promotion, anti-inflammation, and neuroprotection. *J. Surg. Res.* **2013**, *179*, e203–e210.

[60] Tang, S. J.; Reis, G.; Kang, H.; Gingras, A. -C.; Sonenberg, N.; Schuman, E. M. A rapamycin-sensitive signaling pathway contributes to long-term synaptic plasticity in the hippocampus. *Proc. Natl. Acad. Sci. U.S.A.* **2002**, *99*, 467–472.

[61] Ma, T.; Hoeffer, C. A.; Capetillo-Zarate, E.; Yu, F.; Wong, H.; Lin, M. T., et al. Dysregulation of the mTOR pathway mediates impairment of synaptic plasticity in a mouse model of Alzheimer's disease. *PLoS One* **2010**, *5*, e12845.

[62] Fang, Y.; Westbrook, R.; Hill, C.; Boparai, R. K.; Arum, O.; Spong, A., et al. Duration of rapamycin treatment has differential effects on metabolism in mice. *Cell Metab* **2013**, *17*, 456–462.

[63] Lee, J. A.; Gao, F. B. Regulation of Abeta pathology by beclin 1: a protective role for autophagy? *J. Clin. Invest.* **2008**, *118*, 2015–2018.

[64] Nixon, R. A.; Wegiel, J.; Kumar, A.; Yu, W. H.; Peterhoff, C.; Cataldo, A.; Cuervo, A. M. Extensive involvement of autophagy in Alzheimer disease: an immuno-electron microscopy study. *J. Neuropathol. Exp. Neurol.* **2005**, *64*, 113–122.

[65] Boland, B.; Kumar, A.; Lee, S.; Platt, F. M.; Wegiel, J.; Yu, W. H.; Nixon, R. A. Autophagy induction and autophagosome clearance in neurons: relationship to autophagic pathology in Alzheimer' disease. *J. Neurosci.* **2008**, *28*, 6926–6937.

[66] Lipinski, M. M.; Zheng, B.; Lu, T.; Yan, Z.; Py, B. F.; Ng, A., et al. Genome-wide analysis reveals mechanisms modulating autophagy in normal brain aging and in Alzheimer's disease. *Proc. Natl. Acad. Sci. U.S.A.* **2010**, *107*, 14164–14169.

[67] Nixon, R. A. Autophagy, amyloidogenesis and Alzheimer disease. *J. Cell. Sci.* **2007**, *120*, 4081–4091.

[68] Caccamo, A.; Maldonado, M. A.; Majumder, S.; Medina, D. X.; Holbein, W.; Magrí, A.; Oddo, S. Naturally secreted amyloid-β increases mammalian target of rapamycin (mTOR) activity via a PRAS40–mediated mechanism. *J. Biol. Chem.* **2011**, *286*, 8924–8932.

[69] Spilman, P.; Podlutskaya, N.; Hart, M. J.; Debnath, J.; Gorostiza, O.; Bredesen, D., et al. Inhibition of mTOR by rapamycin abolishes cognitive deficits and reduces amyloid-beta levels in a mouse model of Alzheimer's disease. *PLoS One* **2010**, *5*, e9979.

[70] Caccamo, A.; Majumder, S.; Richardson, A.; Strong, R.; Oddo, S. Molecular interplay between mammalian target of rapamycin (mTOR), amyloid-beta, tau: effects on cognitive impairments. *J. Biol. Chem.* **2010**, *285*, 13107–13120.

[71] Xue, Z.; Zhang, S.; Huang, L.; He, Y.; Fang, R.; Fang, Y. Increased expression of beclin-1-dependent autophagy protects against beta-amyloid-induced cell injury in PC12 cells. *J. Mol. Neurosci.* **2013**, *51*, 180–186.

[72] Lafay-Chebassier, C.; Perault-Pochat, M. -C.; Page, G.; Rioux Bilan, A.; Damjanac, M.; Pain, S., et al. The immunosuppressant rapamycin exacerbates neurotoxicity of Abeta peptide. *J. Neurosci. Res.* **2006**, *84*, 1323–1334.

[73] Zhang, S.; Salemi, J.; Hou, H.; Zhu, Y.; Mori, T.; Giunta, B., et al. Rapamycin promotes beta-amyloid production via ADAM-10 inhibition. *Biochem. Biophys. Res. Commun.* **2010**, *398*, 337–341.

[74] Cai, Z.; Zhao, B.; Li, K.; Zhang, L.; Li, C.; Quazi, S. H.; Tan, Y. Mammalian target of rapamycin: a valid therapeutic target through the autophagy pathway for Alzheimer's disease? *J. Neurosci. Res.* **2012**, *90*, 1105–1118.

[75] Meske, V.; Albert, F.; Ohm, T. G. Coupling of mammalian target of rapamycin with phosphoinositide 3-kinase signaling pathway regulates protein phosphatase 2A- and glycogen synthase kinase-3-dependent phosphorylation of Tau. *J. Biol. Chem.* **2008**, *283*, 100–109.

[76] Liu, Y.; Su, Y.; Wang, J.; Sun, S.; Wang, T.; Qiao, X., et al. Rapamycin decreases tau phosphorylation at Ser214 through regulation of cAMP dependent kinase. *Neurochem Int* **2013**, *62*, 458–467.

[77] Ma, Y. Q.; Wu, D. K.; Liu, J. K. mTOR and tau phosphorylated proteins in the hippocampal tissue of rats with type 2 diabetes and Alzheimer's disease. *Mol. Med. Rep.* **2013**, *7*, 623–627.

[78] Tang, Z.; Bereczki, E.; Zhang, H.; Wang, S.; Li, C.; Ji, X., et al. Mammalian target of rapamycin (mTor) mediates tau protein dyshomeostasis. *J. Biol. Chem.* **2013**, *288*, 25556–25570.

[79] Caccamo, A.; Magrì, A.; Medina, D. X.; Wisely, E. V.; Lopez-Aranda, M. F.; Silva, A. J.; Oddo, S. mTOR regulates tau phosphorylation and degradation: implications for Alzheimer's disease and other tauopathies. *Aging Cell* **2013**, *12*, 370–380.

[80] Ozcelik, S.; Fraser, G.; Castets, P.; Schaeffer, V.; Skachokova, Z.; Breu, K., et al. Rapamycin attenuates the progression of tau pathology in P301S tau transgenic mice. *PLoS ONE* **2013**, *8*, e62459.

[81] Kolosova, N. G.; Vitovtov, A. O.; Muraleva, N. A.; Akulov, A. E.; Stefanova, N. A.; Blagosklonny, M. V. Rapamycin suppresses brain aging in senescence-accelerated OXYS rats. *Aging* **2013**, *5*, 474–484.

[82] Gabardi, S.; Baroletti, S. A. Everolimus: a proliferation signal inhibitor with clinical applications in organ transplantation, oncology, and cardiology. *Pharmacotherapy* **2010**, *30*, 1044–1056.

[83] Yao, J. C.; Shah, M. H.; Ito, T.; Lombard Bohas, C.; Wolin, E. M.; Van Cutsem, E., et al. Everolimus for advanced pancreatic neuroendocrine tumors. *N. Engl. J. Med.* **2011**, *364*, 514–523.

[84] Roscic, A.; Baldo, B.; Crochemore, C.; Marcellin, D.; Paganetti, P. Induction of autophagy with catalytic mTOR inhibitors reduces huntingtin aggregates in a neuronal cell model. *J. Neurochem.* **2011**, *119*, 398–407.

[85] Fox, J. H.; Connor, T.; Chopra, V.; Dorsey, K.; Kama, J. A.; Bleckmann, D., et al. The mTOR kinase inhibitor Everolimus decreases S6 kinase phosphorylation but fails to reduce mutant huntingtin levels in brain and is not neuroprotective in the R6/2 mouse model of Huntington's disease. *Mol. Neurodeg.* **2010**, *5*, 26.

[86] Rini, B. I. Temsirolimus, an inhibitor of mammalian target of rapamycin. *Clin. Cancer Res.* **2008**, *14*, 1286–1290.

[87] Hess, G.; Herbrecht, R.; Romaguera, J.; Verhoef, G.; Crump, M.; Gisselbrecht, C., et al. Phase III study to evaluate temsirolimus compared with investigator's choice therapy for the treatment of relapsed or refractory mantle cell lymphoma. *J. Clin. Oncol.* **2009**, *27*, 3822–3829.

[88] Menzies, F. M.; Huebener, J.; Renna, M.; Bonin, M.; Riess, O.; Rubinsztein, D. C. Autophagy induction reduces mutant ataxin-3 levels and toxicity in a mouse model of spinocerebellar ataxia type 3. *Brain* **2010**, *133*, 93–104.

[89] Mita, M.; Sankhala, K.; Abdel-Karim, I.; Mita, A.; Giles, F. Deforolimus (AP23573) a novel mTOR inhibitor in clinical development. *Expert Opin. Investig. Drugs* **2008**, *17*, 1947–1954.

[90] Grube, E.; Buellesfeld, L. BioMatrix Biolimus A9-eluting coronary stent: a next-generation drug-eluting stent for coronary artery disease. *Exp. Rev. Med. Dev.* **2006**, *3*, 731–741.

[91] Burke, S.; Kuntz, R. E.; Schwartz, L. B. Zotarolimus (ABT-578) eluting stents. *Adv. Drug Delivery Rev.* **2006**, *58*, 437–446.

[92] Zoncu, R.; Efeyan, A.; Sabatini, D. M. mTOR: from growth signal integration to cancer, diabetes and ageing. *Nature Rev. Mol. Cell Biol.* **2011**, *12*, 21–35.

[93] Fan, Q. W.; Knight, Z. A.; Goldenberg, D. D.; Yu, W.; Mostov, K. E.; Shokoe, D., et al. A dual PI3 kinase/mTOR inhibitor reveals emergent efficacy in glioblastoma. *Cancer Cell* **2006**, *9*, 341–349.

[94] Raynaud, F. I.; Eccles, S. A.; Patel, S.; Alix, S.; Box, G.; Chuckowree, I., et al. Biological properties of potent inhibitors of class I phosphatidylinositide 3-kinases: from PI-103 through PI-540, PI-620 to the oral agent GDC-0941. *Mol. Cancer Ther.* **2009**, *8*, 1725–1738.

[95] Wallin, J. J.; Edgar, K. A.; Guan, J.; Berry, M.; Prior, W. W.; Lee, L., et al. GDC-0980 is a novel class I PI3K/mTOR kinase inhibitor with robust activity in cancer models driven by the PI3K pathway. *Mol. Cancer Ther.* **2011**, *10*, 2426–2436.

[96] Dolly, S.; Wagner, A. J.; Bendell, J. C.; Yan, Y.; Ware, J. A.; Mazina, K. E., et al. A first-in-human, Phase 1 study to evaluate the dual PI3K/mTOR inhibitor GDC-0980 administered QD in patients with advanced solid tumors or non-Hodgkin's lymphoma. *J. Clin. Oncol.* **2010**, *28*, 3079.

[97] Maira, S. M.; Stauffer, F.; Brueggen, J.; Furet, P.; Schnell, C.; Fritsch, C., et al. Identification and characterization of NVP-BEZ235, a new orally available dual phosphatidylinositol 3-kinase/mammalian target of rapamycin inhibitor with potent in vivo antitumor activity. *Mol. Cancer Ther.* **2008**, *7*, 1851–1863.

[98] Chang, K. Y.; Tsai, S. Y.; Wu, C. M.; Yen, C. J.; Chuang, B. F.; Chang, J. Y. Novel phosphoinositide 3-kinase/mTOR dual inhibitor, NVP-BGT226, displays potent growth-inhibitory activity against human head and neck cancer cells in vitro and in vivo. *Clin. Cancer Res.* **2011**, *17*, 7116–7126.

[99] Burris, H.; Rodon, J.; Sharma, S.; Herbst, R.; Tabernero, J.; Infante, J., et al. First-in-human phase I study of the oral PI3K inhibitor BEZ235 in patients (pts) with advanced solid tumors. In ASCO Annual Meeting. *J. Clin. Oncol.* **2010**, *28* Abstract 3005.

[100] Markman, B.; Tabernero, J.; Krop, I.; Shapiro, G. I.; Siu, L.; Chen, L. C., et al. Phase I safety, pharmacokinetic, and pharmacodynamic study of the oral phosphatidylinositol-3-kinase and mTOR inhibitor BGT226 in patients with advanced solid tumors. *Ann. Oncol.* **2012**, *23*, 2399–2408.

[101] Dai, C.; Zhang, B.; Liu, X.; Ma, S.; Yang, Y.; Yao, Y., et al. Inhibition of PI3K/AKT/mTOR pathway enhances temozolomide-induced cytotoxicity in pituitary adenoma cell lines in vitro and xenografted pituitary adenoma in female nude mice. *Endocrinology* **2013**, *154*, 1247–1259.

[102] Papadopoulos, K. P.; Markman, B.; Tabernero, J.; Patnaik, A.; Heath, E. I.; DeCillis, A., et al. A Phase I dose-escalation study of the safety, pharmacokinetics (PK), and pharmacodynamics (PD) of a novel PI3K inhibitor, XL765, administered orally to patients (pts) with advanced solid tumors. *J. Clin. Oncol.* **2008**, *26*, 3510.

[103] Knight, S. D.; Adams, N. D.; Burgess, J. L.; Chaudhari, A. M.; Darcy, M. G.; Donatelli, C. A., et al. Discovery of GSK2126458, a highly potent inhibitor of PI3K and the mammalian target of rapamycin. *ACS Med. Chem. Lett.* **2010**, *1*, 39–43.

[104] Dehnhardt, C. M.; Venkatesan, A. M.; Delos Santos, E.; Chen, Z.; Santos, O.; Ayral-Kaloustian, S., et al. Lead optimization of N-3-substituted 7-morpholinotriazolopyrimidines as dual phosphoinositide 3-kinase/mammalian target of rapamycin inhibitors: Discovery of PKI-402. *J. Med. Chem.* **2010**, *53*, 798–810.

[105] Mallon, R.; Hollander, I.; Feldberg, L.; Lucas, J.; Soloveva, V.; Venkatesan, A., et al. Antitumor efficacy profile of PKI-402, a dual phosphatidylinositol 3-kinase/mammalian target of rapamycin inhibitor. *Mol. Cancer Ther.* **2010**, *9*, 976–984.

[106] Venkatesan, A. M.; Dehnhardt, C. M.; Delos Santos, E.; Chen, Z.; Dos Santos, O.; Ayral-Kaloustian, S., et al. Bis-Morpholino-1,3,5-triazine derivatives: potent, ATP-competitive phosphatidylinositol-3-kinase (PI3K)/mammalian target of rapamycin (mTOR) inhibitors: discovery of PKI-587 a highly efficacious dual inhibitor. *J. Med. Chem.* **2010**, *53*, 2636–2645.

[107] Mallon, R.; Feldberg, L. R.; Lucas, J.; Chaudhary, I.; Dehnhardt, C.; Delos Santos, E., et al. Antitumor efficacy of PKI-587, a highly potent dual PI3K/mTOR kinase inhibitor. *Clin. Cancer Res.* **2011**, *17*, 3193–3203.

[108] Workman, P.; Raynaud, F. I.; Clarke, P. A.; Te Poele, R.; Eccles, S.; Kelland, L., et al. Pharmacological properties and in vitro and in vivo antitumor activity of the potent and selective PI3 kinase inhibitor PI103. *Eur. J. Cancer* **2004**, *40* 414A.

[109] Garlich, J. R.; De, P.; Dey, N.; Su, J. D.; Peng, X.; Miller, A., et al. A vascular targeted pan phosphoinositide 3-kinase inhibitor prodrug, SF1126, with antitumor and antiangiogenic activity. *Cancer Res* **2008**, *68*, 206–215.

[110] Mahadevan, D.; Chiorean, E. G.; Harris, W. B.; Von Hoff, D. D.; Stejskal-Barnett, A.; Qi, W., et al. Phase I pharmacokinetic and pharmacodynamics study of the pan-PI3K/mTORC vascular targeted prodrug SF1126 in patients with advanced solid tumours and B-cell malignancies. *Eur. J. Cancer* **2012**, *48*, 3319–3327.

[111] Janes, M. R.; Limon, J. J.; So, L.; Chen, J.; Lim, R. J.; Chavez, M. A., et al. Effective and selective targeting of leukemia cells using a TORC1/2 kinase inhibitor. *Nature Med* **2010**, *16*, 205–213.

[112] Apsel, B.; Blair, J. A.; Gonzalez, B.; Nazif, T. M.; Feldman, M. E.; Aizenstein, B., et al. Targeted polypharmacology: discovery of dual inhibitors of tyrosine and phosphoinositide kinases. *Nat. Chem. Biol.* **2008**, *4*, 691–699.

[113] Liu, Q.; Kirubakaran, S.; Hur, W.; Niepel, M.; Westover, K.; Thoreen, C. C., et al. Kinome-wide selectivity profiling of ATP-competitive mammalian target of rapamycin (mTOR) inhibitors and characterization of their binding kinetics. *J. Biol. Chem.* **2012**, *287*, 9742–9752.

[114] Janes, M. R.; Vu, C.; Mallya, S.; Shieh, M. P.; Limon, J. J.; Li, L. -S., et al. Efficacy of the investigational mTOR kinase inhibitor MLN0128/INK128 in models of B-cell acute lymphoblastic leukemia. *Leukemia* **2013**, *27*, 586–594.

[115] Jessen, K.; Wang, S.; Kessler, L.; Guo, X.; Kucharski, J.; Staunton, J., et al. INK128 is a potent and selective TORC1/2 inhibitor with broad oral antitumor activity. *Mol. Cancer Ther.* **2009**, *8* (Suppl. 12) Abstr. B148.

[116] Yu, K.; Toral-Barza, L.; Shi, C.; Zhang, W. G.; Lucas, J.; Shor, B., et al. Biochemical, cellular, and in vivo activity of novel ATPcompetitive and selective inhibitors of the mammalian target of rapamycin. *Cancer Res.* **2009**, *69*, 6232–6240.

[117] Yu, K.; Shi, C.; Toral-Barza, L.; Lucas, J.; Shor, B.; Kim, J. E., et al. Beyond rapalog therapy: preclinical pharmacology and antitumor activity of WYE-125132, an ATP-competitive and specific inhibitor of mTORC1 and mTORC2. *Cancer Res.* **2010**, *70*, 621–631.

[118] García-Martínez, J. M.; Moran, J.; Clarke, R. G.; Gray, A.; Cosulich, S. C.; Chresta, C. M.; Alessi, D. R. Ku–0063794 is a specific inhibitor of the mammalian target of rapamycin (mTOR). *Biochem. J.* **2009**, *421*, 29–42.

[119] Chresta, C. M.; Davies, B. R.; Hickson, I.; Harding, T.; Cosulich, S.; Critchlow, S. E., et al. AZD8055 is a potent, selective, and orally bioavailable ATP-competitive mammalian target of rapamycin kinase inhibitor with in vitro and in vivo antitumor activity. *Cancer Res.* **2010**, *70*, 288–298.

[120] Pike, K. G.; Malagu, K.; Hummersone, M. G.; Menear, K. A.; Duggan, H. M. E.; Gomez, S., et al. Optimization of potent and selective dual mTORC1 and mTORC2 inhibitors: The discovery of AZD8055 and AZD2014. *Bioorg. Med. Chem. Lett.* **2013**, *23*, 1212–1216.

[121] Malagu, K.; Duggan, H.; Menear, K.; Hummersone, M.; Gomez, S.; Bailey, C., et al. The discovery and optimisation of pyrido[2,3-d]pyrimidine-2,4-diamines as potent and selective inhibitors of mTOR kinase. *Bioorg. Med. Chem. Lett.* **2009**, *19*, 5950–5953.

[122] Naing, A.; Aghajanian, C.; Raymond, E.; Olmos, D.; Schwartz, G.; Oelmann, E., et al. Safety, tolerability, pharmacokinetics and pharmacodynamics of AZD8055 in advanced solid tumours and lymphoma. *Br. J. Cancer* **2012**, *107*, 1093–1099.

[123] Guichard, S. M.; Howard, Z.; Heathcote, D.; Roth, M.; Hughes, G.; Curwen, J., et al. AZD2014, a dual mTORC1 and mTORC2 inhibitor is differentiated from allosteric inhibitors of mTORC1 in ER+ breast cancer. *Cancer Res.* **2012**, *72* (Suppl. 1), 817.

[124] Bhagwat, S. V.; Gokhale, P. C.; Crew, A. P.; Cooke, A.; Yao, Y.; Mantis, C., et al. Preclinical characterization of OSI-027, a potent and selective inhibitor of mTORC1 and mTORC2: distinct from rapamycin. *Mol. Cancer Ther.* **2011**, *10*, 1394–1406.

[125] Tan, D. S.; Dumez, H.; Olmos, D.; Sandhu, S. K.; Hoeben, A.; Stephens, A. W., et al. First–in–human phase I study exploring three schedules of OSI–027, a novel small molecule TORC1/TORC2 inhibitor, in patients with advanced solid tumors and lymphoma. *J. Clin. Oncol.* **2010**, *28* (Suppl. 15) Abstr. 3006.

[126] Xue, Q.; Hopkins, B.; Perruzzi, C.; Udayakumar, D.; Sherris, D.; Benjamin, L. E. Palomid 529, a novel small-molecule drug, is a TORC1/TORC2 inhibitor that reduces tumor growth, tumor angiogenesis, and vascular permeability. *Cancer Res.* **2008**, *68*, 9551–9557.

[127] Dalal, M.; Jacobs-El, N.; Nicholson, B.; Tuo, J.; Chew, E.; Chan, C. -C., et al. Subconjunctival Palomid 529 in the treatment of neovascular age-related macular degeneration. *Graefe's Arch. Clin. Exp. Ophthalmol* **May 2013**, .

[128] Liu, Q.; Chang, J. W.; Wang, J.; Kang, S. A.; Thoreen, C. C.; Markhard, A., et al. Discovery of 1-(4-(4-propionylpiperazin-1-yl)-3-(trifluoromethyl)phenyl)-9-(quinolin-3-yl)benzo[h][1,6]naphthyridin-2(1H)-one as a highly potent, selective mammalian target of rapamycin (mTOR) inhibitor for the treatment of cancer. *J. Med. Chem.* **2010**, *53*, 7146–7155.

[129] Liu, Q.; Wang, J.; Kang, S. A.; Thoreen, C. C.; Hur, W.; Ahmed, T., et al. Discovery of 9-(6-aminopyridin-3-yl)-1-(3-(trifluoromethyl)-phenyl)benzo[h][1,6]naphthyridin-2(1H)-one (torin2) as a potent, selective, and orally available mammalian target of rapamycin (mTOR) inhibitor for treatment of cancer. *J. Med. Chem.* **2011**, *54*, 1473–1480.

[130] Thoreen, C. C.; Kang, S. A.; Chang, J. W.; Liu, Q.; Zhang, J.; Gao, Y., et al. An ATP-competitive mammalian target of rapamycin inhibitor reveals rapamycin-resistant functions of mTORC1. *J. Biol. Chem.* **2009**, *284*, 8023–8032.

[131] Liu, Q.; Xu, C.; Kirubakaran, S.; Zhang, X.; Hur, W.; Liu, Y., et al. Characterization of torin2, an ATP-competitive inhibitor of mTOR, ATM, and ATR. *Cancer Res* **2013**, *73*, 2574–2586.

[132] Hanson, K. K.; Ressurreicao, A. S.; Buchholz, K.; Prudencio, M.; Herman-Ornelas, J. D.; Rebelo, M., et al. Torins are potent antimalarials that block replenishment of Plasmodium liver stage parasitophorous vacuole membrane proteins. *Proc. Natl. Acad. Sci. U.S.A.* **2013**, *110*, E2838–E2847.

[133] Zhou, J.; Tan, S. -H.; Nicolas, V.; Bauvy, C.; Yang, N. -D.; Zhang, J., et al. Activation of lysosomal function in the course of autophagy via mTORC1 suppression and autophagosome-lysosome fusion. *Cell Res.* **2013**, *23*, 508–523.

[134] Bachar–Wikstrom, E.; Wikstrom, J. D.; Ariav, Y.; Tirosh, B.; Kaiser, N.; Cerasi, E.; Leibowitz, G. Stimulation of autophagy improves endoplasmic reticulum stress-induced diabetes. *Diabetes* **2013**, *62*, 1227–1237.

[135] Obara, I.; Tochiki, K. K.; Geranton, S. M.; Carr, F. B.; Lumb, B. M.; Liu, Q.; Hunt, S. P. Systemic inhibition of the mammalian target of rapamycin (mTOR) pathway reduces neuropathic pain in mice. *Pain* **2011**, *152*, 2582–2595.

[136] Saiki, S.; Sasazawa, Y.; Imamichi, Y.; Kawajiri, S.; Fujimaki, T.; Tanida, I., et al. Caffeine induces apoptosis by enhancement of autophagy via PI3K/Akt/mTOR/p70S6K inhibition. *Autophagy* **2011**, *7*, 176–187.

[137] Wanke, V.; Cameroni, E.; Uotila, A.; Piccolis, M.; Urban, J.; Loewith, R.; De Virgilio, C. Caffeine extends yeast lifespan by targeting TORC1. *Mol. Microbiol.* **2008**, *69*, 277–285.

[138] Fang, Y.; Vilella-Bach, M.; Bachmann, R.; Flanigan, A.; Chen, J. Phosphatidic acid-mediated mitogenic activation of mTOR signaling. *Science* **2001**, *294*, 1942–1945.

[139] Veverka, V.; Crabbe, T.; Bird, I.; Lennie, G.; Muskett, F. W.; Taylor, R. J.; Carr, M. D. Structural characterization of the interaction of mTOR with phosphatidic acid and a novel class of inhibitor: compelling evidence for a central role of the FRB domain in small molecule-mediated regulation of mTOR. *Oncogene* **2008**, *27*, 585–595.

[140] Morad, S. A. F.; Schmid, M.; Buechele, B.; Siehl, H. -U.; El Gafaary, M.; Lunov, O., et al. A novel semisynthetic inhibitor of the FRB domain of mammalian target of rapamycin blocks proliferation and triggers apoptosis in chemoresistant prostate cancer cells. *Mol. Pharmacol.* **2013**, *83*, 531–541.

[141] Molitoris, J. K.; McColl, K. S.; Swerdlow, S.; Matsuyama, M.; Lam, M.; Finkel, T. H., et al. Glucocorticoid elevation of dexamethasone-induced gene 2 (Dig2/RTP801/REDD1) protein mediates autophagy in lymphocytes. *J. Biol. Chem.* **2011**, *286*, 30181–30189.

[142] Dowling, R. J.; Zakikhani, M.; Fantus, I. G.; Pollak, M.; Sonenberg, N. Metformin inhibits mammalian target of rapamycin-dependent translation initiation in breast cancer cells. *Cancer Res.* **2007**, *67*, 10804–10812.

[143] Zhou, G.; Myers, R.; Li, Y.; Chen, Y.; Shen, X.; Fenyk-Melody, J., et al. Role of AMP-activated protein kinase in mechanism of metformin action. *J. Clin. Invest.* **2001**, *108*, 1167–1174.

[144] Kalender, A.; Selvaraj, A.; Kim, S. Y.; Gulati, P.; Brûlé, S.; Viollet, B., et al. Metformin, independent of AMPK, inhibits mTORC1 in a rag GTPase-dependent manner. *Cell Metab* **2010**, *11*, 390–401.

[145] Ben Sahra, I.; Regazzetti, C.; Robert, G.; Laurent, K.; Le Marchand-Brustel, Y.; Auberger, P., et al. Metformin, independent of AMPK, induces mTOR inhibition and cell-cycle arrest through REDD1. *Cancer Res.* **2011**, *71*, 4366–4372.

[146] Shi, W. -Y.; Xiao, D.; Wang, L.; Dong, L. -H.; Yan, Z. -X.; Shen, Z. -X., et al. Therapeutic metformin/AMPK activation blocked lymphoma cell growth via inhibition of mTOR pathway and induction of autophagy. *Cell Death Dis.* **2012**, *3*, e275.

[147] Soares, H. P.; Ni, Y.; Kisfalvi, K.; Sinnett-Smith, J.; Rozengurt, E. Different patterns of Akt and ERK feedback activation in response to rapamycin, active-site mTOR inhibitors and metformin in pancreatic cancer cells. *PLoS One* **2013**, *8*, e57289.

[148] Din, F. V. N.; Valanciute, A.; Houde, V. P.; Zibrova, D.; Green, K. A.; Sakamoto, K., et al. Aspirin inhibits mTOR signaling, activates AMP-activated protein kinase, and induces autophagy in colorectal cancer cells. *Gastroenterology* **2012**, *142*, 1504–1515.

[149] Strong, R.; Miller, R. A.; Astle, C. M.; Floyd, R. A.; Flurkey, K.; Hensley, K. L., et al. Nordihydroguaiaretic acid and aspirin increase lifespan of genetically heterogeneous male mice. *Aging Cell* **2008**, *7*, 641–650.

[150] Rothwell, P. M.; Fowkes, F. G.; Belch, J. F.; Ogawa, H.; Warlow, C. P.; Meade, T. W. Effect of daily aspirin on long-term risk of death due to cancer: analysis of individual patient data from randomised trials. *Lancet* **2011**, *377*, 31–41.

[151] Vara, D.; Salazar, M.; Olea-Herrero, N.; Guzman, M.; Velasco, G.; Díaz-Laviada, I. Antitumoral action of cannabinoids on hepatocellular carcinoma: role of AMPK-dependent activation of autophagy. *Cell Death Differ.* **2011**, *18*, 1099–1111.

[152] Pallauch, K.; Rimbach, G. Autophagy, polyphenols and healthy ageing. *Ageing Res. Rev.* **2013**, *12*, 237–252.

[153] Zhang, J.; Chiu, J. F.; Zhang, H.; Qi, T.; Tang, Q.; Ma, K., et al. Autophagic cell death induced by resveratrol depends on the Ca^{2+}/AMPK/mTOR pathway in A549 cells. *Biochem. Pharmacol.* **2013**, *86*, 317–328.

[154] Vingtdeux, V.; Chandakkar, P.; Zhao, H.; d'Abramo, C.; Davies, P.; Marambaud, P. Novel synthetic small-molecule activators of AMPK as enhancers of autophagy and amyloid-β peptide degradation. *FASEB J.* **2011**, *25*, 219–231.

[155] Gurusamy, N.; Lekli, I.; Mukherjee, S.; Ray, D.; Ahsan, K.; Gherghiceanu, M., et al. Cardioprotection by resveratrol: a novel mechanism via autophagy involving the mTORC2 pathway. *Cardiovascular Res.* **2010**, *86*, 103–112.

[156] Oh, W. J.; Jacinto, E. mTOR complex 2 signaling and functions. *Cell Cycle* **2011**, *10*, 2305–2316.

[157] Ofran, Y.; Rost, B. Protein protein interaction hotspots carved into sequences. *PLoS Comput. Biol.* **2007**, *3*, e119.

[158] Lee, J. A.; Uhlik, M. T.; Moxham, C. M.; Tomandl, D.; Sall, D. J. Modern phenotypic drug discovery is a viable, neoclassic pharma strategy. *J. Med. Chem.* **2012**, *55*, 4527–4538.

[159] Mahe, E.; Morelon, E.; Lechaton, S.; Sand, K. H.; Mansouri, R.; Ducasse, M. F., et al. Cutaneous adverse events in renal transplant recipients receiving sirolimus-based therapy. *Transplantation* **2005**, *79*, 476–482.

[160] Gyurus, E.; Kaposztas, Z.; Kahan, B. D. Sirolimus therapy predisposes to new-onset diabetes mellitus after renal transplantation: a long-term analysis of various treatment regimens. *Transplant Proc.* **2011**, *43*, 1583–1592.

[161] Garelick, M. G.; Kennedy, B. K. TOR on the brain. *Exp. Gerontol.* **2011**, *46*, 155–163.

[162] Bové, J.; Martínez-Vicente, M.; Vila, M. Fighting neurodegeneration with rapamycin: mechanistic insights. *Nat. Rev. Neurosci.* **2011**, *12*, 437–452.

[163] Apel, A.; Zentgraf, H.; Buechler, M. W.; Herr, I. Autophagy—a double-edged sword in oncology. *Autophagy* **2009**, *125*, 991–995.

[164] Kingston, D. G. I. Taxol, a molecule for all seasons. *Chem. Commun.* **2001**, 867–880.

[165] Dunn, C.; Croom, K. F. Everolimus: a review of its use in renal and cardiac transplantation. *Drugs* **2006**, *66*, 547–570.

[166] Bennet, J. W.; Chung, K. -T. Alexander Fleming and the discovery of penicillin. *Adv. Appl. Microbiol.* **2001**, *49*, 163–184.

[167] Altmann, K. -H.; Buchner, J.; Kessler, H.; Diederich, F.; Krautler, B.; Lippard, S., et al. The state of the art of chemical biology. *ChemBioChem* **2009**, *10*, 16–29.

[168] Cong, F.; Cheung, A. K.; Huang, S. -M. A. Chemical genetics-based target identification in drug discovery. *Annu. Rev. Pharmacol. Toxicol.* **2012**, *52*, 57–78.

[169] Klionsky, D. J.; Abdalla, F. C.; Abeliovich, H.; Abraham, R. T.; Acevedo-Arozena, A.; Adeli, K., et al. Guidelines for the use and interpretation of assays for monitoring autophagy. *Autophagy* **2012**, *8*, 445–544.

[170] Ichimura, Y.; Imamura, Y.; Emoto, K.; Umeda, M.; Noda, T.; Ohsumi, Y. In vivo and in vitro reconstitution of Atg8 conjugation essential for autophagy. *J. Biol. Chem.* **2004**, *279*, 40584–40592.

[171] Tsien, R. The green fluorescent protein. *Annu. Rev. Biochem.* **1998**, *67*, 509–544.

[172] Zhang, L.; Yu, J.; Pan, H.; Hu, P.; Hao, Y.; Cai, W., et al. Small molecule regulators of autophagy identified by an image-based high-throughput screen. *Proc. Natl. Acad. Sci. U.S.A.* **2007**, *104*, 19023–19028.

[173] Baek, K. -H.; Park, J.; Shin, I. Autophagy-regulating small molecules and their therapeutic applications. *Chem. Soc. Rev.* **2012**, *41*, 3245–3263.

[174] Sarkar, S. Chemical screening platforms for autophagy drug discovery to identify therapeutic candidates for Huntington's disease and other neurodegenerative disorders. *Drug Discov. Today: Technol.* **2013**, *10*, e137–e144.

[175] Kimura, S.; Noda, T.; Yoshimori, T. Dissection of the autophagosome maturation process by a novel reporter protein, tandem fluorescent-tagged LC3. *Autophagy* **2007**, *3*, 452–460.

[176] Hailey, D. W.; Lippincott-Schwartz, J. Using photoactivatable proteins to monitor autophagosome lifetime. *Methods Enzymol.* **2009**, *452*, 25–45.

[177] Farkas, T.; Hoyer-Hansen, M.; Jaattela, M. Identification of novel autophagy regulators by a luciferase-based assay for the kinetics of autophagic flux. *Autophagy* **2009**, *5*, 1018–1025.

[178] Hundeshagen, P.; Hamacher-Brady, A.; Eils, R.; Brady, N. R. Concurrent detection of autolysosome formation and lysosomal degradation by flow cytometry in a high-content screen for inducers of autophagy. *BMC Biology* **2011**, *9*, 38.

[179] Farkas, T.; Daugaard, M.; Jaattela, M. Identification of small molecule inhibitors of phosphatidylinositol 3-kinase and autophagy. *J. Biol. Chem.* **2011**, *286*, 38904–38912.

[180] Sarkar, S.; Perlstein, E. O.; Imarisio, S.; Pineau, S.; Cordenier, A.; Maglathlin, R. L., et al. Small molecules enhance autophagy and reduce toxicity in Huntington's disease models. *Nat. Chem. Biol.* **2007**, *3*, 331–338.

[181] Lee, S.; Kim, E.; Park, S. B. Discovery of autophagy modulators through the construction of a high-content screening platform via monitoring of lipid droplets. *Chem. Sci.* **2013**, *4*, 3282–3287.

[182] Williams, A.; Sarkar, S.; Cuddon, P.; Ttofi, E. K.; Saiki, S.; Siddiqi, F. H., et al. Novel targets for Huntington's disease in an mTOR-independent autophagy pathway. *Nat. Chem. Biol.* **2008**, *4*, 295–305.

[183] Balgi, A. D.; Fonseca, B. D.; Donohue, E.; Tsang, T. C. F.; Lajoie, P.; Proud, C. G., et al. Screen for chemical modulators of autophagy reveals novel therapeutic inhibitors of mTORC1 signaling. *PLoS ONE* **2009**, *4*, e7124.

[184] Singh, B. N.; Kumar, D.; Shankar, S.; Srivastava, R. K. Rottlerin induces autophagy which leads to apoptotic cell death through inhibition of PI3K/Akt/mTOR pathway in human pancreatic cancer stem cells. *Biochem. Pharmacol.* **2012**, *84*, 1154–1163.

[185] Ozpolat, B.; Akar, U.; Mehta, K.; Lopez-Berenstein, G. PKCδ and tissue transglutaminase are novel inhibitors of autophagy in pancreatic cancer cells. *Autophagy* **2007**, *3*, 480–483.

[186] Maioli, E.; Torricelli, C.; Valacchi, G. Rottlerin and cancer: novel evidence and mechanisms. *Sci. World J.* **2012**, 350826.

[187] Fonseca, B. D.; Diering, G. H.; Bidinosti, M. A.; Dalal, K.; Alain, T.; Balgi, A. D., et al. Structure-activity analysis of niclosamide reveals potential role for cytoplasmic pH in control of mammalian target of rapamycin complex 1 (mTORC1) signalling. *J. Biol. Chem.* **2012**, *287*, 17530–17545.

[188] Gies, E.; Wilde, I.; Winget, J. M.; Brack, M.; Rotblat, B.; Arias Novoa, C., et al. Niclosamide prevents the formation of large ubiquitin-containing aggregates caused by proteasome inhibition. *PLoS ONE* **2010**, *5*, e14410.

[189] Xia, H. -g.; Zhang, L.; Chen, G.; Zhang, T.; Liu, J.; Jin, M., et al. Control of basal autophagy by calpain1 mediated cleavage of ATG5. *Autophagy* **2010**, *6*, 61–66.

[190] Indelicato, M.; Pucci, B.; Schito, L.; Reali, V.; Aventaggiato, M.; Mazzarino, M. C., et al. Role of hypoxia and autophagy in MDA-MB-231 invasiveness. *J. Cell. Physiol.* **2010**, *223*, 359–368.

[191] Tsvetkov, A. S.; Miller, J.; Arrasate, M.; Wong, J. S.; Pleiss, M. A.; Finkbeiner, S. A small-molecule scaffold induces autophagy in primary neurons and protects against toxicity in a Huntington disease model. *Proc. Natl. Acad. Sci. U.S.A.* **2010**, *107*, 16982–16987.

[192] Chen, G.; Xia, H.; Cai, Y.; Ma, D.; Yuan, J.; Yuan, C. Synthesis and SAR study of diphenylbutylpiperidines as cell autophagy inducers. *Bioorg. Med. Chem. Lett.* **2011**, *21*, 234–239.

[193] Chen, G.; Xia, H.; Cai, Y.; Ma, D.; Yuan, J.; Yuan, C. Diphenylbutylpiperidine-based cell autophagy inducers: Design, synthesis and SAR studies. *Med. Chem. Commun.* **2011**, *2*, 315–320.

[194] Shen, D.; Coleman, J.; Chan, E.; Nicholson, T. P.; Dai, L.; Sheppard, P. W.; Patton, W. F. Novel cell- and tissue-based assays for detecting misfolded and aggregated protein accumulation within aggresomes and inclusion bodies. *Cell. Biochem. Biophys.* **2011**, *60*, 173–185.

[195] Tian, Y.; Bustos, V.; Flajolet, M.; Greengard, P. A small-molecule enhancer of autophagy decreases levels of Aβ and APP-CTF *via* Atg5-dependent autophagy pathway. *FASEB J.* **2011**, *25*, 1934–1942.

[196] Rose, C.; Menzies, F. M.; Renna, M.; Acevedo-Arozena, A.; Corrochano, S.; Sadiq, O., et al. Rilmenidine attenuates toxicity of polyglutamine expansions in a mouse model of Huntington's disease. *Hum. Mol. Genet.* **2010**, *19*, 2144–2153.

[197] Jo, Y. K.; Park, S. J.; Shin, J. H.; Kim, Y.; Hwang, J. J.; Cho, D. -H.; Kim, J. C. ARP101, a selective MMP-2 inhibitor, induces autophagy-associated cell death in cancer cells. *Bioch. Biophys. Res. Commun.* **2011**, *404*, 1039–1043.

[198] Shin, J. H.; Park, S. J.; Kim, E. S.; Jo, Y. K.; Hong, J.; Cho, D. H. Sertindole, a potent antagonist at dopamine D2 receptors, induces autophagy by increasing reactive oxygen species in SH-SY5Y neuroblastoma cells. *Biol. Pharm. Bull.* **2012**, *35*, 1069–1075.

[199] Chu, C.; Zhang, X.; Ma, W.; Li, L.; Wang, W.; Shang, L.; Fu, P. Induction of autophagy by a novel small molecule improves Aβ pathology and ameliorates cognitive deficits. *PLoS ONE* **2013**, *8*, e65367.

[200] Shen, S.; Niso-Santano, M.; Adjemian, S.; Takehara, T.; Ahmad Malik, S.; Minoux, H., et al. Cytoplasmic STAT3 represses autophagy by inhibiting PKR activity. *Mol. Cell* **2012**, *48*, 667–680.

[201] Shen, S.; Kepp, O.; Michaud, M.; Martins, I.; Minoux, H.; Metivier, D., et al. Association and dissociation of autophagy, apoptosis and necrosis by systematic chemical study. *Oncogene* **2011**, *30*, 4544–4556.

[202] Niso-Santano, M.; Shen, S.; Adjemian, S.; Malik, S. A.; Marino, G.; Lachkar, S., et al. Direct interaction between STAT3 and EIF2AK2 controls fatty acid-induced autophagy. *Autophagy* **2013**, *9*, 415–417.

[203] Choi, I. -K.; Cho, Y. S.; Jung, H. J.; Kwon, H. J. Autophagonizer, a novel synthetic small molecule, induces autophagic cell death. *Biochem. Biophys. Res. Commun.* **2010**, *393*, 849–854.

[204] Ryu, H. -W.; Oh, W. K.; Jang, I. -S.; Park, J. Amurensin G induces autophagy and attenuates cellular toxicities in a rotenone model of Parkinson's disease. *Biochem. Biophys. Res. Commun.* **2013**, *433*, 121–126.

[205] Høyer-Hansen, M.; Bastholm, L.; Szyniarowski, P.; Campanella, M.; Szabadkai, G.; Farkas, T., et al. Control of macroautophagy by calcium, calmodulin-dependent kinase kinase-beta, and Bcl-2. *Mol. Cell* **2007**, *25*, 193–205.

[206] Wang, Y.; Qiu, Q.; Shen, J. -J.; Li, D. -D.; Jiang, X. -J.; Si, S. -Y., et al. Cardiac glycosides induce autophagy in human non-small cell lung cancer cells through regulation of dual signaling pathways. *Int. J. Biochem. Cell Biol.* **2012**, *44*, 1813–1824.

Targeting Selective Autophagy of Insoluble Protein Aggregates
The Sniper's Philosophy

5.1 AGGREPHAGY-MEDIATED DEGRADATION OF PROTEIN AGGREGATES

Selective autophagy processes are needed, either in particular cell types or in specific environmental conditions, when a specific cell component must be disposed of, while the cell remains viable.

When the refolding chaperone system, the ubiquitin–proteasome system (UPS), and chaperone-mediated autophagy (CMA) fail to clean up damaged proteins, their aggregation is a common event [1]. Any misfolded, insoluble protein aggregate needs eventually to be disposed of in a cell. Neurons are a privileged location for selective protein autophagy/aggrephagy [2,3]. Rather than to engage in bulk-aspecific autophagy, they often need to deploy selective autophagic processes targeting only large protein aggregates, which are more and more abundant in aging neurons [4,5].

K63 ubiquitination of soluble or insoluble species leads to small K63-ubiquitinated aggregates that bind to aggrephagy transporters. In neuronal cells, aggrephagy transporters take K63-polyUBQ protein aggregates to a number of local microtubule nucleation points [6], forming large insoluble inclusion bodies named aggresomes [2]. Aggrephagy scaffolds [7] and aggrephagy receptors [8,9] contribute to the recognition and loading of ubiquitinated protein aggregates. They promote their transportation to aggresomes and the subsequent recruitment of aggresomes onto autophagosome (AP) structures. The genesis and growth of aggrephagy-specific APs follows the same path described for macro-autophagy (MA, Chapter 4, biology-oriented companion book [10]). Similarly, the fate of protein aggregate-specific APs is identical to their MA counterparts. In fact, they merge with lysosomes (LSs) and cause degradation of protein aggregates in the resulting autolysosomes.

Chemical Modulators of Protein Misfolding and Neurodegenerative Disease. http://dx.doi.org/10.1016/B978-0-12-801944-3.00005-9

Histone deacetylase 6 (HDAC6) and p62-sequestosome 1 are the chosen targets to rescue impaired aggrephagy processes in neurodegenerative diseases (NDDs). They are extensively covered in the biology-oriented companion book [10]. The former is the most relevant aggrephagy transporter, while the latter is the most studied aggrephagy receptor. The next sections describe small molecule modulators acting on each of the selected targets.

5.2 p62

Treasure troves are usually well hidden, and treasure hunters need patience, planning, and technical skills to unravel their secrets. The same is true for p62, as p62-directed modulators (i.e., biologically active compounds exerting their activity primarily through selective modulation of one or more functions of p62) are largely unknown. p62-targeted autophagy modulators may cause different outcomes in neurodegeneration. Autophagy induction in *Huntington disease (HD)* and *Parkinson's disease (PD)* would promote the clearance of neurotoxic protein aggregates, while other diseases, such as *Alzheimer's disease (AD)* and *frontotemporal dementia (FTD)*, show defects in late autophagy and accumulation of APs [11]. An increase of p62, and the subsequent increase of APs, would accelerate the autophagic clearance of huntingtin and α-synuclein in the former scenario. Conversely, the generation of additional APs in a stalled autophagy process in AD could even be neurotoxic [12]. Although p62-targeted autophagy inducers may seem appropriate only for HD- and PD-like NDDs, one could conceive their use in combination with agents that target late autophagy defects in cargo turnover/AP transport observed in AD [13] (see also HDAC6, section 5.3).

Selective autophagy may be modulated through p62 either directly (hot spots determining the interaction of p62 with protein partners—unknown small molecule modulators) or indirectly (e.g., post-translational modifications (PTMs) of p62 affecting its behavior in autophagy—known small molecule modulators).

Homodimerization of p62 through its N-terminal Phox and Bem 1p (PB1) domain is required for aggrephagy [14]. p62 is a type AB PB1 protein, containing both an acidic OPCA (OPR/PC/AID) type A motif and a basic type B cluster [15]. Type A/type B homodimerization of two p62 copies starts the formation of p62 inclusion bodies. As p62 heterodimerizes with other type A and type B PB1-carrying proteins [15,16], there is a competition between homo- and heterodimerization of p62. The binding competition, and consequently the concentration of p62-containing homo- and heterodimers, depends on the concentration of PB1-containing proteins and on their relative binding affinities for p62 [17]. In fact, type A and type B motifs are highly conserved among PB1 proteins, but protein-specific PB1–PB1 interactions are known [17]. Nuclear magnetic resonance (NMR) studies show that type AB PB1 proteins may homodimerize (p62) or not (atypical protein kinase C, aPKC) due to different local electrostatic properties [18]. Any means to reduce p62 heterodimerization and to increase p62 homodimerization (sequence-specific interaction between rationally designed small molecules and PB1 motifs of p62-interacting proteins) should increase p62-containing aggregate formation and stimulate autophagy [12]. Conversely, selective inhibitors of the homodimerization between p62 PB1 domains may reduce the formation of protein aggregates, and decrease the excessive AP burden in pathologies such as AD [11,19]. Target-specific modulators of type A–type B PB1 interactions may arise from further structural studies, which are badly needed, but could also be targeted through high-throughput screening (HTS) campaigns using published assay formats [20]. Their identification and biological profiling should help unravel the role of p62 as a scaffolding protein through its PB1 domain, and may lead to useful scaffolds for medicinal chemistry optimization.

p62 dimerizes through its ubiquitin-associated (UBA) domain, and UBA-connected p62 dimers are unable to bind ubiquitin (UBQ) [21]. UBA-driven p62 dimerization is a negative regulator of p62 activity, as UBA-connected p62 dimers are usually inactive. This is particularly true in selective autophagy, as the interaction of polyUBQ proteins and p62 is inhibited, preventing aggresome formation and disposal of polyUBQ protein by aggrephagy. UBA-connected p62 dimers inhibit p62/UBA–UBQ binding at high p62/UBA concentrations, while p62/UBA–UBQ binding is stronger at lower p62/UBA concentrations [22]. The deletion of a four amino acid (AA) sequence at the C-terminus of the UBA domain (AAs 435–438, YSKH, **5.1**, Figure 5.1) is sufficient to prevent UBA dimerization [23]. Additionally, some clinically observed p62 mutations in the UBA domain modulate its UBA-driven dimerization, leading to *Paget's disease of bone (PDB)* [24]. A mechanistic hypothesis [21] suggests that p62 exists as inactive PB1- and UBA-connected oligomeric species, where multiple PB1 domains polymerize (aggrephagy-inducing effect) and UBA domains dimerize in pairs (p62 inactivation, aggrephagy-inhibiting effect). The PB1 multi-domain architecture is relatively stable, and leads to aggregation and eventually aggrephagy when K63-bound polyUBQ-protein species and APs provide (with a largely unknown mechanism) enough binding affinity to stabilize isolated UBA domains/to minimize UBA dimerization [22]. Is it possible to inhibit UBA dimerization while preserving the affinity of p62 UBA domains for UBQ? Is it possible to rationally design p62 UBA-selective small molecules, targeted against the tetrapeptide **5.1**, that do not interfere with other UBA-containing proteins? Does inhibition of UBA-driven p62 dimerization cause major side effects? Further structural studies, and the synthesis of selective chemical tools, should tell us if the UBA domain of

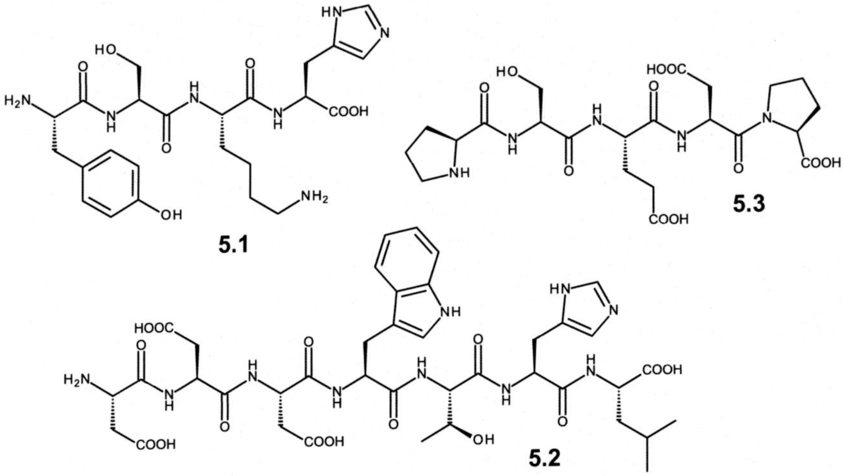

FIGURE 5.1 Peptidic sequences of p62 determining its role in aggrephagy: chemical structures, 5.1–5.3.

p62 is a validated target in drug discovery, and in impaired aggrephagy-dependent NDDs in particular.

The light chain 3 (LC3)-interacting region (LIR) domain of p62 interacts with LC3 isoforms and complexes, and influences aggrephagy [25,26]. AP elongation, and aggrephagy progression, depend on the LC3 pathway. LC3 is cleaved by the protease autophagy-targeted gene 4B (Atg4B), the resulting LC3-I isoform is activated by Atg7 and conjugated with phosphatidylethanolamine (PE) by Atg3. Then, the LC3-II PE conjugate associates with sites on the pre-autophagosomal membrane and promotes their enlargement. p62, through its LIR domain, binds both to LC3-I and LC3-II [27]. As to the former, p62 binds either to free LC3-I or to LC3-I:Atg7 complexes, while LC3-I complexed either with Atg3 or with Atg4B does not bind to p62. As to membrane-bound LC3-II, its binding affinity for p62 is much higher [27]. Thus, in physiological conditions p62 preferentially binds to membrane-bound LC3-II–PE, and contributes to basal autophagy without disturbing the lipidation of LC3-I. When autophagy/aggrephagy is stimulated and p62 levels are increased, LC3-I–p62 binding starts to occur. A negative feedback/reduction of lipidated LC3-II levels may stem from lower levels of free LC3-I, especially when autophagy is impaired (e.g., aging, NDDs). Autophagy/aggrephagy impairment and accumulation of p62 aggregates, thus, may exacerbate neurotoxicity [27]. The interaction between the DDD-containing 11-mer LIR sequence (and in particular the 7-mer DDDWTHL sequence, 5.2) and the hydrophobic C-terminus and basic N-terminus regions of LC3 is characterized by X-ray [26] and NMR [28].

Rational design of either LIR-mimetics or LC3-mimetics seems to be possible. The interaction of an LIR mimetic with LC3-II should cause p62 accumulation coupled with autophagy impairment in physiological/basal autophagy conditions, and the exacerbation of neurotoxicity in pathological/impaired autophagy/neurodegeneration cellular models. Conversely, peptidomimetics mimicking the LC3 region and interacting with LIR may bind the excess of p62 and restore/stimulate the autophagic flux by increasing the levels of free LC3-I. The validation of these hypotheses and their translation into hits and leads against p62/aggrephagy-determined neurodegeneration depends on the rational design and synthesis of such peptidomimetics. Alternatively, modulators of the LIR–LC3 interaction could be identified through HTS campaigns on chemical collections, providing that reliable assay formats are set up.

Other protein–protein interactions (PPIs) involving p62 may modulate aggrephagy or even autophagy. p62 interacts with the mammalian target of rapamycin complex 1 (mTORC1) through raptor, and activates it upon amino acid stimulation (but not upon insulin stimulation), inhibiting autophagy and cell growth [29]. p62 acts as a scaffolding protein to bind raptor and ras-related GTPases (Rag GTPases), and to promote the formation of active Rag/raptor heterodimers. Such interaction causes the localization of mTORC1 to the lysosomal compartment. Oligomerization of p62 may lead to local activation of the mTORC1 complex and to autophagy inhibition [29]. Targeting the p62-raptor PPI requires the identification of their binding epitopes. The binding spot on p62 seems to be located between the tumor necrosis factor (TNF) receptor-associated factor 6 (TRAF6) binding site (TBS) and the zinc finger motif (ZZ) (AAs 167–230) [29]. Such a sequence must be narrowed down to a few AAs become useful for structure-based design of small molecule inducers of autophagy.

The K63-ubiquitinating E3 enzyme TRAF6 binds to p62 (TBS domain), and is required with p62 for AA-dependent stimulation of mTORC1 [30]. Upon AA stimulation (but not upon insulin stimulation) TRAF6 binds to p62, translocates to the LSs with mTORC1 and connects polyUBQ chains to mTORC1 via K63 binding. Thus, TRAF6–p62 binding promotes mTORC1 activation and autophagy inhibition in AA-rich environments [30]. p62 binds to other proteins, possibly to activate the former towards local K63-mediated polyubiquitination of the latter. PB1-bound mitogen-activated protein kinase kinase kinase 3 (MEKK3) is a substrate for p62-bound TRAF6 E3 UBQ activity, leading to the regulation of localized nuclear factor kappa-light-chain-enhancer of activated B cells (NF-κB) [15]. A large number of putative TRAF6–p62 binders/K63 polyUBQ substrates are known [31]. TRAF6 interacts with the scaffolding protein-activating molecule in beclin 1-regulated autophagy (AMBRA1) to promote K63-mediated polyubiquitination. Following autophagy induction, the TRAF6–AMBRA1 PPI promotes K63-mediated polyubiquitination and activation of UNC-51-like kinase 1 (Ulk1), a component of mTORC1 with a positive role in autophagy regulation [32]. Expression of the TBS domain of p62 (AAs 228–254) disrupts the TRAF6–p62 PPI, while single point mutations on the TRAF6 binding-consensus sequence PSEDP (AAs 228–232, **5.3**, Figure 5.1) reduce the affinity of the TBS domain for TRAF6 [30]. Thus, PSEDP-inspired molecules may bind TRAF6, prevent its docking onto p62 and subsequent mTORC1 ubiquitination/activation. Do such molecules promote autophagy only under AA stimulation (when TRAF6 decorates mTORC1 with K63-bound polyUBQ chains), or also under pathological conditions characterized by aggrephagy impairment (i.e., AD and other NDDs)? How would a PSEDP mimetic influence the poly-ubiquitination of MEKK3 and other p62-bound substrates by TRAF6, or the autophagy-stimulating TRAF6–AMBRA1 PPI? How would a PSEDP mimetic influence abnormal p62 levels and p62 aggregates observed in neurodegeneration? Would the contiguity between raptor and TRAF6 binding sites on p62 allow the design of small molecules modulating both p62–TRAF6 and p62–raptor PPIs? We should learn more about these PPIs, and their putative relevance in neurodegeneration, in the near future.

Kelch-like ECH-associated protein 1 (Keap1) is an adaptor protein that binds under basal conditions to the transcription factor NF-E2-related factor 2 (Nrf2) and to the Cullin3-Roc1 E3 ligase, promoting ubiquitination of the former by the latter protein. PolyUBQ Nrf2 is then degraded by the UPS [33]. Oxidation of Cys residues in Keap1 in oxidative stress conditions disrupts Keap1–Nrf2 complexes, prevents ubiquitination and UPS degradation of Nrf2, and transiently activates the transcription of a battery of Nrf2-dependent genes encoding for an antioxidant response [34]. p62 binds to Keap1 and competes with Nrf2, so that the p62–Keap1 PPI frees Nrf2 and prevents its degradation. The p62–Keap1 PPI leads to persistent Nrf2 activation when p62 levels are high [35]. Keap1 is sequestered in p62 inclusion bodies/aggregates when autophagy is impaired (loss of Atg7) and p62 accumulates. Nrf2 is stably protected from Keap1-driven UPS degradation as a result [35]. Keap1 is degraded by autophagy in p62 inclusion bodies [36], while Keap1 accumulates in autophagy-deficient cells [37]. p62-dependent autophagic degradation of Keap1 is promoted by the interaction of p62 and Keap1 with sestrins, proteins involved in the defense from reactive oxygen species (ROS), possibly due to their effect on Nrf2 activation [38]. Surprisingly, the *p62* promoter contains an antioxidant response element (ARE), which makes it susceptible to increased expression by Nrf2 [36]. A positive feedback loop leads to increased levels of Nrf2 (p62-dependent, disruption of Keap1–Nrf2 PPI) and p62 (Nrf2-dependent, increased transcription of the *p62* promoter) [36]. Persistent p62-dependent Nrf2 activation (non-canonical Nrf2 activation) caused by autophagy impairment (Atg7 knockout (KO)) produces severe liver injuries [35], the development of hepatocellular carcinoma [39], and airway hyper-responsiveness in the lung [40]. A double Atg7–Nrf2 KO mouse ameliorates the liver pathology, while liver injury is worsened in the double Atg7–Keap1 KO mouse. Triple Atg7–Keap1–Nrf2 KO mice rescue most liver injuries, proving the central role of persistent Nrf2 activation in the pathology. Conversely, triple Atg7–Keap1–p62 KO mice do not rescue liver injuries, proving that p62 accumulation *per se* is not toxic for the liver [41]. AD shows age-dependent autophagy impairment and p62 accumulation [42,43], which should translate into Keap1 sequestration and persistent Nrf2 activation. The Keap1–p62 PPI is detected in control human brains, and Keap1 is sequestered into p62-containing protein aggregates in the brains of NDD patients (i.e., NFT tangles in AD, Lewy bodies in PD and *dementia with Lewy bodies (DLB)*, neuronal cytoplasmic inclusions in *multiple system atrophy (MSA)*) [44]. mRNA levels of Nrf2-activated genes are higher in the same clinical samples [44]. Small molecule inhibitors of the p62–Keap1–Nrf2 PPIs could be a promising strategy to restore autophagy/aggrephagy, and possibly to protect/rescue diseased tissues and organs in neurodegeneration.

The Keap1-interacting region (KIR) domain of p62, involved in the Keap1–p62 PPI, consists of a 14-mer residue (AAs 346–359). It interacts with the Keap1 kelch repeat and C-terminus region (DC domain, AAs 315–624) in a 1:1 stoichiometry, and with a moderate μM affinity [35]. The X-ray structure of the p62 KIR–Keap1 DC complex shows that p62 competes with Nrf2 for Keap1 binding, as most of the residues in Keap1 DC bind either p62 or Nrf2. The Keap1-interacting portion on KIR is narrowed down to the DPSTGEL 7-mer sequence (AAs 349–355, **5.4**, Figure 5.2) [35].

The Keap1–Nrf2 complex is made of one Nrf2 and two Keap1 molecules (or a Keap1 homodimer) [35]. It is established through a primary/potent ETGE binding motif (**5.5**, $\approx$20 nM affinity measured by isothermal titration calorimetry (ITC)) and a secondary/weaker DLG binding motif (**5.6**, $\approx$1 μM affinity/ITC) on Nrf2. The former is structurally similar to the DPSTGEL sequence on KIR p62, but the affinity of the latter is similar to the p62 KIR–Keap1 DC PPI [35]. Thus, under basal conditions p62 competes with Nrf2 for binding to the secondary DLG site, while Nrf2 remains bound to Keap1 through the EDGE motif. PolyUBQ functionalization of Nrf2 is nevertheless prevented, so that accumulation of p62 and of Keap1 in insoluble aggregates in autophagy-impaired samples frees Nrf2 and activates Nrf2-dependent transcription [35].

The p62–Keap1 interaction further interferes with autophagy/aggrephagy. Oligomerization-deficient ΔPB1 p62 mutants cannot bind simultaneously to Keap1 and to LC3 isoforms, due to the proximity of the KIR and LIR domains in p62. Accordingly, the rate of p62 degradation in cells overexpressing both p62 and Keap1 is significantly diminished [36]. As p62

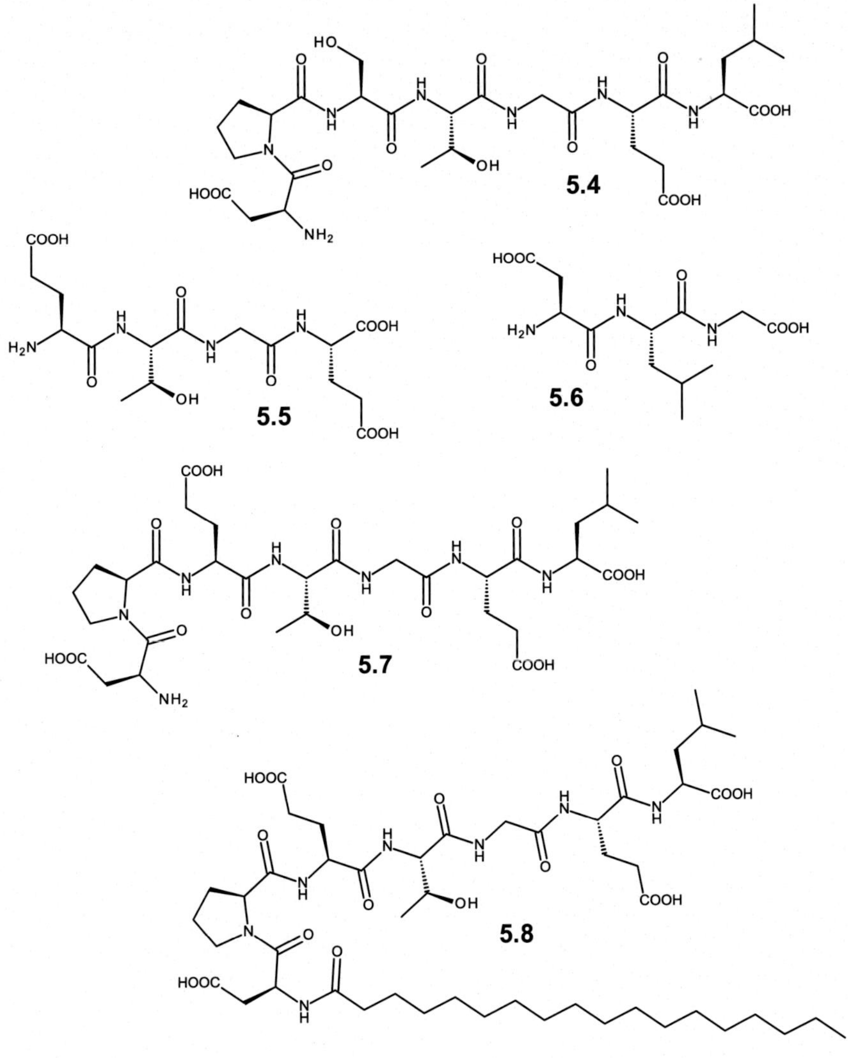

FIGURE 5.2   Peptidic sequences of p62 and Nrf2 determining their role in aggrephagy: chemical structures, **5.4–5.8**.

must oligomerize to be active in aggrephagy/autophagy, and Keap1 and LC3 can bind different p62 copies in a p62 oligomer, it is unclear if the mutual exclusion of LC3 and Keap1 on an isolated p62 monomer has pathological consequences in general, and in autophagy impairment/NDDs in particular.

Although rationally designed inhibitors of the p62–Keap1 PPI are not known, a preliminary SAR around peptide oligomers is available. In particular, the heptamer **5.7** shows an ≈300-fold affinity increase for Keap1 due to a single amino acid substitution with respect to the p62 7-mer sequence **5.4** (E to S in position 3) [45]. The stearyl heptamer **5.8** (Figure 5.2) shows moderate cellular potency due to its membrane permeability [46]. These peptides should be useful to identify drug-like inhibitors of the p62–Keap1 PPI with structure-based drug design.

The activity of p62 in autophagy/aggrephagy-impaired environments may be indirectly modulated, by acting on its phosphorylation pattern. Two papers describe aggrephagy-relevant site-selective phosphorylation of p62 and may have therapeutic implications. The phosphorylation pattern of p62 in Neuro2a cells under basal conditions, in the presence of a proteasome inhibitor and in impaired autophagy conditions, clarifies the autophagy-dependent phosphorylation status of the S403/pS403 residue in the UBA domain of p62 [47]. pS403 is observed in Atg5 KO, autophagy-impaired mice in UBQ- and p62-positive inclusion bodies. The introduction of a pseudo-phosphorylated S403E mutant p62 positively modulates autophagy, increases p62 affinity for both K63- and K48-linked poly-UBQ proteins, and increases polyUBQ protein incorporation in APs. pS403 p62 bound to polyUBQ proteins is resistant to dephosphorylation [47].

Casein kinase 2 (CK2) is an S403-phosphorylating kinase that increases pS403 levels in cells and accelerates p62 turnover (a sign of promoted autophagy). Overexpression of CK2 reduces the aggregation and neurotoxicity of mutant polyQ aggregates through autophagic clearance [47]. Conversely, CK2 inhibition by small interference RNA (siRNA) leads to cancer cell death *via* autophagy induction through mTOR inhibition and down-regulation of raptor expression [48].

Direct activation of CK2, or of any other pS403-phosphorylating kinase, with small molecule modulators is not reported. Conversely, CK2 inhibition may be useful in NDDs where autophagy is stalled at late stages, APs are not processed further, and p62-containing aggregates accumulate (e.g., AD). CK2 inhibitors are covered in section 2.4.3 (see Figures 2.13 and 2.14), as inhibitors of the activation of the co-chaperone cdc37 and inhibitors of the stabilization of hyperphosphorylated (HP), misfolded tau by the Hsp90–Cdc37 complex. The multi-targeting profile of such inhibitors suggests caution, due to the large (>1000) panel of CK2 substrates, whose regulation impacts on many cellular functions. Nonetheless, CK2 inhibition could lead to therapeutic effects against neurodegeneration in general and tauopathies in particular. CK2 inhibition should impact on tau HP, should negatively regulate the interaction of Hsp90 with cdc37 and FKBP52, and should reduce amyloid neurotoxicity. The use of clinically tested CK2 inhibitor CX-4945 (**2.81a**, Figure 2.14) in neurodegeneration-tauopathy animal models characterized by late stage autophagy impairment (possibly determining its impact on p62 phosphorylation and accumulation) is recommended, although it may lead to a complex phenotypic outcome.

Rab proteins are well-known membrane trafficking regulators in autophagy. Rab8, in particular, affects autophagic killing of mycobacteria and interacts with optineurin (OPN), an LIR-containing autophagy receptor [49]. The kinase TANK-binding kinase 1 (TBK1) interacts with OPN, most likely through Rab proteins, and regulates xenophagy through phosphorylation of OPN at the S177 residue [50]. pS177 OPN increases its binding affinity to LC3 isoforms and promotes autophagic clearance of *Salmonella typhimurium* [50]. TBK1 plays a similar, autophagy-promoting role on p62. TBK1 binds OPN and co-localizes on APs, where it affects, *inter alia*, p62 clearance [49]. TBK1 phosphorylates the S403 residue on p62 and promotes its autophagic degradation by increasing its affinity for LC3 proteins. This leads to TBK1- and p62-dependent autophagic clearance of *Mycobacterium tuberculosis* [49], but should also lead to the effects described for CK2-driven p62 S403 phosphorylation on aggrephagy. TBK1-driven regulation of OPN and p62 may provide an overall stronger impact on autophagic clearance. Interestingly, there is evidence of TBK1 activation through K63-bound ubiquitination of TBK1 by the K63-directing TRAF6 E3 ligase [51,52], linking two p62-connected key players in autophagy/aggrephagy activation.

TBK1 is involved in the innate immune response [53], in antimicrobial [54] and anticancer research [55] due to its effects on autophagy. Its mechanism of action, its substrate specificity, and role in the regulation of signaling networks are known [56]. The X-ray structures of human [57] and mouse [58] C-truncation forms, and of full length (FL) human TBK1 [51], are published. Such structures, either as such or complexed with small molecule inhibitors, are useful tools for rational drug design efforts.

Although TBK1 is a validated target in several pathologies, only a few TBK1 selective inhibitors are known. One of them is effective in a therapeutic animal model, another is marketed in Japan and the USA.

The indolinone SU6668 (**5.9**, Figure 5.3) is an anti-angiogenetic inhibitor of receptor tyrosine kinases [59]. A proteomic study identifies TBK1 (low μM inhibition), Aurora A and B as cellular targets of SU6668 [60]. SU6668 is effective at low μM concentration in suppressing TBK1-dependent poly(I:C)-induced interferon β (IFN-β) induction and regulated on activation, normal T cell expressed and secreted (RANTES) biosynthesis [60]. Structural optimization of the indolinone chemotype towards selective TBK1 inhibition is not reported.

The aminopyrimidine BX795 (**5.10**) and its structural analogues are potent ($IC_{50} \approx 100$ nM) 3-phosphoinositide-dependent kinase 1 (PDK1) inhibitors identified through a screening/structural optimization program [61]. Their evaluation against a larger kinase panel highlights stronger potency against ≈10 kinases, including TBK1 ($IC_{50} = 6$ nM) [62]. BX795 is a useful chemical tool for the characterization of the mechanism of action of TBK1, including its activation and its regulation in cells [62]. Its limited selectivity,

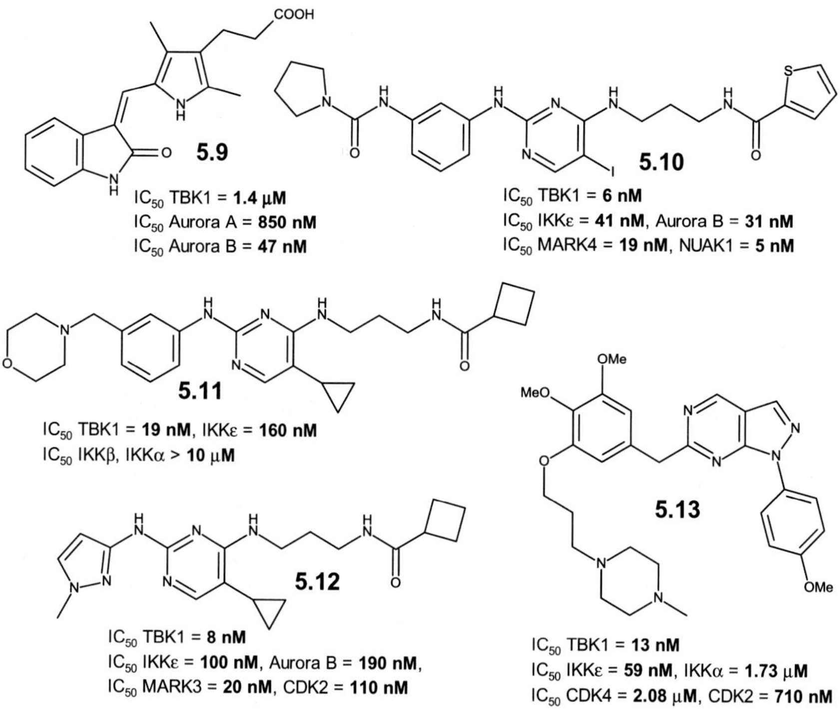

FIGURE 5.3　Small molecule TBK1/IKKε inhibitors: chemical structures, **5.9–5.13**.

off-target effects on c-jun N-terminal kinase (JNK) and p38 pathways, and poor adsorption/distribution/metabolism/excretion/toxicity (ADMET) properties do not allow its further development. MRT67307 (**5.11**) is a more TBK1/inhibitor of NF-κB kinase ε (IKKε)-selective aminopyrimidine, devoid of major side effects [63]. TBK1/IKKε-dependent regulation mechanisms preventing inflammatory and autoimmune diseases are now understood due to MRT67307 [63]. A drug-like set of aminopyrimidines includes pyrazole-substituted **5.12** [64]. The compound has good selectivity and reasonable ADMET properties. *In vivo* testing (i.p. route) confirms the potential of TBK1 inhibitors in inflammation, but highlights a sub-optimal pharmacokinetic (PK) profile and some toxicity at higher dosages [64].

The 6-aminopyrazolopyrimidine **5.13** (Figure 5.3) originates from an HTS campaign targeted against TBK1/IKKε on an ≈250K compound collection [65]. The compound is potent, selective, and cell permeable. Its efficacy as an anticancer agent depends on inhibition of TBK1-promoted pathological Akt activation in TBK1-dependent cancer cell lines [65]. The azabenzimidazole chemotype, and in particular compound **5.14** (Figure 5.4), is identified through HTS of a kinase-focused library, and successive structural

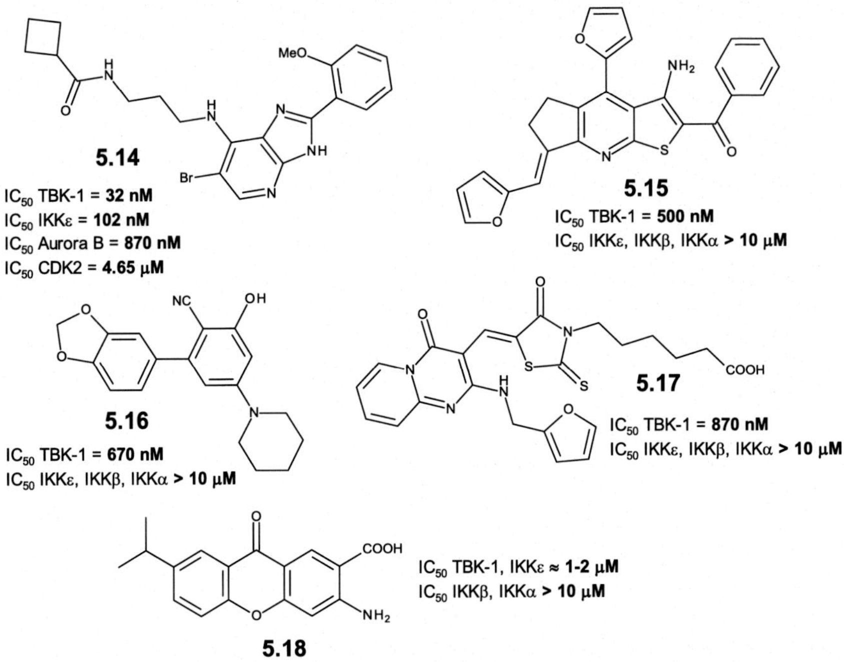

5.14

IC$_{50}$ TBK-1 = 32 nM
IC$_{50}$ IKKε = 102 nM
IC$_{50}$ Aurora B = 870 nM
IC$_{50}$ CDK2 = 4.65 μM

5.15

IC$_{50}$ TBK-1 = 500 nM
IC$_{50}$ IKKε, IKKβ, IKKα > 10 μM

5.16

IC$_{50}$ TBK-1 = 670 nM
IC$_{50}$ IKKε, IKKβ, IKKα > 10 μM

5.17

IC$_{50}$ TBK-1 = 870 nM
IC$_{50}$ IKKε, IKKβ, IKKα > 10 μM

IC$_{50}$ TBK-1, IKKε ≈ 1-2 μM
IC$_{50}$ IKKβ, IKKα > 10 μM

5.18

FIGURE 5.4   Small molecule TBK1/IKKε inhibitors: chemical structures, **5.14–5.18**.

optimization [66]. The compound is potent and selective *vs.* other kinases, and is active in cellular assays. A recently reported HTS assay format to screen for selective TBK1 inhibitors, based on an optimized TBK1 phosphorylation substrate, is the gateway to hits (i.e., the pyrimidine **5.15**, the pyridopyrimidine **5.16**, and the tricyclic pyridothiophene **5.17**) with unprecedented selectivity *vs.* IKKε [67]. Their structural optimization may lead to drug-like, *in vivo* active, truly selective TBK1 inhibitors.

Interestingly, an HTS campaign on a ≈150K compound collection identifies amlexanox (**5.18**, Figure 5.4) as a moderately potent, selective, ATP-competitive TBK1/IKKε inhibitor [68]. Amlexanox is a chromenopyridine marketed against asthma and allergic rhinitis in Japan, and against aphthous ulcers in the USA. Its mechanism of action is unknown. Orally administered amlexanox shows TBK1-related *in vivo* efficacy in mice models of obesity, liver disease, and inflammation [68]. It remains to be seen if TBK1/IKKε inhibition is the main factor to determine the therapeutic effects of amlexanox, as the compound has pharmacological effects on non-kinase targets [69].

TBK1 inhibitors are not yet tested in models of neurodegeneration where autophagy impairment and p62 accumulation play a role. Either amlenaxol—providing that no off-target effects appear in neurons—or structurally optimized compounds from TBK1-selective compounds **5.9**–**5.11** could be suitable candidates for testing. The wealth of available structural information (i.e., published X-ray structures of complexes between TBK1 constructs and BX795, MRT67307, SU6668 [51,57,58]) and the availability of reliable and selective HTS assay formats [67] should facilitate the identification of potent and selective TBK1-targeted leads.

Phosphorylation plays a major role in the Keap1–p62 PPI. The tetrameric primary Keap1-binding site on Nrf2 binds efficiently to Keap1 ($\approx$20 nM affinity, ETGE sequence, **5.5**, Figure 5.2). The heptamer peptide corresponding to the KIR domain on p62 contains a similar motif, but binds less efficiently to Keap1 ($\approx$1 μM affinity, DPSTGEL sequence, **5.4**, AAs 349–355 in p62). Substitution of Ser351 with Glu (DPETGEL sequence, **5.7**, a stable phosphorylated-like mutant) mirrors the primary Keap1-binding site on Nrf2 in terms of binding strength ($\approx$100 nM affinity). Modelization of the interaction between DPpSTGEL–p62 and Keap1 shows the formation of two additional hydrogen bonds with respect to DPSTGEL–p62, due to the phosphate group [70]. The X-ray structure of the Keap1 DC domain complexed with the 14-meric 346–359 KIR peptide sequence of p62 confirms the predicted stronger interaction of pS351-containing peptides with Keap1. ITC determines an $\approx$30-fold increase of binding affinity for pSer351 KIR with respect to its non-phosphorylated counterpart [70]. Experiments with mutant S351E full length p62 show that Nrf2 activation/translation is permanently inhibited.

Using a pSer351 p62-specific antibody, a low level of pSer351 is observed under basal conditions. Once oxidative stress is applied to cells, pSer351 p62 levels increase both as soluble species and in p62 aggregates, and Keap1 co-localizes with pSer351 p62 [70]. The aggregates are eventually disposed of through autophagy, but continue to accumulate (and to switch the Ser351/pSer351 p62 equilibrium towards the phosphoisoform) when autophagy is impaired at a late stage/post-AP formation. Thus, inhibition of phosphorylation on Ser351 of p62 may be useful to prevent p62 accumulation, Keap1 degradation and toxic, persistent Nrf2 activation in NDDs such as AD.

mTORC1 kinase phosphorylates p62 on Ser351 [70]. Cellular experiments show abnormally high pSer351 p62 levels and persistent Nrf2 activation in autophagy-deficient livers and in human hepatocarcinoma cells (HCC) [70]. Rapamycin and torin1 (see section 4.2, respectively **4.1a**, Figure 4.1, and **4.20**, Figure 4.5) inhibit phosphorylation of Ser351 and Nrf2-dependent gene transcription in murine embryonic fibroblasts (MEFs), but they only partially rescue autophagy impaired conditions (i.e., valinomycin treatment/impaired mitophagy, *Salmonella typhimurium* invasion/xenophagy) [70]. Other unknown p351-phosphorylating kinases may compensate mTORC1 activity in the presence of its inhibitors. Nevertheless, mTORC1 inhibitors (described in detail in section 4.2, and shown in

Figures 4.1 to 4.6) should provide synergistic therapeutic benefits due to their effects on autophagy activation, on persistent Nfr2 activation, and on pSer351 levels in neuronal samples under physiological and pathological/ neurodegeneration-mimicking conditions.

5.3 HDAC6

Histone deacetylase 6 (HDAC6) is a deceivingly simple target, due to the focus on its enzymatic activity and on its inhibitors. Inhibitors of HDAC6 deacetylase activity are extensively covered in this section, together with unconventional HDAC6 modulators. The latter compounds affect the interactions of proteins with non-catalytic domains of HDAC6, that should also be the object of future drug discovery efforts.

Reviews describing structures and SARs for small molecule inhibitors of HDAC deacetylase enzymes are available [71,72]. The main indication for HDAC inhibitors is cancer [73,74] but other clinical indications are supported by preclinical evidence [75]. Vorinostat/Zolinza® [76] and romidepsin/Istodax® [77] (respectively the hydroxamate SAHA-**5.19** and the cyclic depsipeptide FK228-**5.20**, Figure 5.5) are broad spectrum HDAC inhibitors, acting on all HDAC isoforms. They are both approved for the treatment of cutaneous T-cell lymphoma [78], and romidepsin also for the treatment of peripheral T-cell lymphoma [79]. Orally available hydroxamates panobinostat [80], abexinostat [81], and givinostat [82] (respectively LBH-589-**5.21**, PCI-24781-**5.22**, and ITF2357-**5.23**) are second generation broad HDAC inhibitors undergoing clinical evaluation. Panobinostat, either alone or in combination, is in Phase III against multiple myeloma, Hodgkin's and T-cell lymphomas, and in Phase II against non-Hodgkin's lymphoma, myelodysplastic syndromes, soft tissue sarcoma, breast, renal, prostate, pancreas, thyroid, and malignant brain cancers [83]. Abexinostat, either alone or in combination, is in Phase I/II against lymphomas and multiple myeloma, and in Phase I against sarcoma and metastatic solid tumors [84]. Givinostat, either alone or in combination, is in Phase II against myoproliferative diseases, chronic lymphocytic leukemia, and multiple myeloma, and in Phase I/II against Hodgkin's lymphoma [85]. Broad spectrum HDAC inhibitors show dose-limiting toxic effects, including gastrointestinal disturbances (nausea, vomiting, diarrhea, fatigue) [78], thrombocytopenia [86], and human ether-a-go-go related gene (hERG)-derived QT prolongation and cardiac arrhythmias [87]. Such side effects would not be tolerated in patients suffering from chronic diseases, including NDDs.

The aminobenzamides entinostat [88] and mocetinostat [89] (respectively MS-275-**5.24** and MGCD0103-**5.25**, Figure 5.5) are class I-specific HDAC inhibitors acting on HDAC 1, 2, 3, and 8, currently in clinical trials. Entinostat is a sub-μM HDAC1 inhibitor, with ≈five-fold selectivity

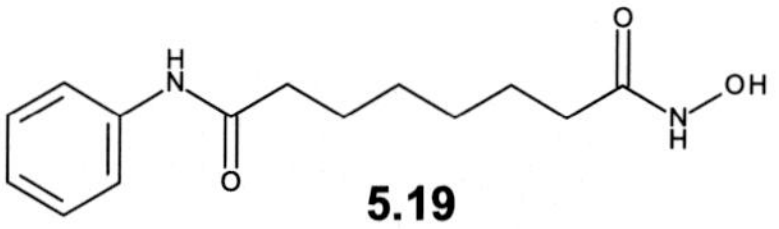

5.19

IC_{50} HDAC6 = **21 nM**
IC_{50} HDAC1 = **25 nM**
IC_{50} HDAC3 = **37 nM**
IC_{50} HDAC8 = **1.2 μM**

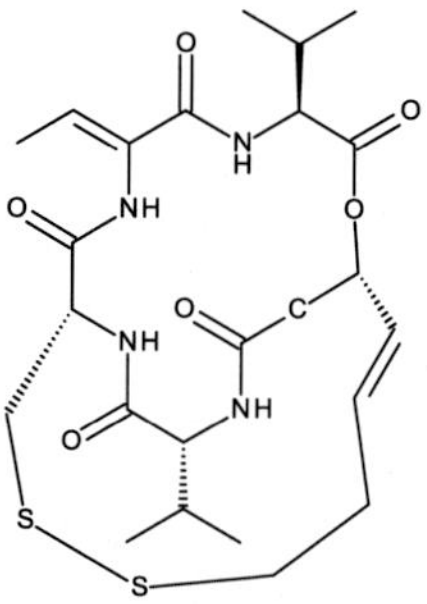

5.20 IC_{50} HDAC6 = **14 μM**
IC_{50} HDAC1 = **36 nM**

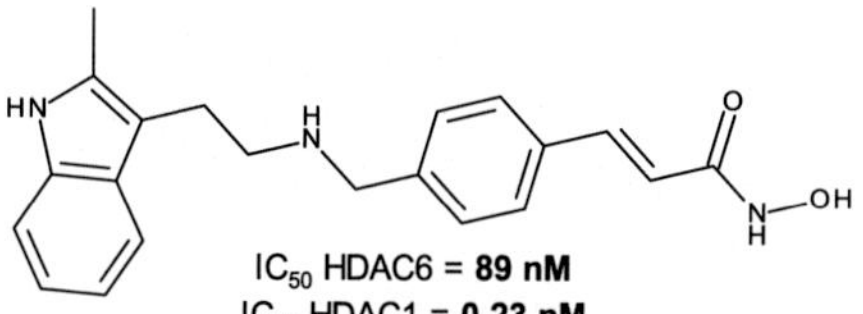

5.21

IC_{50} HDAC6 = **89 nM**
IC_{50} HDAC1 = **0.23 nM**
IC_{50} HDAC8 = **283 nM**

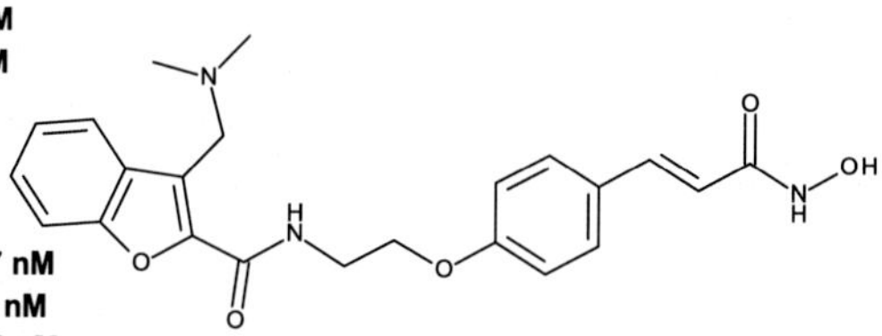

5.22

IC_{50} HDAC6 = **17 nM**
IC_{50} HDAC1 = **7 nM**
IC_{50} HDAC3 = **8.2 nM**
IC_{50} HDAC8 = **280 nM**

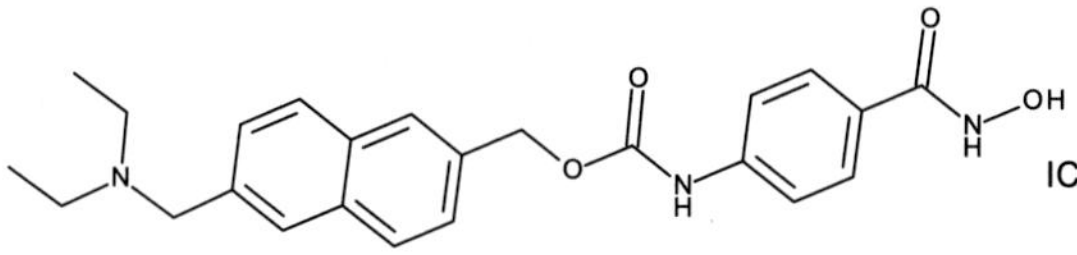

5.23

IC_{50} HDAC1 = **10 nM**

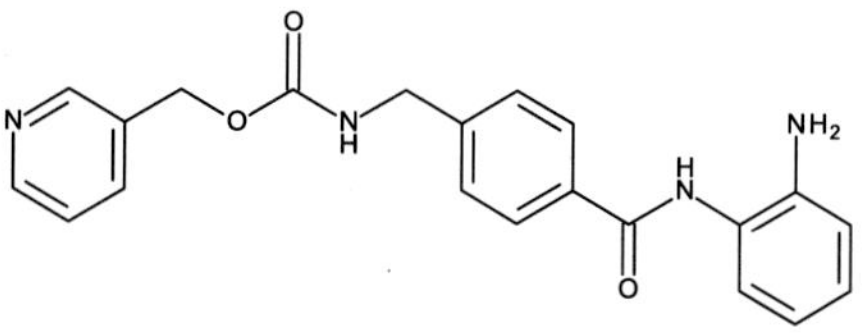

5.24

IC_{50} HDAC1 = **510 nM**
IC_{50} HDAC3 = **1.7 μM**

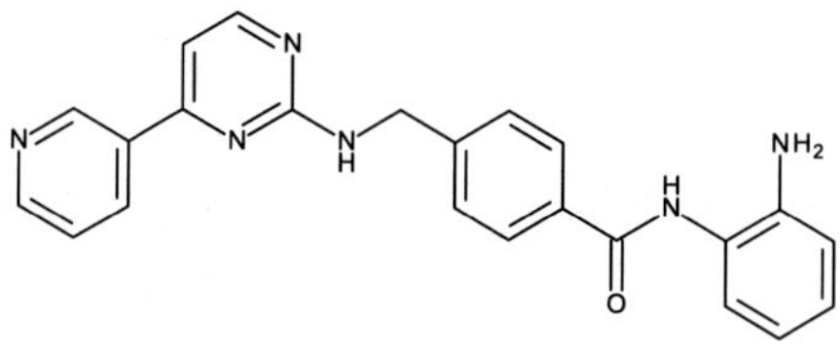

5.25

IC_{50} HDAC1 = **150 nM**
IC_{50} HDAC3 = **1.66 μM**

FIGURE 5.5 Small molecule HDAC hydroxamate and benzamide inhibitors: chemical structures of unselective (**5.19–5.23**) and class I selective (**5.24–5.25**) compounds.

vs. HDAC9, ≈10-fold *vs.* HDAC2, ≈four-fold *vs.* HDAC3, and no measurable activity against other HDAC isoforms [90]. It is evaluated in Phase II against acute and chronic myeloid leukemia, myelodysplastic syndromes, Hodgkin's lymphoma, melanoma, breast, colon, and non-small cell lung cancer [91]. Mocetinostat is a nM inhibitor of HDAC1 and -2, with ≈15-fold selectivity *vs.* HDAC3 and no measurable activity against other HDAC isoforms [90]. It is evaluated in Phase II against acute myeloid and chronic lymphoid leukemia, Hodgkin's lymphoma, and myelodysplastic syndromes [92]. A detailed survey of the toxicological and metabolic profile of clinically tested HDAC inhibitors is available [93].

Class I-selective HDAC inhibitors for non-oncology indications [94,95]—and neurodegeneration in particular [96,97]—are supported by a scientific rationale. Even focused inhibition of class I HDACs, though, leads to sub-optimal toxicity profiles that should be acceptable only in acute, oncology-like indications [98]. Four hundred and fourteen clinical trials employing HDAC inhibitors were documented in late 2013 [99]. The vast majority of them are targeted against oncological diseases. Around 25 trials for non-oncology indications are targeted against chronic obstructive pulmonary disease (COPD), human immunodeficiency virus 1 (HIV-1), and graft *vs.* host disease. A few trials target NDDs but use old aspecific compounds such as lithium carbonate, vitamin B3, and sodium valproate [99].

Class II-specific HDAC inhibitors (acting on HDAC 4, 5, 6, 7, and 9) seem to be better tolerated when compared with class I HDAC inhibitors [100], although clinical toxicology data on them are still scarce. In particular, genetic ablation of HDAC6 in mice does not interfere with their development, and does not seem to affect their lives [101]. HDAC6-specific inhibitors are known since the discovery of tubacin in 2003 [102]. Their slow progress towards clinical testing will eventually determine their usefulness in chronic diseases [100].

Tubacin (**5.26**, Figure 5.6) is the result of a phenotypic screen [102] aimed to discover tubulin-selective deacetylase agents among a trichostatin A (TSA)/trapoxin-inspired, ≈8000-member 1,3-dioxane hydroxamate library [103]. Tubacin does not act on histone deacetylation-dependent cellular mechanisms, and interacts specifically with the C-terminal deacetylase domain of HDAC6. Tubacin does not act on cell cycle progression and on gene expression, as it happens with broad HDAC inhibitors [102]. Tubacin is used as a chemical tool to decouple tubulin/microtubule (MT) deacetylation from chromatin deacetylation in physiological and pathological environments.

Tubacin inhibits cell growth and, when used in combination with the UPS inhibitor bortezomib, prevents resistance in multiple myeloma [104], enhances apoptosis in Burkitt lymphoma [105] and endoplasmic reticulum (ER) stress-induced apoptosis in breast cancer cells [106]. Tubacin protects prostate cancer LNCaP cells from hydrogen peroxide-induced death at low μM concentration *via* peroxiredoxin (Prx) acetylation [107]. The combination of tubacin and DNA damaging agents (etoposide, doxorubicin) leads to increased cytotoxicity, most likely through induction of apoptosis, in LNCaP cells [108]. Increased cytotoxicity in acute lymphoblastic leukemia results from polyUBQ protein accumulation [109]. HDAC6/cortactin-dependent migration and invasion mechanisms observed in

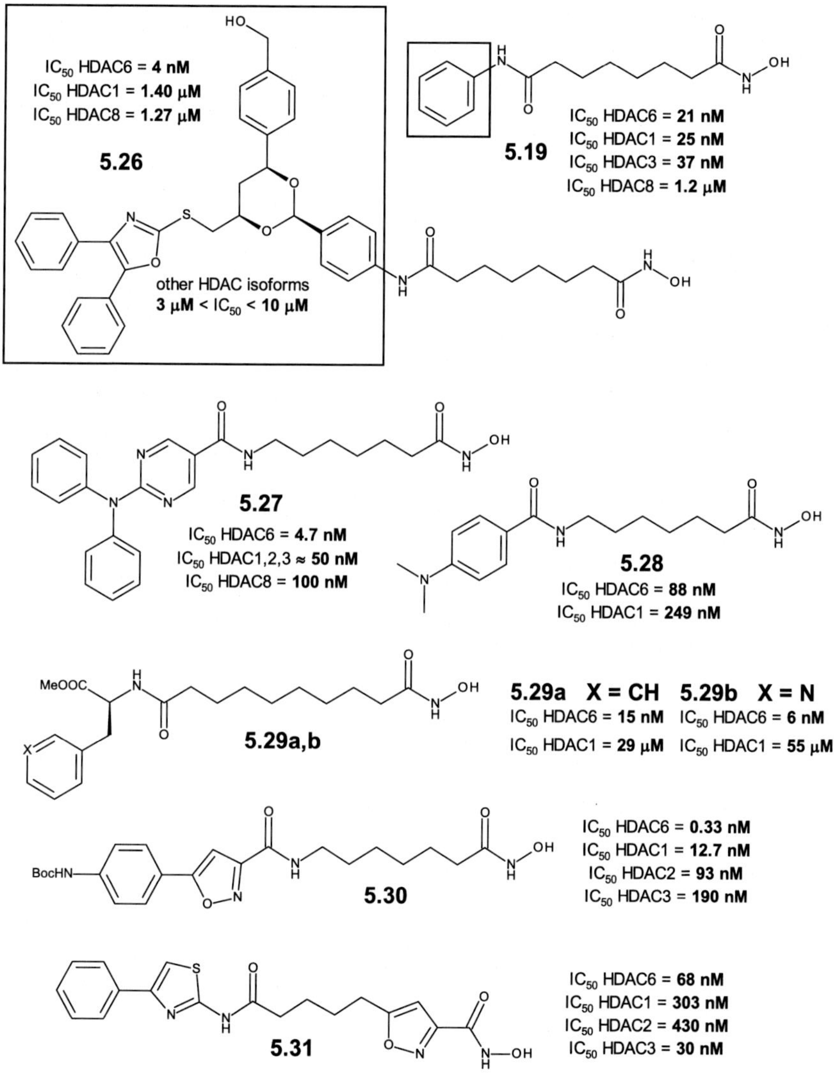

FIGURE 5.6 Small molecule HDAC6 hydroxamate inhibitors with varying selectivity: chemical structures, **5.26–5.31**.

bladder cancer may be attenuated by tubacin [110]. Tubacin causes down-regulation of FADD-like interleukin-1 beta-converting enzyme (FLICE) inhibitory protein (FLIP) through deacetylation of Ku70 [111] and desta-bilization of the Ku70–FLIP complex in colon cancer cells [112].

Tubacin is useful in non-oncological indications. Tubacin reverses the progression of established colitis *in vivo* in a forkhead box P3 (Foxp3)$^{+}$-reg-ulatory T cells (Tregs)-dependent manner [113]. Class I-specific entinostat is not immunosuppressant [114]. The combination of tubacin with either rapamycin (low dose) or the Hsp90 inhibitor tanespimycin shows long-term protection against allograft rejection in a murine model, possibly through Hsp90 regulation by HDAC6 [113]. Tubacin inhibits HDAC6 in PDK1-mutated renal epithelial cells, contrasting HDAC6 up-regulation and preventing HDAC6-mediated increased transport and degradation of epidermal growth factor receptor (EGFR), with putative usefulness against polycystic renal disease [115]. HDAC6 inhibition by tubacin influences hepatic glucocorticoid-induced gluconeogenesis by controlling nuclear translocation of glucocorticoid receptors (GRs), and may address glucocorticoid-induced diabetes [116]. As to NDDs, tubacin restores physiological levels of acetylated α-tubulin and alleviates axonal transport deficiencies in a transgenic (TG) mouse model of Charcot–Marie–Tooth (CMT) disease caused by mutations in the heat shock protein B1 (HspB1)/Hsp27 chaperone protein [117]. It stimulates kinesin-1-mediated mitochondrial transport in hippocampal neurons, with potential usefulness against abnormal mitochondrial transport in AD and in other NDDs [118]. Conversely, tubacin does not influence the interaction between HDAC6 and tau, although it reduces tau phosphorylation on Thr231 in cortical neurons [119]. Interestingly, tau and tubacin similarly inhibit α-tubulin deacetylation [120]. Tubacin treatment of a *Drosophila* model rescues tau-induced MT defects, possibly through increased acetylation, recruitment of motor proteins, and MT transport stimulation [121].

HDAC isoform-non-selective hydroxamate vorinostat/SAHA can be made HDAC6-selective by connecting the zinc-binding hydroxamate to a bulky, p-(2,5-dioxanyl)phenyl-based cap portion in tubacin (compare the boxed cap region of **5.19** and **5.26**, Figure 5.6) through the same C6-alkylamido linker. The selectivity of tubacin for HDAC6 can be rationalized by docking it into a homology model of HDAC6 [122].

Hydroxamates bearing various linker/cap regions and possessing varying HDAC6 selectivity are reported in the literature. Rocilinostat (ACY-1215, **5.27**) [123] is a moderately HDAC6-selective hydroxamate containing an inverted C6-alkylamido linker and a diphenylaminopyrimidine cap region. It is evaluated in Phase I/II clinical trials in combination with UPS inhibitors against multiple myeloma [124]. The rocilinostat–bortezomib combination in a mouse multiple myeloma model shows a significant increase in polyUBQ proteins and a good tolerability, with synergistic efficacy compared to the compounds tested as standalone treatments [123].

The hydroxamate M344 (**5.28**) contains an inverted C6-alkylamido linker and a dimethylaminophenyl cap region [125]. Its HDAC6 selectivity is marginal *vs.* HDAC2 [126]. Data about other HDAC isoforms are missing, but its use as an HDAC6-selective inhibitor in non-oncological cellular models is documented. Treatment with M344 increases the levels of the survival motor neuron 2 (SMN2) protein that modulates the severity of spinal muscular atrophy (SMA) in fibroblasts of SMA patients [127]. Broad spectrum HDAC inhibitors cause a lower SMN2 level increase and are more toxic to human fibroblasts [127]. M344 treatment enhances the clearance of tau in primary neuronal cultures by inhibiting Hsp90-mediated tau clearance in a C-terminus of Hsc70 interacting protein (CHIP)-dependent manner [128]. M344 causes HDAC6 inhibition-dependent increase in MT acetylation, reduction of p23 levels *via* Hsp90 complexation and clearance, synergism with the Hsp90 inhibitor tanespimycin, and enhancement of inducible Hsp70 activity by HSF1-mediated gene transcription [128]. M344 reactivates HIV-1 gene expression in latently infected Jurkat cells through HDAC inhibition and NF-κB activation [129]. M344 causes an HDAC6-independent, possibly HDAC3-dependent [130], increase in histone H3 and H4 acetylation, which leads to the reactivation of HIV gene expression and clearance. Indirect mechanisms are postulated to explain the inhibitory and activating effects of M344 observed respectively on HDAC6 and NF-κB [129].

The hydroxamate ST80 (**5.29a**) [131] contains a longer C8-alkylamido linker, and its cap is an (S)-Phe methyl ester. A structurally similar analogue (**5.29b**) contains a 3-pyridine in place of the phenyl ring cap [132]. ST80 is ≈40-fold more potent on HDAC6 than on HDAC1, while data about the inhibition of other HDAC isoforms are missing [133]. Compound **5.29b** shows slightly lower potency and selectivity [132]. ST80 produces a moderate dose-dependent growth arrest of ERBB2-overexpressing breast cancer cells by accelerating the decay of mature ERBB2 mRNA [133]. Its effect is HDAC6-specific, as it reduces ERBB2 stability by inhibiting the molecular interaction of HDAC6 with the transcript stability factor human antigen R (HuR), while it does not repress the *ERBB2* promoter function as pan-HDAC inhibitors do. ST80 causes cell growth reduction in leukemia cell lines and in primary patient blasts [134]. The activity of ST80 is connected to increased α-acetylated tubulin levels and altered MT growth velocity, while histone acetylation is marginally increased. Combined administration of ST80 and bortezomib is suggested to kill cancer cells hitting two protein degradation pathways [134,135]. Interestingly, a substantial population of rhabdomyosarcoma (RDS) cells becomes resistant to ST80/bortezomib co-treatment by up-regulating Bcl-2-associated athanogene 3 (BAG-3) mRNA levels and activating BAG-3-dependent aggrephagy/protein aggregate disposal [136]. As BAG-3 promotes autophagic clearance when complexed with Hsp70 and CHIP in ageing and/or protein aggregate-rich tissues [137], and interacts with p62 [9], combined proteasome/HDAC6 inhibition may be useful in NDDs.

The hydroxamate **5.30** contains an inverted C6-alkylamido linker, and its cap is built around a 3,5-disubstituted isoxazole [138]. It is a sub-nM HDAC6 inhibitor with sub-μM potency against several class I HDACs, so that its remarkable cytotoxicity against a panel of cancer cells may be attributed to HDAC1-3 inhibition [138]. Interestingly, the incorporation of an isoxazole in the linker region (e.g., **5.31**, Figure 5.6) causes both a decrease in HDAC6 inhibition and a loss of selectivity *vs.* other HDAC isoforms [139].

The hydroxamate **5.32** (Figure 5.7) contains a C6-alkylamido linker and an *m*-triazolylphenyl cap region [140]. Compound **5.32** and close structural analogues show limited selectivity against other HDAC isoforms and moderate cytotoxicity against cancer cells. Compound **5.32** is a potent inhibitor of *Plasmodium falciparum* (50% growth inhibition at ≈20 nM), with a comparable antimalarial activity with potent broad spectrum HDAC inhibitors, although its toxicity profile is more acceptable [109]. Recently reported hydroxamates include the unsaturated linker-, sulfonamide cap-containing C1A (**5.33**), endowed with good *in vivo* PK and efficacy against colon tumors [141]; the unsaturated linker-, furylamine cap-containing **5.34**, showing good solubility [142]; the arylalkene linker-containing, phenyl-capped dual HDAC6–HDAC8 inhibitor **5.35**, capable of cellular α-tubulin acetylation without any effect on histone deacetylation [143]; the phenylmethyl linker-containing, dihydroquinoxalinone-capped chiral compound **5.36**, which shows moderate selectivity for HDAC6 and cellular activity (α-tubulin acetylation) [144]; and the linker-less pyrroline compound **5.37** (Figure 5.7), which achieves good HDAC6 selectivity within an extremely small structure [145].

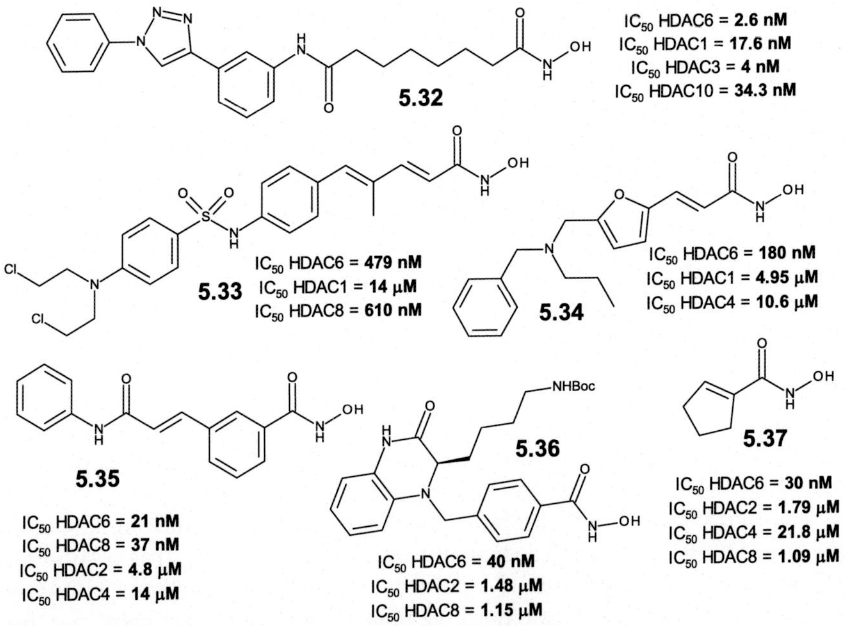

5.32
IC$_{50}$ HDAC6 = 2.6 nM
IC$_{50}$ HDAC1 = 17.6 nM
IC$_{50}$ HDAC3 = 4 nM
IC$_{50}$ HDAC10 = 34.3 nM

5.33
IC$_{50}$ HDAC6 = 479 nM
IC$_{50}$ HDAC1 = 14 μM
IC$_{50}$ HDAC8 = 610 nM

5.34
IC$_{50}$ HDAC6 = 180 nM
IC$_{50}$ HDAC1 = 4.95 μM
IC$_{50}$ HDAC4 = 10.6 μM

5.35
IC$_{50}$ HDAC6 = 21 nM
IC$_{50}$ HDAC8 = 37 nM
IC$_{50}$ HDAC2 = 4.8 μM
IC$_{50}$ HDAC4 = 14 μM

5.36
IC$_{50}$ HDAC6 = 40 nM
IC$_{50}$ HDAC2 = 1.48 μM
IC$_{50}$ HDAC8 = 1.15 μM

5.37
IC$_{50}$ HDAC6 = 30 nM
IC$_{50}$ HDAC2 = 1.79 μM
IC$_{50}$ HDAC4 = 21.8 μM
IC$_{50}$ HDAC8 = 1.09 μM

FIGURE 5.7　Small molecule HDAC6 hydroxamate inhibitors with varying selectivity: chemical structures, **5.32–5.37**.

Homology models of HDAC1 and HDAC6 are used to identify a more drug-like, HDAC6-selective hydroxamate inhibitor [146]. A rigid, short linker (a phenylmethyl group) and a bulky, tricyclic cap (a tetrahydropyridoindole) determine the structure of tubastatin (**5.38a**, Figure 5.8). The selectivity and drug-like properties of tubastatin make it a suitable compound for further mechanistic elucidation of the role of HDAC6 in physiological and pathological processes. Tubastatin is exceptionally selective against all HDAC isoforms, showing moderate activity only against HDAC8. It is much more synthetically accessible than tubacin, and has a suitable ADMET profile to be exploited in models of chronic diseases [146].

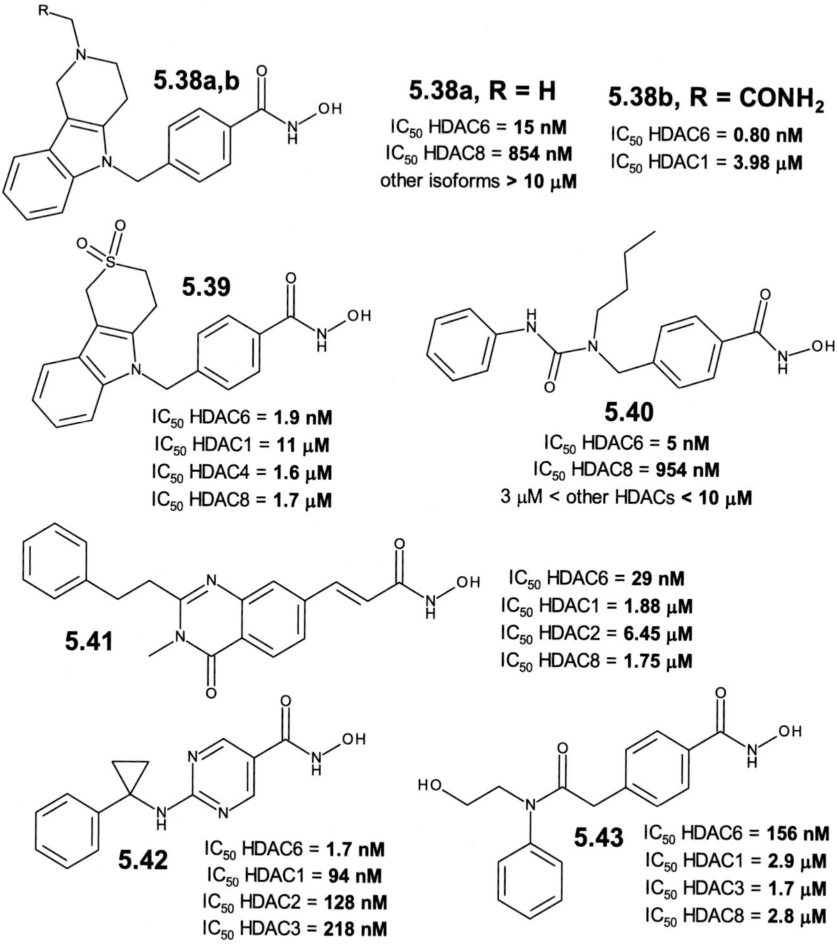

FIGURE 5.8 Small molecule selective HDAC6 hydroxamate inhibitors: chemical structures, **5.38a–5.43**.

Arsenite-induced oxidative stress causes the association of HDAC6 with ribosomes, and its involvement in stress-induced protein translation [147]. Tubastatin treatment in HaCaT keratinocytes prevents the association of Nrf2 mRNA with ribosomes, the increase of Nrf2 protein levels and the transcription of a battery of Nrf2-dependent genes encoding for an antioxidant response. Deacetylation of ribosomal proteins and regulation of the shuttling process of Nrf2 mRNA to microsomes are proposed as HDAC6-mediated processes in the oxidative/arsenite stress-induced response [147]. Overexpression of HDAC6 in cholangiocytes (epithelial cells lining the biliary tree) and in cholangiocarcinoma cells leads to the reduction of multisensory primary cilia, increasing the cell proliferation rate [148]. Tubastatin treatment restores the basal levels of primary cilia in cholangiocytes, and significantly decreases tumor growth in a cholangiocarcinoma animal model [148]. Tubastatin treatment of breast adenocarcinoma MCF-7 cells causes the increased localization of HDAC6 on MTs and on the polymeric tubulin fraction [149]. Increased MT–HDAC6 binding in the presence of tubastatin stabilizes the MT/polymeric tubulin structure and reduces the rate of MT assembly and growth [149]. As HDAC6 ablation does not influence the rate of MT growth, a "capping" function of MT binding-capable, catalytically inactive HDAC6 (possibly at MT ends, to explain the slower MT growth [150]) may influence MT dynamics together with HDAC6-dependent α-tubulin acetylation.

Tubastatin shows efficacy in cellular models of cystic fibrosis (CF) by restoration of physiological α-tubulin acetylation and reduction of NF-κB activation [151]. Tubastatin acts on Tregs, and on associated suppression of T cell-dependent immune responses *in vitro* and *in vivo* (progression of established cholitis) [113]. Tubastatin exerts moderate anti-inflammatory and anti-rheumatic effects in animal models at 30 mg/kg i.p. (Freund's complete adjuvant (FCA)-induced rat model of inflammation, collagen-induced arthritis (CIA) DBA1 mouse model), possibly by inhibiting the release of pro-inflammatory cytokines interleukin 6 (IL-6) and TNF-α, and of chemokine nitric oxide (NO) [152]. Tubastatin protects against atrial remodeling at 1 mg/kg in a dog model of atrial fibrillation (AF), likely due to HDAC6-dependent α-tubulin deacetylation and MT disruption [152]. Increased HDAC6 levels and deacetylase activity are observed in clinical samples from patients suffering from permanent AF [153]. Broad spectrum HDAC inhibitors act similarly to tubastatin in CF-arthritis-inflammation-AF, but they show significant adverse events *in vivo*. Tubacin has a lower efficacy because it is much less bioavailable than tubastatin.

Mitochondrial transport is influenced by α-tubulin acetylation, and Aβ peptides reduce mitochondrial transport by MTs in neurons. Treatment with tubastatin restores basal mitochondrial velocity and increases mitochondrial length [154], probably by restoring motor protein binding to MTs and increasing their dynamics. An HTS campaign targeted against enhancers of progranulin as putative FTD treatments identifies broad spectrum HDAC inhibitors such as SAHA, and HDAC inhibitors with limited selectivity such as M344 as progranulin enhancers with limited toxicity at higher dosages [155]. HDAC6-selective tubastatin and tubacin do not enhance progranulin expression, pointing towards other HDAC isoforms as potential FTD targets.

The tubastatin analogue **5.38b** bearing a 2-methylamido substituent is the result of a structural optimization program [156]. Its selectivity *vs.* HDAC1 is an unprecedented ≈5000, and its ADMET profile is further improved compared to tubastatin. The efficacy of **5.38b** in cellular assays related to Tregs-dependent suppression of T cell-dependent immune responses is similar to tubastatin [156]. The sulfone analogue of tubastatin (tubathian A, **5.39**) is as potent and selective *in vitro* as its parent compound, and may represent another structural avenue for further optimization of HDAC6-selective inhibitors [157].

The hydroxamate nexturastat (**5.40**) contains the same linker as tubastatin, and an N-butyl-N′-phenylurea-based cap region [158]. Nexturastat is extremely selective against HDAC6, does not deacetylate histones up to µM potency and shows slightly decreased cellular potency (low µM) against B16 melanoma cells when compared with tubastatin [158]. The last data may indicate that cytotoxicity stems from histone deacetylation, while α-tubulin deacetylation and MT transport regulation happens at low nM concentrations and do not harm cells.

The hydroxamate ACY-738 (**5.41**) contains a rigid pyrimidine linker and a cycloprophylphenyl cap region, is potent against HDAC6 and moderately selective *vs.* HDACs1-3 [159]. ACY-738 is brain permeable and shows a reasonable PK profile, providing increased α-tubulin acetylation in the brain up to 4 hours after a single 5 mg/kg injection (tubastatin does not increase α-tubulin acetylation in the brain in similar conditions) without affecting histone acetylation levels. *In vivo* testing of ACY-738 in mice displays rapid psychoactive properties upon acute and chronic administration (50 mg/kg-acute; 5 mg/kg, 10 days-chronic), including acute exploration-enhancing effects in novel environments, anxiolytic-like effects and acute antidepressant-like anti-immobility effects. These effects can be justified by HDAC6 inhibition in the central nervous system (CNS) [159].

The hydroxamate **5.42** contains a small alkene linker and a disubstituted quinazolin-4-one cap region, and is endowed with good HDAC6 selectivity [160]. It causes substantial neurite outgrowth and increased synaptic activity in undifferentiated neuronal cells, and it induces the expression of growth-associated protein 43 (GAP-43). Its cellular selectivity is limited, as it acts both on α-tubulin and histone acetylation in neuronal cells [160]. Interestingly, the zinc-binding hydroxamic acid group prevents zinc-mediated Aβ aggregation *in vitro*, and is effective *in vivo* on learning-impaired mice bearing Zn-mediated Aβ hippocampal lesions (10 mg/kg, i.p., 30 days, complete learning recovery). Compound **5.42** appears to be safe *in vitro* and *in vivo* [160].

The hydroxamate HPOB (**5.43**, Figure 5.8) contains a phenylmethyl linker and a small N-phenyl, N-hydroxyethyl cap region [161]. HPOB is potent and HDAC6-selective, increases α-tubulin acetylation, inhibits the growth of various tumor cells, and shows synergistic effects in combination with anticancer agents. HPOB does not interfere with UBQ binding by HDAC6 and does not influence HDAC6-dependent autophagy [161].

HDAC6-targeted inhibitors may contain zinc-binding groups different from hydroxamates. Thiol-containing HDAC6 inhibitors are exemplified by the cycloheptyl-capped, C6-alkyl-linked thiol **5.44** (Figure 5.9), endowed with good potency on HDAC6 and moderate isoform selectivity [162]. Mercaptoamide-containing HDAC6 inhibitors are exemplified by moderately selective, C5-alkyl linker-containing DMA-PB (**5.45**) [163]. DMA-PB promotes survival and regeneration of cortical neurons, protecting them against oxidative stress-induced neurodegeneration [164]. DMA-PB inhibits microglial transformation and reduces neuronal degradation when administered shortly after traumatic brain injury (TBI) in rats [165]. A naphthalene replacement of the dimethylaminophenyl cap in **5.45** and the same C5-alkyl-linker constitute the mercaptoamide **5.46** [166]. This compound is neuroprotective *in vivo* against AD in triple TG APP-tau-PS1 mice (50 mg/kg, i.p., once daily, 4 weeks), significantly decreasing soluble and insoluble

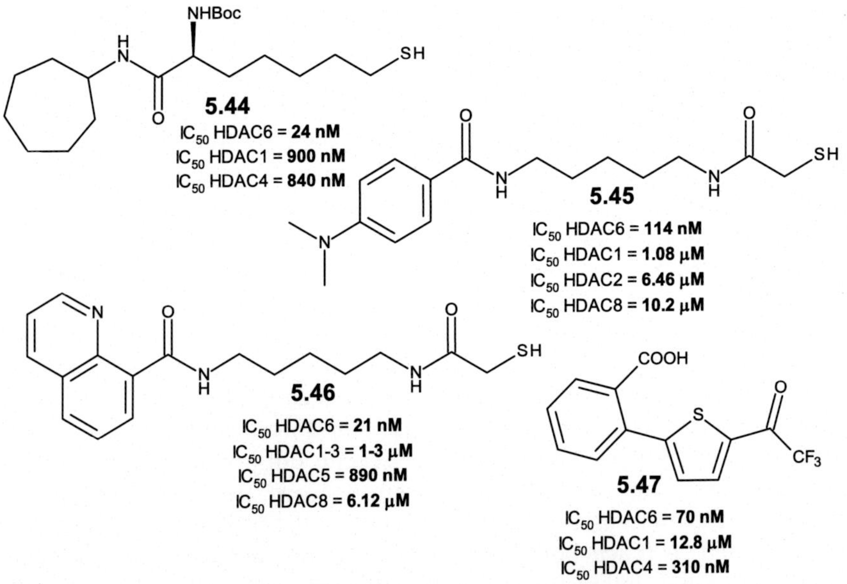

FIGURE 5.9   Small molecule HDAC6 inhibitors with zinc-binding motifs different from hydroxamates: chemical structures, **5.44–5.47**.

$A\beta_{40}$ levels, reducing tau phosphorylation at the Thr181 epitope, and improving learning and memory of treated mice [166]. The limited selectivity of **5.46** *vs.* other HDAC isoforms makes it unlikely that its AD effects are exclusively due to HDAC6 inhibition. The trifluoroacetylthiophene **5.47** (Figure 5.9) shows potent inhibition of HDAC6 and HDAC4, causes α-tubulin acetylation and is ineffective on histone acetylation in colon cancer HCT116 cells [167].

PTMs of HDAC6 indirectly regulate its deacetylase activity, and some of its PPIs. CK2 phosphorylates the dynein motor binding (DMB) domain of HDAC6 at the Ser458 residue [168]. CK2 inhibition should decrease HDAC enzymatic activity and prevent aggresome formation, possibly through physical hindrance of the HDAC6–dynein interaction [168]. CK2 inhibitors are covered in section 2.4.3 (Figures 2.13 and 2.14) as putative activation inhibitors of the co-chaperone cdc37 and indirect destabilizers of HP misfolded tau through inhibition of the Hsp90–cdc37 complex. They are mentioned in section 5.2, as CK2-mediated phosphorylation of the Ser403 residue in the UBA domain of p62 has a strong aggrephagy-promoting effect [47], and CK2 inhibitors may be useful in the presence of accumulated protein aggregates/stalled aggrephagy. Thus, the careful evaluation of CK2 inhibitors should clarify the relevance of each of the three mentioned molecular mechanisms (and maybe others) as a possibly effective and non-toxic chronic treatment in NDDs. Indirect HDAC6 regulation may also derive from rho-associated coiled-coil kinase (ROCK) and/or cyclin-dependent kinase 1 (Cdk-1) inhibition, through their phosphorylation of HDAC6-binding tubulin polymerization-promoting protein 1 (TPPP1) [169,170].

Several HDAC6-involving PPIs, which are not exploited as sources of new leads against NDDs yet, appear to be totally or partially deacetylase-independent, and connected with aggrephagy/autophagy and neurodegeneration. HDAC6 recognizes polyUBQ-containing protein aggregates by interacting with the C-terminal Gly–Gly sequence of free UBQ [171]. Free UBQ is released from ubiquitinated proteins by the deubiquitinase (DUB) ataxin-3, and remains exposed onto the aggregate surface. HDAC6 binding to exposed G75–G76–UBQ chains activates the transport of UBQ-containing microaggregates to aggresomes. The X-ray structure of the complex between UBQ and HDAC6 highlights the network of interactions between the Gly–Gly dipeptide and the deep hydrophobic pocket in the C-terminus of HDAC6, and the gatekeeper role of Arg1155–Tyr1156 residues [171]. Non-peptidic mimetics of the C-terminal UBQ sequence could be useful tools to study the influence of HDAC6 in the accumulation of insoluble protein aggregates, and may become new leads for *in vivo* studies in animal models of NDDs.

The AP–LS fusion is an essential step in autophagy, which is impaired in AD, leading to an accumulation of APs and autophagy stalling [12,19]. Stimulation of AP–LS fusion is an attractive therapeutic option, but small molecule activators of such autophagy step are unknown.

HDAC6–cortactin binding and deacetylation promotes cortactin-dependent F-actin remodeling, and stimulates AP–LS fusion in aggrephagy [172]. Conversely, starvation-induced autophagy is not HDAC6-dependent, providing desirable *inter*-autophagy selectivity. The HDAC6–cortactin PPI takes place between the central repeat region of cortactin and the catalytic DD1–DBM–DD2 region of HDAC6 [173]. This interaction should be narrowed to much smaller epitopes in both proteins to be useful for the rational design of small molecule stimulators (a difficult task *per se*). The setup of an HTS assay format targeted towards stimulators of the HDAC6–cortactin PPI could alternatively lead, *via* screening of large compound collections, to the same result.

An inhibitory function of tubacin-treated, deacetylase-impaired HDAC6 on MT growth and dynamics in B16F1 melanoma cells is due to a prolonged stay of non-functional HDAC6 on the MT tips, physically interfering with the addition of new tubulin subunits and with the detachment of tubulin subunits [150]. Tubastatin causes an increased localization of HDAC6 on MTs, stabilizes the MT/polymeric tubulin structure and reduces the rate of MT assembly and growth in MCF-7 cells [149]. Cells containing siRNA-silenced HDAC6 do not show impaired MT dynamics. MT capping/binding by HDAC6 must happen through either one of the catalytic DD domains, as DD1- and DD2-mutated HDAC6 still reduces MT dynamics (binds to MT tips), while mutations on both HDAC6 catalytic domains prevent the effects on MT dynamics (does not bind to MT tips) [150]. Is HDAC6 capping of MTs/impairment of MT dynamics achievable without inhibition of the deacetylase activity of HDAC6? Is it possible to identify (either rationally, or through suitable HTS assay formats) HDAC6 deacetylase inhibitors devoid of MT capping effects/impairment of MT dynamics? The capping effect should be further elucidated, and smaller interacting regions between mutated/inhibited HDAC6 and α-tubulin should be identified to answer these questions.

5.4 RECAP

This chapter deals with small molecule modulators of neuropathological alterations related to selective autophagy/aggrephagy-related protein misfolding and aggregation in general, and to tau and/or tau-connected events in particular. Two molecular targets are arbitrarily chosen, and discussed in detail in the companion biology book [10]. Forty-nine compounds/scaffolds acting on p62 and HDAC6 are reported in Figures 5.1 to 5.9, and are briefly summarized in Table 5.1. The chemical core of each scaffold/compound is structurally defined; its molecular target is mentioned; the developing laboratory (either public or private) is listed; and the development status—according to publicly available information—is finally provided.

TABLE 5.1   Compounds **5.1–5.47** Chemical Class, Target, Developing Organization, Development Status

Number	Chemical cpd./class	Target	Organization	Dev. status
5.1	YSKH peptide	p62	–	–
5.2	DDDWTHL	p62	–	–
5.3	PSEDP peptide	p62	–	–
5.4	DSPTGEL peptide	p62	–	–
5.5	ETGE peptide	p62	–	–
5.6	DLG peptide	p62	–	–
5.7	DPETGEL peptide	p62	–	–
5.8	Stearyl-NH-DPETGEL peptide amide	p62	–	–
5.9	Indolinones, SU6668	TBK1, other kinase inhibitor	Sugen/Pfizer	Ph II
5.10	Aminopyrimidines, BX795	TBK1, other kinase inhibitor	Novartis	Ph I
5.11	Aminopyrimidines, MRT67307	TBK1-IKKε inhibitors	Dundee University	LO
5.12	Aminopyrimidines	TBK1-IKKε inhibitors	Dundee University	PE
5.13	6-Aminopyrazolopyrimidines	TBK1-IKKε inhibitors	SouthWestern Medical Center, Dallas, TX	DD
5.14	Azabenzimidazoles	TBK1-IKKε inhibitors	AstraZeneca	LO
5.15	Pyrimidines	TBK1 inhibitors	University of North Carolina	DD
5.16	Pyridopyrimidines	TBK1 inhibitors	University of North Carolina	DD
5.17	Tricyclic pyridothiophenes	TBK1 inhibitors	University of North Carolina	DD
5.18	Amlexanox	TBK1-IKKε inhibitor	University of Michigan	MKTD
5.19	Hydroxamic acids, SAHA	Broad HDAC inhibitors	Merck Sharp & Dohme	MKTD
5.20	Cyclic depsipeptides, romidepsin, FK-228	Broad HDAC inhibitor	Celgene	MKTD

(Continued)

TABLE 5.1 Compounds **5.1–5.47** Chemical Class, Target, Developing Organization, Development Status (*cont.*)

Number	Chemical cpd./class	Target	Organization	Dev. status
5.21	Panobinostat, LBH-589	Broad HDAC inhibitor	Novartis	Ph III
5.22	Abexinostat, PCI-24781	Broad HDAC inhibitor	Pharmacyclics	Ph II
5.23	Givinostat, ITF2357	Broad HDAC inhibitor	Italfarmaco	Ph II
5.24	Entinostat, MS-275	Class I-specific HDAC inhibitor	Syndax Pharmac.	Ph II
5.25	Mocetinostat, MGCD0103	Class I-specific HDAC inhibitor	MethylGene	Ph I
5.26	Tubacin	HDAC6 inhibitor	Harvard University	PE
5.27	Rocilinostat, ACY-1215	HDAC6 inhibitor	Acetylon Pharmac.	Ph I/II
5.28	M344	Broad HDAC inhibitor	Munster University, Germany	PE
5.29a,b	ST80 (a)	HDAC6 inhibitor	Munster University, Germany	LO
5.30, 5.31	Isoxazole hydoxamates	Class I-specific HDAC inhibitors	University of Illinois	LO
5.32	Triazolylphenyl hydroxamates	Broad HDAC inhibitors	University of Illinois	LO
5.33	C1A	Broad HDAC inhibitor	Imperial College, UK	LO
5.34	Furylamine hydroxamates	Broad HDAC inhibitors	China Pharm. University	DD
5.35	Arylalkene hydroxamates	HDAC6–HDAC8 inhibitor	Broad Institute, Boston, USA	LO
5.36	Dihydroquinoxalinone hydroxamates	HDAC6 inhibitor	MethylGene	LO
5.37	Pyrroline hydroxamates	HDAC6 inhibitor	Broad Institute, Boston, USA	DD
5.38a,b	Tubastatin (a)	HDAC6 inhibitor	University of Illinois	PE
5.39	Tubathian A	HDAC6 inhibitor	University of Ghent, Belgium	LO
5.40	Nexturastat	HDAC6 inhibitor	University of Illinois	DD

TABLE 5.1   Compounds **5.1–5.47** Chemical Class, Target, Developing Organization, Development Status (*cont.*)

Number	Chemical cpd./class	Target	Organization	Dev. status
5.41	ACY-738	HDAC6 inhibitor	University of Pennsylvania	LO
5.42	Quinazolin-4-one hydroxamates	HDAC6 inhibitor	Taiwan University	PE
5.43	HPOB	HDAC6 inhibitor	Sloan Kettering Cancer Center, USA	LO
5.44	Thiols	Broad HDAC inhibitors	Nagoya University	DD
5.45, 5.46	Mercaptoamides, DMA-CB	Broad HDAC inhibitors	University of Illinois	PE
5.47	Trifluoroacetylthio-phenes	Broad HDAC inhibitors	Angeletti, Rome, Italy	DD

Not progressed, NP; early discovery, DD; lead optimization, LO; preclinical evaluation, PE; clinical Phase I–II–III, Ph I–Ph III; marketed, MKTD; food supplement, FS.

References

[1] Tyedmers, J.; Moegk, A.; Bukau, B. Cellular strategies for controlling protein aggregation. *Nat. Rev. Mol. Cell Biol.* **2010**, *11*, 777–788.

[2] Lamark, T.; Johansen, T. Aggrephagy: selective disposal of protein aggregates by macroautophagy. *Int. J. Cell Biol.* **2012**, 736905.

[3] Reggiori, F.; Komatsu, M.; Finley, K.; Simonsen, A. Selective types of autophagy. *Int. J. Cell Biol.* **2012**, 219625.

[4] Wong, E.; Cuervo, A. M. Autophagy gone awry in neurodegenerative diseases. *Nat. Neurosci. Rev.* **2010**, *13*, 805–811.

[5] Cuervo, A. M.; Bergamini, E.; Brunk, U. T.; Droge, W.; French, M.; Terman, A. Autophagy and aging: the importance of maintaining "clean" cells. *Autophagy* **2005**, *1*, 131–140.

[6] McNaught, K. S.; Shashidharan, P.; Perl, D. P.; Jenner, P.; Olanow, C. W. Aggresome-related biogenesis of Lewy bodies. *Eur. J. Neurosci.* **2002**, *16*, 2136–2148.

[7] Filimonenko, M.; Isakson, P.; Finley, K. D.; Anderson, M.; Jeong, H.; Melia, T. J., et al. The selective macroautophagic degradation of aggregated proteins requires the PI3P-binding protein Alfy. *Mol. Cell* **2010**, *38*, 265–279.

[8] Shin, J. p62 and the sequestosome, a novel mechanism for protein metabolism. *Arch. Pharm. Res.* **1998**, *21*, 629–633.

[9] Gamerdinger, M.; Hajieva, P.; Kaya, A. M.; Wolfrum, U.; Hartl, F. U.; Behl, C. Protein quality control during aging involves recruitment of the macroautophagy pathway by BAG3. *EMBO J.* **2009**, *28*, 889–901.

[10] Seneci, P. *Molecular targets in protein misfolding and neurodegenerative disease.* Academic Press, **2014**, 278 pages.

[11] He, L. -Q.; Lu, J. -H.; Yue, Z. -Y. Autophagy in ageing and ageing-associated diseases. *Acta Pharmacol. Sin.* **2013**, *34*, 605–611.

[12] Nixon, R. A.; Yang, D. S. Autophagy failure in Alzheimer's disease—locating the primary defect. *Neurobiol. Dis.* **2011**, *43*, 38–45.

[13] Boland, B.; Kumar, A.; Lee, S.; Platt, F. M.; Wegiel, J.; Huang Yu, W.; Nixon, R. A. Autophagy induction and autophagosome clearance in neurons: relationship to autophagic pathology in Alzheimer's disease. *J. Neurosci.* **2008**, *28*, 6926–6937.

[14] Bjørkøy, G.; Lamark, T.; Brech, A.; Outzen, H.; Perander, M.; Oeervatn, A., et al. p62/SQSTM1 forms protein aggregates degraded by autophagy and has a protective effect on huntingtin-induced cell death. *J. Cell Biol.* **2005**, *171*, 603–614.

[15] Nakamura, K.; Kimple, A. J.; Siderovski, D. P.; Johnson, G. L. PB1 domain interaction of p62/sequestosome 1 and MEKK3 regulates NF-κB activation. *J. Biol. Chem.* **2010**, *285*, 2077–2089.

[16] Sanz, L.; Diaz-Meco, M. T.; Nakano, H.; Moscat, J. The atypical PKC-interacting protein p62 channels NF-κB activation by the IL-1-TRAF6 pathway. *EMBO J.* **2000**, *19*, 1576–1586.

[17] Moscat, J.; Diaz-Meco, M. T.; Albert, A.; Campuzano, S. Cell signaling and function organized by PB1 domain interactions. *Mol. Cell* **2006**, *23*, 631–640.

[18] Saio, T.; Yokochi, M.; Inagaki, F. The NMR structure of the p62 PB1 domain, a key protein in autophagy and NF-*k*B signaling pathway. *J. Biomol. NMR* **2009**, *45*, 335–341.

[19] Nixon, R. A.; Wegiel, J.; Kumar, A.; Yu, W. H.; Peterhoff, C.; Cataldo, A.; Cuervo, A. M. Extensive involvement of autophagy in Alzheimer disease: an immuno-electron microscopy study. *J. Neuropathol. Exp. Neurol.* **2005**, *64*, 113–122.

[20] Nakamura, K.; Zawistowski, J. S.; Hughes, M. A.; Sexton, J. Z.; Yeh, L. -A.; Johnson, G. L.; Scott, J. E. Homogeneous time-resolved fluorescence resonance energy transfer assay for measurement of Phox/Bem1p (PB1) domain heterodimerization. *J. Biomol. Screen.* **2008**, *13*, 396–405.

[21] Long, J.; Garner, T. P.; Pandya, M. J.; Craven, C. J.; Chen, P.; Shaw, B., et al. Dimerisation of the UBA domain of p62 inhibits ubiquitin binding and regulates NFκB signalling. *J. Mol. Biol.* **2010**, *396*, 178–194.

[22] Searle, M. S.; Garner, T. P.; Strachan, J.; Long, J.; Adlington, J.; Cavey, J. R., et al. Structural insights into specificity and diversity in mechanisms of ubiquitin recognition by ubiquitin-binding domains. *Biochem. Soc. Trans.* **2012**, *40*, 404–408.

[23] Isogai, S.; Morimoto, D.; Arita, K.; Unzai, S.; Tenno, T.; Hasegawa, J., et al. Crystal structure of the ubiquitin-associated (UBA) domain of p62 and its interaction with ubiquitin. *J. Biol. Chem.* **2011**, *286*, 31864–31874.

[24] Garner, T. P.; Long, J.; Layfield, R.; Searle, M. S. Impact of p62/SQSTM1 UBA domain mutations linked to Paget's disease of bone on ubiquitin recognition. *Biochemistry* **2011**, *50*, 4665–4674.

[25] Pankiv, S.; Clausen, T. H.; Lamark, T.; Brech, A.; Bruun, J. A.; Outzen, H., et al. p62/SQSTM1 binds directly to Atg8/LC3 to facilitate degradation of ubiquitinated protein aggregates by autophagy. *J. Biol. Chem.* **2007**, *282*, 24131–24145.

[26] Ichimura, Y.; Kumanomidou, T.; Sou, Y. S.; Mizushima, T.; Ezaki, J.; Ueno, T., et al. Structural basis for sorting mechanism of p62 in selective autophagy. *J. Biol. Chem.* **2008**, *283*, 22847–22857.

[27] Gao, W.; Chen, Z.; Wang, W.; Stang, M. T. E1-like activating enzyme Atg7 is preferentially sequestered into p62 aggregates via its interaction with LC3-I. *PLoS ONE* **2013**, *8*, e73229.

[28] Noda, N. N.; Kumeta, H.; Nakatogawa, H.; Satoo, K.; Adachi, W.; Ishii, J., et al. Structural basis of target recognition by Atg8/LC3 during selective autophagy. *Genes Cells* **2008**, *13*, 1211–1218.

[29] Duran, A.; Amanchy, R.; Linares, J. F.; Joshi, J.; Abu-Baker, S.; Porollo, A., et al. p62 is a key regulator of nutrient sensing in the mTORC1 pathway. *Mol. Cell* **2011**, *44*, 134–146.

[30] Linares, J. F.; Duran, A.; Yajima, T.; Pasparakis, M.; Moscat, J.; Diaz-Meco, M. T. K63 polyubiquitination and activation of mTOR by the p62-TRAF6 complex in nutrient-activated cells. *Mol. Cell* **2013**, *51*, 283–296.

[31] Jadhav, T. S.; Wooten, M. W.; Wooten, M. C. Mining the TRAF6/p62 interactome for a selective ubiquitination motif. *BMC Proc.* **2011**, *5*, S4.

[32] Nazio, F.; Strappazzon, F.; Antonioli, M.; Bielli, P.; Cianfanelli, V.; Bordi, M., et al. mTOR inhibits autophagy by controlling ULK1 ubiquitination, self-association and function through AMBRA1 and TRAF6. *Nat. Cell Biol.* **2013**, *15*, 406–416.

[33] Furukawa, M.; Xiong, Y. BTB protein Keap1 targets antioxidant transcription factor Nrf2 for ubiquitination by the Cullin 3-Roc1 ligase. *Mol. Cell. Biol.* **2005**, *25*, 162–171.

[34] Lau, A.; Wang, X. -J.; Zhao, F.; Villeneuve, N. F.; Wu, T., et al. A noncanonical mechanism of Nrf2 activation by autophagy deficiency: direct interaction between keap1 and p62. *Mol. Cell. Biol.* **2010**, *30*, 3275–3285.

[35] Komatsu, M.; Kurokawa, H.; Waguri, S.; Taguchi, K.; Kobayashi, A.; Ichimura, Y., et al. The selective autophagy substrate p62 activates the stress responsive transcription factor Nrf2 through inactivation of Keap1. *Nat. Cell Biol.* **2010**, *12*, 213–223.

[36] Jain, A.; Lamark, T.; Sjøttem, E.; Bowitz Larsen, K.; Atesoh Awuh, J.; Øvervatn, A., et al. *p62/SQSTM1* is a target gene for transcription factor NRF2 and creates a positive feedback loop by inducing antioxidant response element-driven gene transcription. *J. Biol. Chem.* **2010**, *285*, 22576–22591.

[37] Fan, W.; Tang, Z.; Chen, D.; Moughon, D.; Ding, X.; Chen, S., et al. Keap1 facilitates p62-mediated ubiquitin aggregate clearance via autophagy. *Autophagy* **2010**, *6*, 614–621.

[38] Bae, S. H.; Sung, S. H.; Oh, S. Y.; Lim, J. M.; Lee, S. K.; Park, Y. N., et al. Sestrins activate Nrf2 by promoting p62-dependent autophagic degradation of Keap1 and prevent oxidative liver damage. *Cell Metab.* **2013**, *17*, 73–84.

[39] Inami, Y.; Waguri, S.; Sakamoto, A.; Kouno, T.; Nakada, K.; Hino, O., et al. Persistent activation of Nrf2 through p62 in hepatocellular carcinoma cells. *J. Cell Biol.* **2011**, *193*, 275–284.

[40] Inoue, D.; Kubo, H.; Taguchi, K.; Suzuki, T.; Komatsu, M.; Motohashi, H.; Yamamoto, M. Inducible disruption of autophagy in the lung causes airway hyper-responsiveness. *Biochem. Biophys. Res. Commun.* **2011**, *405*, 13–18.

[41] Taguchi, K.; Fujikawa, N.; Komatsu, M.; Ishii, T.; Unno, M.; Akaike, T., et al. Keap1 degradation by autophagy for the maintenance of redox homeostasis. *Proc. Natl. Acad. Sci. U.S.A.* **2012**, *109*, 13561–13566.

[42] Pickford, F.; Masliah, E.; Britschgi, M.; Lucin, K.; Narasimhan, R.; Jaeger, P. A., et al. The autophagy-related protein beclin 1 shows reduced expression in early Alzheimer disease and regulates amyloid beta accumulation in mice. *J. Clin. Invest.* **2008**, *118*, 2190–2199.

[43] Nixon, R. A.; Cataldo, A. M.; Mathews, P. M. The endosomal-lysosomal system of neurons in Alzheimer's disease pathogenesis: a review. *Neurochem. Res.* **2000**, *25*, 1161–1172.

[44] Tanji, K.; Maruyama, A.; Odagiri, S.; Mori, F.; Itoh, K.; Kakita, A., et al. Keap1 is localized in neuronal and glial cytoplasmic inclusions in various neurodegenerative diseases. *J. Neuropathol Exp. Neurol.* **2012**, *72*, 18–28.

[45] Hancock, R.; Bertrand, H. C.; Tsujita, T.; Naz, S.; El-Bakry, A.; Laoruchupong, J., et al. Peptide inhibitors of the Keap1–Nrf2 protein–protein interaction. *Free Radic. Biol. Med.* **2012**, *52*, 444–451.

[46] Hancock, R.; Schaap, M.; Pfister, H.; Wells, G. Peptide inhibitors of the Keap1–Nrf2 protein–protein interaction with improved binding and cellular activity. *Org. Biomol. Chem.* **2013**, *11*, 3553–3557.

[47] Matsumoto, G.; Wada, K.; Okuno, M.; Kurosawa, M.; Nukina, N. Serine 403 phosphorylation of p62/SQSTM1 regulates selective autophagic clearance of ubiquitinated proteins. *Mol. Cell* **2011**, *44*, 279–289.

[48] Olsen, B. B.; Svenstrup, T. H.; Guerra, B. Downregulation of protein kinase CK2 induces autophagic cell death through modulation of the mTOR and MAPK signaling pathways in human glioblastoma cells. *Int. J. Oncol.* **2012**, *41*, 1967–1976.

[49] Pilli, M.; Arko-Mensah, J.; Ponpuak, M.; Roberts, E.; Master, S.; Mandell, M. A., et al. TBK1 promotes autophagy mediated antimicrobial defense by controlling autophagosome maturation. *Immunity* **2012**, *37*, 223–234.

[50] Wild, P.; Farhan, H.; McEwan, D. G.; Wagner, S.; Rogov, V. V.; Brady, N. R., et al. Phosphorylation of the autophagy receptor optineurin restricts Salmonella growth. *Science* **2011**, *333*, 228–233.

[51] Tu, D.; Zhu, Z.; Zhou, A. Y.; Yun, C. H.; Lee, K. E.; Toms, A. V., et al. Structure and ubiquitination-dependent activation of TANK-binding kinase 1. *Cell Rep.* **2013**, *3*, 747–758.

[52] Qu, F.; Gao, H.; Zhu, S.; Shi, P.; Zhang, Y.; Liu, Y., et al. TRAF6-dependent Act1 phosphorylation by the IkB kinase-related kinases suppresses interleukin-17-induced NF-kB activation. *Mol. Cell. Biol.* **2012**, *32*, 3925–3937.

[53] Clément, J. -F.; Meloche, S.; Servant, M. J. The IKK-related kinases: from innate immunity to oncogenesis. *Cell Res.* **2008**, *18*, 889–899.

[54] Randow, F. How cells deploy ubiquitin and autophagy to defend their cytosol from bacterial invasion. *Autophagy* **2011**, *7*, 304–309.

[55] Newman, A. C.; Scholefield, C. L.; Kemp, A. J.; Newman, M.; Mciver, E. G.; Kamal, A.; Wilkinson, S. TBK1 kinase addiction in lung cancer cells is mediated via autophagy of Tax1bp1/Ndp52 and non-canonical NF-κB signalling. *PLoS ONE* **2012**, *7*, e50672.

[56] Helgason, E.; Phung, Q. T.; Dueber, E. C. Recent insights into the complexity of Tank-binding kinase 1 signaling networks: The emerging role of cellular localization in the activation and substrate specificity of TBK1. *FEBS Lett.* **2013**, *587*, 1230–1237.

[57] Shu, C.; Sankaran, B.; Chaton, C. T.; Herr, A. B.; Mishra, A.; Peng, J.; Li, P. Structural insights into the functions of TBK1 in innate antimicrobial immunity. *Structure* **2013**, *21*, 1137–1148.

[58] Larabi, A.; Devos, J. M.; Ng, S. -L.; Nanao, M. H.; Round, A.; Maniatis, T.; Panne, D. Crystal structure and mechanism of activation of TANK-binding kinase 1. *Cell Rep.* **2013**, *3*, 734–746.

[59] Laird, A. D.; Vajkoczy, P.; Shawver, L. K.; Thurnher, A.; Liang, C.; Mohammadi, M., et al. SU6668 is a potent antiangiogenic and antitumor agent that induces regression of established tumors. *Cancer Res.* **2000**, *60*, 4152–4162.

[60] Godl, K.; Gruss, O. J.; Eickhoff, J.; Wissing, J.; Blencke, S.; Weber, M., et al. Proteomic characterization of the angiogenesis inhibitor SU6668 reveals multiple impacts on cellular kinase signalling. *Cancer Res.* **2005**, *65*, 6919–6926.

[61] Feldman, R. I.; Wu, J. M.; Polokoff, M. A.; Kochanny, M. J.; Dinter, H.; Zhu, D., et al. Novel small molecule inhibitors of 3-phosphoinositide-dependent kinase-1. *J. Biol. Chem.* **2005**, *280*, 19867–19874.

[62] Clark, K.; Plater, L.; Peggie, M.; Cohen, P. Use of the pharmacological inhibitor BX795 to study the regulation and physiological roles of TBK1 and IκB kinase ε: a distinct upstream kinase mediates Ser-172 phosphorylation and activation. *J. Biol. Chem.* **2009**, *284*, 14136–14146.

[63] Clark, K.; Peggie, M.; Plater, L.; Sorcek, R. J.; Young, E. R. R.; Madwed, J. D.; Hough, J.; McIver, E. G.; Cohen, P. Novel cross-talk within the IKK family controls innate immunity. *Biochem. J.* **2011**, 93–104.

[64] McIver, E. G.; Bryans, J.; Birchall, K.; Chugh, J.; Drake, T.; Lewis, S. J., et al. Synthesis and structure–activity relationships of a novel series of pyrimidines as potent inhibitors of TBK1/IKKe kinases. *Bioorg. Med. Chem. Lett.* **2012**, *22*, 7169–7173.

[65] Ou, Y. -H.; Torres, M.; Ram, R.; Formstecher, E.; Roland, C.; Cheng, T., et al. TBK1 directly engages Akt/PKB survival signaling to support oncogenic transformation. *Mol. Cell* **2011**, *41*, 458–470.

[66] Wang, T.; Block, M. A.; Cowen, S.; Davies, A. M.; Devereaux, E.; Gingipalli, L., et al. Discovery of azabenzimidazole derivatives as potent, selective inhibitors of TBK1/IKKe kinases. *Bioorg. Med. Chem. Lett.* **2012**, *22*, 2063–2069.

[67] Hutti, J. E.; Porter, M. A.; Cheely, A. W.; Cantley, L. C.; Wang, X.; Kireev, D., et al. Development of a high-throughput assay for identifying inhibitors of TBK1 and IKKε. *PLoS ONE* **2012**, *7*, e41494.

[68] Reilly, S. M.; Chiang, S. H.; Decker, S. J.; Chang, L.; Uhm, M.; Larsen, M. J., et al. An inhibitor of the protein kinases TBK1 and IKK-ε improves obesity-related metabolic dysfunctions in mice. *Nat. Med.* **2013**, *19*, 313–321.

[69] Makino, H.; Saijo, T.; Ashida, Y.; Kuriki, H.; Maki, Y. Mechanism of action of an antiallergic agent, amlexanox (AA-673), in inhibiting histamine release from mast cells. Acceleration of cAMP generation and inhibition of phosphodiesterase. *Int. Arch. Allergy Appl. Immunol.* **1987**, *8*, 66–71.

[70] Ichimura, Y.; Waguri, S.; Sou, Y. -S.; Kageyama, S.; Hasegawa, J.; Ishimura, R., et al. Phosphorylation of p62 activates the Keap1-Nrf2 pathway during selective autophagy. *Mol. Cell* **2013**, *51*, 618–631.

[71] Paris, M.; Porcelloni, M.; Binaschi, M.; Fattori, D. Histone deacetylase inhibitors: from bench to clinic. *J. Med. Chem.* **2008**, *51*, 1505–1529.

[72] Pontiki, E.; Hadjipavlou-Litina, D. Histone deacetylase inhibitors (HDACIs). Structure-activity relationships: History and new QSAR perspectives. *Med. Res. Rev.* **2012**, *32*, 1–165.

[73] Li, J.; Li, G.; Xu, W. Histone deacetylase inhibitors: an attractive strategy for cancer therapy. *Curr. Med. Chem.* **2013**, *20*, 1858–1886.

[74] Giannini, G.; Cabri, W.; Fattorusso, C.; Rodriquez, M. Histone deacetylase inhibitors in the treatment of cancer: overview and perspectives. *Fut. Med. Chem.* **2012**, *4*, 1439–1460.

[75] Sangshetti, J. N.; Sakle, N. S.; Dehghan, M. H. G.; Shinde, D. B. Histone deacetylases as targets for multiple diseases. *Mini-Rev. Med. Chem.* **2013**, *13*, 1005–1026.

[76] Kelly, W. K.; Marks, P. A. Drug insight: histone deacetylase inhibitors—development of the new targeted anticancer agent suberoylanilide hydroxamic acid. *Nat. Clin. Pract. Oncol.* **2005**, *2*, 150–157.

[77] Harrison, S. J.; Bishton, M.; Bates, S. E.; Grant, S.; Piekarz, R. L.; Johnstone, R. W., et al. A focus on the preclinical development and clinical status of the histone deacetylase inhibitor, romidepsin (depsipeptide, Istodax). *Epigenomics* **2012**, *4*, 571–589.

[78] Rangwala, S.; Duvic, M.; Zhang, C. Trends in the treatment of cutaneous T-cell lymphoma—critical evaluation and perspectives on vorinostat. *Blood Lymph. Canc.: Targets Ther.* **2012**, *2*, 17–27.

[79] McGraw, A. L. Romidepsin for the treatment of T-cell lymphomas. *Am. J. Health-System Pharm.* **2013**, *70*, 1115–1122.

[80] Khot, A.; Dickinson, M.; Prince, H. M. Panobinostat in lymphoid and myeloid malignancies. *Exp. Opin. Invest. Drugs* **2013**, *22*, 1211–1223.

[81] Banuelos, C. A.; Banáth, J. P.; MacPhail, S. H.; Zhao, J.; Reitsema, T.; Olive, P. L. Radiosensitization by the histone deacetylase inhibitor PCI-24781. *Clin. Cancer Res.* **2007**, *13*, 6816–6826.

[82] Rambaldi, A.; Dellacasa, C. M.; Finazzi, G.; Carobbio, A.; Ferrari, M. L.; Guglielmelli, P., et al. A pilot study of the histone-deacetylase inhibitor givinostat in patients with JAK2V617F positive chronic myeloproliferative neoplasms. *Br. J. Haematol.* **2010**, *150*, 446–455.

[83] http://clinicaltrials.gov/ct2/results?intr=lbh-589&pg=5, October 26th, 2013.

[84] http://clinicaltrials.gov/ct2/results?term=pci-24781&Search=Search, October 26th, 2013.

[85] http://clinicaltrials.gov/ct2/results?term=itf2357&Search=Search, October 26th, 2013.

[86] Bishton, M. J.; Harrison, S. J.; Martin, B. P.; McLaughlin, N.; James, C.; Josefsson, E. C., et al. Deciphering the molecular and biological processes that mediate histone deacetylase inhibitor-induced thrombocytopenia. *Blood* **2011**, *117*, 3658–3668.

[87] Shultz, M. D.; Cao, X.; Chen, C. H.; Cho, Y. S.; Davis, N. R.; Eckman, J., et al. Optimization of the in vitro cardiac safety of hydroxamate-based histone deacetylase inhibitors. *J. Med. Chem.* **2011**, *54*, 4572–4592.

[88] Knipstein, J.; Gore, L. Entinostat for treatment of solid tumors and hematologic malignancies. *Exp. Opin. Invest. Drugs* **2011**, *20*, 1455–1467.

[89] Boumber, Y.; Younes, A.; Garcia-Manero, G. Mocetinostat (MGCD0103): a review of an isotype-specific histone deacetylase inhibitor. *Exp. Opin. Invest. Drugs* **2011**, *20*, 823–829.

[90] Bertrand, P. Inside HDAC with HDAC inhibitors. *Eur. J. Med. Chem.* **2010**, *45*, 2095–2116.

[91] http://clinicaltrials.gov/ct2/results?term=ms-275&Search=Search, October 26th, 2013.

[92] http://clinicaltrials.gov/ct2/results?term=mgcd0103&Search=Search, October 26th, 2013.

[93] Fraczek, J.; Vanhaecke, T.; Rogiers, V. Toxicological and metabolic considerations for histone deacetylase inhibitors. *Exp. Opin. Drug Metab. & Toxicol.* **2013**, *9*, 441–457.

[94] Liu, N.; He, S.; Ma, L.; Ponnusamy, M.; Tang, J.; Tolbert, E., et al. Blocking the class I histone deacetylase ameliorates renal fibrosis and inhibits renal fibroblast activation via modulating TGF-beta and EGFR signaling. *PLoS ONE* **2013**, *8*, e54001.

[95] Denk, F.; Huang, W.; Sidders, B.; Bithell, A.; Crow, M.; Grist, J., et al. HDAC inhibitors attenuate the development of hypersensitivity in models of neuropathic pain. *Pain* **2013**, *154*, 1668–1679.

[96] Welberg, L. Neurodegenerative disorders: HDAC2 is the one. *Nat. Rev. Drug Discov.* **2009**, *8*, 538–539.

[97] Jia, H.; Pallos, J.; Jacques, V.; Lau, A.; Tang, B.; Cooper, A., et al. Histone deacetylase (HDAC) inhibitors targeting HDAC3 and HDAC1 ameliorate polyglutamine-elicited phenotypes in model systems of Huntington's disease. *Neurobiol. Dis.* **2012**, *46*, 351–361.

[98] Best, J. D.; Carey, N. Epigenetic therapies for non-oncology indications. *Drug Discovery Today* **2010**, *15*, 1008–1014.

[99] http://clinicaltrials.gov/ct2/results?term=hdac&Search=Search, October 26th, 2013.

[100] Kalin, J. H.; Bergman, J. A. Development and therapeutic implications of selective histone deacetylase 6 inhibitors. *J. Med. Chem.* **2013**, *56*, 6297–6313.

[101] Zhang, Y.; Kwon, S.; Yamaguchi, T.; Cubizolles, F.; Rousseaux, S.; Kneissel, M., et al. Mice lacking histone deacetylase 6 have hyperacetylated tubulin but are viable and develop normally. *Mol. Cell. Biol.* **2008**, *28*, 1688–1701.

[102] Haggarty, S. J.; Koeller, K. M.; Wong, J. C.; Grozinger, C. M.; Schreiber, S. L. Domain-selective small-molecule inhibitor of histone deacetylase 6 (HDAC6)-mediated tubulin deacetylation. *Proc. Natl. Acad. Sci. U.S.A.* **2003**, *100*, 4389–4394.

[103] Sternson, S. M.; Wong, J. C.; Grozinger, C. M.; Schreiber, S. L. Synthesis of 7200 small molecules based on a substructural analysis of the histone deacetylase inhibitors trichostatin and trapoxin. *Org. Lett.* **2001**, *3*, 4239–4242.

[104] Hideshima, T.; Bradner, J. E.; Wong, J.; Chauhan, D.; Richardson, P.; Schreiber, S. L.; Anderson, K. C. Small-molecule inhibition of proteasome and aggresome function induces synergistic antitumor activity in multiple myeloma. *Proc. Natl. Acad. Sci. U.S.A.* **2005**, *102*, 8567–8572.

[105] Kawada, J.; Zou, P.; Mazitschek, R.; Bradner, J. E.; Cohen, J. I. Tubacin kills Epstein–Barr virus (EBV)–Burkitt lymphoma cells by inducing reactive oxygen species and EBV lymphoblastoid cells by inducing apoptosis. *J. Biol. Chem.* **2009**, *284*, 17102–17109.

[106] Komatsu, S.; Moriya, S.; Che, X. -F.; Yokoyama, T.; Kohno, N.; Miyazawa, K. Combined treatment with SAHA, bortezomib, and clarithromycin for concomitant targeting of aggresome formation and intracellular proteolytic pathways enhances ER stress-mediated cell death in breast cancer cells. *Biochem. Biophys. Res. Commun.* **2013**, *437*, 41–47.

[107] Parmigiani, R. B.; Xu, W. S.; Venta-Perez, G.; Erdjument-Bromage, H.; Yaneva, M.; Tempst, P.; Marks, P. A. HDAC6 is a specific deacetylase of peroxiredoxins and is involved in redox regulation. *Proc. Natl. Acad. Sci. U.S.A.* **2008**, *105*, 9633–9638.

[108] Namdar, M.; Perez, G.; Ngo, L.; Marks, P. A. Selective inhibition of histone deacety-lase 6 (HDAC6) induces DNA damage and sensitizes transformed cells to anticancer agents. *Proc. Natl. Acad. Sci. U.S.A.* **2010**, *107*, 20003–20008.

[109] Aldana-Masangkay, G. I.; Rodriguez-Gonzalez, A.; Lin, T.; Ikeda, A. K.; Hsieh, Y. T.; Kim, Y. M., et al. Tubacin suppresses proliferation and induces apoptosis of acute lym-phoblastic leukemia cells. *Leuk. Lymphoma* **2011**, *52*, 1544–1555.

[110] Zuo, Q.; Wu, W.; Li, X.; Zhao, L.; Chen, W. HDAC6 and SIRT2 promote bladder cancer cell migration and invasion by targeting cortactin. *Oncol. Rep.* **2012**, *27*, 819–824.

[111] Wickstrom, S. A.; Masoumi, K. C.; Khochbin, S.; Fassler, R.; Massoumi, R. CYLD nega-tively regulates cell-cycle progression by inactivating HDAC6 and increasing the lev-els of acetylated tubulin. *EMBO J.* **2010**, *29*, 131–144.

[112] Kerr, E.; Holohan, C.; McLaughlin, K. M.; Majkut, J.; Dolan, S.; Redmond, K., et al. Identification of an acetylation-dependent Ku70/FLIP complex that regulates FLIP ex-pression and HDAC inhibitor-induced apoptosis. *Cell Death Differ.* **2012**, *19*, 1317–1327.

[113] de Zoeten, E. F.; Wang, L.; Butler, K.; Beier, U. H.; Akimova, T.; Sai, H., et al. Histone deacetylase 6 and heat shock protein 90 control the functions of Foxp3$^+$ T-regulatory cells. *Mol. Cell. Biol.* **2011**, *31*, 2066–2078.

[114] Akimova, T.; Beier, U. H.; Liu, Y.; Wang, L.; Hancock, W. W. Histone/protein deacety-lases and T-cell immune responses. *Blood* **2012**, *119*, 2443–2451.

[115] Liu, W.; Fan, L. X.; Zhou, X.; Sweeney, W. E., Jr.; Avner, E. D.; Li, X. HDAC6 regulates epidermal growth factor receptor (EGFR) endocytic trafficking and degradation in re-nal epithelial cells. *PLoS ONE* **2012**, *7*, e49418.

[116] Winkler, R.; Benz, V.; Clemenz, M.; Bloch, M.; Foryst-Ludwig, A.; Wardat, S., et al. His-tone deacetylase 6 (HDAC6) is an essential modifier of glucocorticoid-induced hepatic gluconeogenesis. *Diabetes* **2012**, *61*, 513–523.

[117] d'Ydewalle, C.; Krishnan, J.; Chiheb, D. M.; Van Damme, P.; Irobi, J.; Kozikowski, A. P., et al. HDAC6 inhibitors reverse axonal loss in a mouse model of mutant HSPB1-induced Charcot-Marie-Tooth disease. *Nat. Med.* **2011**, *17*, 968–974.

[118] Chen, S.; Owens, G. C.; Makarenkova, H.; Edelman, D. B. HDAC6 regulates mitochon-drial transport in hippocampal neurons. *PLoS ONE* **2010**, *5*, e10848.

[119] Ding, H.; Dolan, P. J.; Johnson, G. V. Histone deacetylase 6 interacts with the microtu-bule-associated protein tau. *J. Neurochem.* **2008**, *106*, 2119–2130.

[120] Perez, M.; Santa-Maria, I.; Gomez de Barreda, E.; Zhu, X.; Cuadros, R.; Cabrero, J. R., et al. Tau—an inhibitor of deacetylase HDAC6 function. *J. Neurochem.* **2009**, *109*, 1756–1766.

[121] Xiong, Y.; Zhao, K.; Wu, J.; Xu, Z.; Jin, S.; Zhang, Y. Q. HDAC6 mutations rescue human tau-induced microtubule defects in Drosophila. *Proc. Natl. Acad. Sci. U.S.A.* **2013**, *110*, 4604–4609.

[122] Estiu, G.; Greenberg, E.; Harrison, C. B.; Kwiatkowski, N. P.; Mazitschek, R.; Bradner, J. E.; Wiest, O. Structural origin of selectivity in class II-selective histone deacetylase inhibitors. *J. Med. Chem.* **2008**, *51*, 2898–2906.

[123] Santo, L.; Hideshima, T.; Kung, A. L.; Tseng, J. C.; Tamang, D.; Yang, M., et al. Preclini-cal activity, pharmacodynamic, and pharmacokinetic properties of a selective HDAC6 inhibitor, ACY-1215, in combination with bortezomib in multiple myeloma. *Blood* **2012**, *119*, 2579–2589.

[124] http://clinicaltrials.gov/ct2/results/displayOpt?flds=a&flds=b&flds=f&submit_fld_opt=on&=Update+Display&term=acy-1215&show_flds=Y, October 26th, 2013.

[125] Jung, M.; Brosch, G.; Koelle, D.; Scherf, H.; Gerhauser, C.; Loidl, P. Amide analogues of trichostatin A as inhibitors of histone deacetylase and inducers of terminal cell dif-ferentiation. *J. Med. Chem.* **1999**, *42*, 4669–4679.

[126] Heltweg, B.; Dequiedt, F.; Marshall, B. L.; Brauch, C.; Yoshida, M.; Nishino, N., et al. Subtype selective substrates for histone deacetylases. *J. Med. Chem.* **2004**, *47*, 5235–5243.

[127] Riessland, M.; Brichta, L.; Hahnen, E.; Wirth, B. The benzamide M344, a novel histone deacetylase inhibitor, significantly increases SMN2 RNA/protein levels in spinal muscular atrophy cells. *Hum. Genet.* **2006**, *120*, 101–110.

[128] Cook, C.; Gendron, T. F.; Scheffel, K.; Carlomagno, Y.; Dunmore, J.; Deture, M.; Petrucelli, L. Loss of HDAC6, a novel CHIP substrate, alleviates abnormal tau accumulation. *Hum. Mol. Genet.* **2012**, *21*, 2936–2945.

[129] Ying, H.; Zhang, Y.; Zhou, X.; Qu, X.; Wang, P.; Liu, S., et al. Selective histone deacetylase inhibitor M344 intervenes in HIV-1 latency through increasing histone acetylation and activation of NF-kappaB. *PLoS ONE* **2012**, *7*, e48832.

[130] Huber, K.; Doyon, G.; Plaks, J.; Fyne, E.; Mellors, J. W.; Sluis-Cremer, N. Inhibitors of histone deacetylases: correlation between isoform specificity and reactivation of HIV type 1 (HIV-1) from latently infected cells. *J. Biol. Chem.* **2011**, *286*, 22118–22211.

[131] Jung, M.; Hoffmann, K.; Brosch, G.; Loidl, P. Analogues of trichostatin A, trapoxin B as histone deacetylase inhibitors. *Bioorg. Med. Chem. Lett.* **1997**, *7*, 1655–1658.

[132] Schaefer, S.; Saunders, L.; Schlimme, S.; Valkov, V.; Wagner, J. M.; Kratz, F., et al. Pyridylalanine-containing hydroxamic acids as selective HDAC6 inhibitors. *ChemMedChem* **2009**, *4*, 283–290.

[133] Scott, G. K.; Marx, C.; Berger, C. E.; Saunders, L. R.; Verdin, E.; Schäfer, S., et al. Destabilization of ERBB2 transcripts by targeting 3′ untranslated region messenger RNA associated HuR and histone deacetylase-6. *Mol. Cancer Res.* **2008**, *6*, 1250–1258.

[134] Hackanson, B.; Rimmele, L.; Benkißer, M.; Abdelkarim, M.; Fliegauf, M.; Jung, M.; Lübbert, M. HDAC6 as a target for antileukemic drugs in acute myeloid leukemia. *Leukemia Res.* **2012**, *36*, 1055–1062.

[135] Wu, W. K.; Sakamoto, K. M.; Milani, M.; Aldana-Masankgay, G.; Fan, D.; Wu, K., et al. Macroautophagy modulates cellular response to proteasome inhibitors in cancer therapy. *Drug Resist. Updat.* **2010**, *13*, 87–92.

[136] Rapino, F.; Jung, M.; Fulda, S. BAG3 induction is required to mitigate proteotoxicity via selective autophagy following inhibition of constitutive protein degradation pathways. *Oncogene* **2014**, *33*, 1713–1724.

[137] Gamerdinger, M.; Carra, S.; Behl, C. Emerging roles of molecular chaperones and co-chaperones in selective autophagy: focus on BAG proteins. *J. Mol. Med.* **2011**, *89*, 1175–1182.

[138] Kozikowski, A. P.; Tapadar, S.; Luchini, D. N.; Kim, K. H.; Billadeau, D. D. Use of the nitrile oxide cycloaddition (NOC) reaction for molecular probe generation: a new class of enzyme selective histone deacetylase inhibitors (HDACIs) showing picomolar activity at HDAC6. *J. Med. Chem.* **2008**, *51*, 4370–4373.

[139] Tapadar, S.; He, R.; Luchini, D. N.; Billadeau, D. D.; Kozikowski, A. P. Isoxazole moiety in the linker region of HDAC inhibitors adjacent to the Zn-chelating group: effects on HDAC biology and antiproliferative activity. *Bioorg. Med. Chem. Lett.* **2009**, *19*, 3023–3026.

[140] Chen, Y.; Lopez-Sanchez, M.; Savoy, D. N.; Billadeau, D. D.; Dow, G. S.; Kozikowski, A. P. A series of potent and selective, triazolylphenyl-based histone deacetylases inhibitors with activity against pancreatic cancer cells and *Plasmodium falciparum*. *J. Med. Chem.* **2008**, *51*, 3437–3448.

[141] Kaliszczak, M.; Trousil, S.; Åberg, O.; Perumal, M.; Nguyen, Q. -D.; Aboagye, E. O. A novel small molecule hydroxamate preferentially inhibits HDAC6 activity and tumour growth. *Br. J. Cancer* **2013**, *108*, 342–350.

[142] Feng, T.; Wang, H.; Su, H.; Lu, H.; Yu, L.; Zhang, X., et al. Novel N-hydroxyfurylacrylamide-based histone deacetylase (HDAC) inhibitors with branched CAP group (Part 2). *Bioorg. Med. Chem.* **2013**, *21*, 5339–5354.

[143] Olson, D. E.; Wagner, F. F.; Kaya, T.; Gale, J. P.; Aidoud, N.; Davoine, E. L., et al. Discovery of the first histone deacetylase 6/8 dual inhibitors. *J. Med. Chem.* **2013**, *56*, 4816–4820.

[144] Smil, D. V.; Manku, S.; Chantigny, Y. A.; Leit, S.; Wahhab, A.; Yan, T. P., et al. Novel HDAC6 isoform selective chiral small molecule histone deacetylase inhibitors. *Bioorg. Med. Chem. Lett.* **2009**, *19*, 688–692.

[145] Wagner, F. F.; Olson, D. E.; Gale, J. P.; Kaya, T.; Weïwer, M.; Aidoud, N., et al. Potent and selective inhibition of histone deacetylase 6 (HDAC6) does not require a surface-binding motif. *J. Med. Chem.* **2013**, *56*, 1772–1776.

[146] Butler, K. V.; Kalin, J.; Brochier, C.; Vistoli, G.; Langley, B.; Kozikowski, A. P. Rational design and simple chemistry yield a superior, neuroprotective HDAC6 inhibitor, tubastatin A. *J. Am. Chem. Soc.* **2010**, *132*, 10842–10846.

[147] Kappeler, K. V.; Zhang, J.; Dinh, T. N.; Strom, J. G.; Chen, Q. M. Histone deacetylase 6 associates with ribosomes and regulates de novo protein translation during arsenite stress. *Toxicol. Sci.* **2012**, *127*, 246–255.

[148] Gradilone, S. A.; Radtke, B. N.; Bogert, P. S.; Huang, B. Q.; Gajdos, G. B.; LaRusso, N. F. HDAC6 inhibition restores ciliary expression and decreases tumor growth. *Cancer Res.* **2013**, *73*, 2259–2270.

[149] Asthana, J.; Kapoor, S.; Mohan, R.; Panda, D. Inhibition of HDAC6 deacetylase activity increases its binding with microtubules and suppresses microtubule dynamic instability in MCF-7 cells. *J. Biol. Chem.* **2013**, *288*, 22516–22526.

[150] Zilberman, Y.; Ballestrem, C.; Carramusa, L.; Mazitschek, R.; Khochbin, S.; Bershadsky, A. Regulation of microtubule dynamics by inhibition of the tubulin deacetylase HDAC6. *J. Cell Sci.* **2009**, *122*, 3531–3541.

[151] Rymut, S. M.; Harker, A.; Corey, D. A.; Burgess, J. D.; Sun, H.; Clancy, J. P.; Kelley, T. J. Reduced microtubule acetylation in cystic fibrosis epithelial cells. *Am. J. Physiol. Lung Cell. Mol. Physiol.* **2013**, *305*, L419–L431.

[152] Vishwakarma, S.; Iyer, L. R.; Muley, M.; Singh, P. K.; Shastry, A.; Saxena, A., et al. Tubastatin, a selective histone deacetylase 6 inhibitor shows anti-inflammatory and anti-rheumatic effects. *Internat. Immunopharm.* **2013**, *16*, 72–78.

[153] Zhang, D.; Wu, C. -T.; Qi, X. -Y.; Meijering, R. A. M.; Hoogstra-Berends, F.; Tadevosyan, A., et al. Activation of histone deacetylase-6 (HDAC6) induces contractile dysfunction through derailment of α-tubulin proteostasis in experimental and human atrial fibrillation. *Circulation* **2014**, *129*, 346–358.

[154] Kim, C.; Choi, H.; Jung, E. S.; Lee, W.; Oh, S.; Jeon, N. L.; Mook-Jung, I. HDAC6 inhibitor blocks amyloid beta-induced impairment of mitochondrial transport in hippocampal neurons. *PLoS ONE* **2012**, *7*, e42983.

[155] Cenik, B.; Sephton, C. F.; Dewey, C. M.; Xian, X.; Wei, S.; Yu, K., et al. Suberoylanilide hydroxamic acid (vorinostat) up-regulates progranulin transcription. *J. Biol. Chem.* **2011**, *286*, 16101–16108.

[156] Kalin, J. H.; Butler, K. V.; Akimova, T.; Hancock, W. W.; Kozikowski, A. P. Second-generation histone deacetylase 6 inhibitors enhance the immunosuppressive effects of Foxp3+ T-regulatory cells. *J. Med. Chem.* **2012**, *55*, 639–651.

[157] De Vreese, R.; Verhaeghe, T.; Desmet, T.; D'hooghe, M. Potent and selective HDAC6 inhibitory activity of N-(4-hydroxycarbamoylbenzyl)-1,2,4,9-tetrahydro-3-thia-9-azafluorenes as novel sulfur analogues of tubastatin A. *Chem. Commun.* **2013**, *49*, 3775–3777.

[158] Bergman, J. A.; Woan, K.; Perez-Villarroel, P.; Villagra, A.; Sotomayor, E. M.; Kozikowski, A. P. Selective histone deacetylase 6 inhibitors bearing substituted urea linkers inhibit melanoma cell growth. *J. Med. Chem.* **2012**, *55*, 9891–9899.

[159] Jochems, J.; Boulden, J.; Lee, B. G.; Blendy, J. A.; Jarpe, M.; Mazitschek, R., et al. Antidepressant-like properties of novel HDAC6-selective inhibitors with improved brain bioavailability. *Neuropsychopharmacology* **2013**, 1–12.

[160] Yu, C. -W.; Chang, P. -T.; Hsin, L. -W.; Chern, J. -W. Quinazolin-4-one derivatives as selective histone deacetylase-6 inhibitors for the treatment of Alzheimer's disease. *J. Med. Chem.* **2013**, *56*, 6775–6791.

[161] Lee, J. -H.; Mahendran, A.; Yao, Y.; Ngo, L.; Venta-Perez, G.; Choy, M. L., et al. Development of a histone deacetylase 6 inhibitor and its biological effects. *Proc. Natl. Acad. Sci. U.S.A.* **2013**, *110*, 15704–15709.

[162] Itoh, Y.; Suzuki, T.; Kouketsu, A.; Suzuki, N.; Maeda, S.; Yoshida, M., et al. Design, synthesis, structure-selectivity relationship, and effect on human cancer cells of a novel series of histone deacetylase 6-selective inhibitors. *J. Med. Chem.* **2007**, *50*, 5425–5438.

[163] Kozikowski, A. P.; Chen, Y.; Gaysin, A.; Chen, B.; D'Annibale, M. A.; Suto, C. M.; Langley, B. C. Functional differences in epigenetic modulators: superiority of mercaptoacetamide-based histone deacetylase inhibitors relative to hydroxamates in cortical neuron neuroprotection studies. *J. Med. Chem.* **2007**, *50*, 3054–3061.

[164] Rivieccio, M. A.; Brochier, C.; Willis, D. E.; Walker, B. A.; D'Annibale, M. A.; McLaughlin, K., et al. HDAC6 is a target for protection and regeneration following injury in the nervous system. *Proc. Natl. Acad. Sci. U.S.A.* **2009**, *106*, 19599–19604.

[165] Zhang, B.; West, E. J.; Van, K. C.; Gurkoff, G. G.; Zhou, J.; Zhang, X. M., et al. HDAC inhibitor increases histone H3 acetylation and reduces microglia inflammatory response following traumatic brain injury in rats. *Brain Res.* **2008**, *1226*, 181–191.

[166] Sung, Y. M.; Lee, T.; Yoon, H.; DiBattista, A. M.; Song, J. M.; Sohn, Y., et al. Mercaptoacetamide-based class II HDAC inhibitor lowers Aβ levels and improves learning and memory in a mouse model of Alzheimer's disease. *Exp. Neurol.* **2013**, *239*, 192–201.

[167] Ontoria, J. M.; Altamura, S.; Di Marco, A.; Ferrigno, F.; Laufer, R.; Muraglia, E., et al. Identification of novel, selective, and stable inhibitors of class II histone deacetylases. Validation studies of the inhibition of the enzymatic activity of HDAC4 by small molecules as a novel approach for cancer therapy. *J. Med. Chem.* **2009**, *52*, 6782–6789.

[168] Watabe, M.; Nakaki, T. Protein kinase CK2 regulates the formation and clearance of aggresomes in response to stress. *J. Cell. Sci.* **2011**, *124*, 1519–1532.

[169] Lehotzky, A.; Tirián, L.; Tökési, N.; Lénárt, P.; Szabó, B.; Kovács, J.; Ovádi, J. Dynamic targeting of microtubules by TPPP/p25 affects cell survival. *J. Cell Sci.* **2004**, *117*, 6249–6259.

[170] Schofield, A. V.; Steel, R.; Bernard, O. Rho-associated coiled coil kinase (ROCK) protein controls microtubule dynamics in a novel signaling pathway that regulates cell migration. *J. Biol. Chem.* **2012**, *287*, 43620–43629.

[171] Ouyang, H.; Ali, Y. O.; Ravichandran, M.; Dong, A.; Qiu, W.; MacKenzie, F., et al. Protein aggregates are recruited to aggresome by histone deacetylase 6 via unanchored ubiquitin C termini. *J. Biol. Chem.* **2012**, *287*, 2317–2327.

[172] Lee, J. Y.; Koga, H.; Kawaguchi, Y.; Tang, W.; Wong, E.; Gao, Y. S., et al. HDAC6 controls autophagosome maturation essential for ubiquitin-selective quality-control autophagy. *EMBO J.* **2010**, *29*, 969–980.

[173] Zhang, X.; Yuan, Z.; Zhang, Y.; Yong, S.; Salas-Burgos, A.; Koomen, J., et al. HDAC6 modulates cell motility by altering the acetylation level of cortactin. *Mol. Cell* **2007**, *27*, 197–213.

6

Targeting Assembly and Disassembly of Protein Aggregates
A Raggle-taggle Bunch with High Hopes

6.1 DISORDERED PROTEIN AGGREGATES AND ORDERED AMYLOID FIBRILS

Protein aggregation [1,2] is influenced by physicochemical and biological factors. Any protein tends to aggregate, depending on the arrangement of its polypeptide regions. *Ordered aggregates* are nm-long (un)branched amyloid fibrils, arranged in a cross-β-sheet structure [3]. Around 30 human amyloid proteins form ordered aggregates *in vivo*, leading to chronic amyloidosis [4]. *Disordered aggregates* [5] are the result of acute cellular stimuli (i.e., stress-caused denaturation, lack of assembly partners). The relocation of disordered aggregates and amyloid structures in specific organelles [6] drives their rescuing/disposal through protein quality control (PQC) pathways, described in details in Chapters 2 to 5 of the biology-oriented companion book [7].

Many neurodegenerative diseases (NDDs) result from the saturation and impairment of the PQC machinery. The transition from a soluble, functional protein to misfolded, aggregation-prone species, and eventually to ordered or disordered aggregates, is promoted by *protein-dependent* (structural features, genetic or post-translational modifications/PTMs) and *protein-independent factors* (cellular, environmental stimuli) [8,9]. *Aging* is the most common cause of protein aggregation [10], with a continuous progression determined by subtle, chronic, pathological changes in cellular components. A "vicious circle" scenario gradually increases the amount of protein aggregates, stressing the capacity of the PQC machinery and causing further misfolding and aggregation.

Chemical Modulators of Protein Misfolding and Neurodegenerative Disease. http://dx.doi.org/10.1016/B978-0-12-801944-3.00006-0

Native proteins aggregate in a concentration-dependent manner [11]. Their aggregation is reversible in its early phases. It becomes irreversible when covalent modifications stabilize the aggregate [12], or when a large number of molecular interactions take place in an ordered amyloid structure [13]. Key events include aggregation-prone conformational activation of monomers, kinetically disfavored aggregation of monomers into aggregation nuclei, and aggregate growth to form mature amyloid fibrils [14].

The formation of aggregation nuclei is promoted by *secondary nucleation events* [15]. They include the fragmentation of mature fibrils, resulting in seeds for secondary nucleation [16], and the creation of new reactive ends/fibril branches that grow into aggregation nuclei, and are then released to fuel the aggregation cycle [17].

The end results of aggregation, i.e., *insoluble aggregates*, were initially considered to be the toxic species in proteinopathies [18,19]. The consensus is now for *non-fibrillary, soluble oligomers* to be the most neurotoxic species [13,20]. The interaction between aggregation-prone protein species and *biological membranes* has a major impact on the cyto/neurotoxicity of aggregates, and on the integrity of membranes [21,22].

A detailed understanding of molecular mechanisms leading to protein aggregation is the gateway to immunotherapeutics/biologicals [23] or small molecule aggregation inhibitors [24]. The biology-oriented companion book [7] contains a detailed description of the steps leading to ordered/amyloid and disordered aggregates. Immunotherapeutics/biologicals are not dealt with here, while small molecule aggregation inhibitors are thoroughly presented. Tau is selected as a privileged aggregation-prone protein target. The toxicity of each intermediate on the path to aggregation, and the consequences of blocking a specific molecular reaction/process with a small molecule (i.e., preventing the formation of a potentially toxic intermediate, or causing the accumulation of an even more toxic precursor [25]), are discussed in the next sections.

6.2 INTERFERING WITH (NEURO)TOXIC TAU SPECIES IN THE AGGREGATION PROCESS

Aggregation steps leading to neurotoxic species must be validated and targeted, while steps leading to non-toxic or even neuroprotective species should not be disturbed. As it happens for other amyloidogenic proteins, the obvious suspects (neurofibrillary tangles (NFTs), insoluble tau aggregates) were long believed to be the major determinants of neurotoxic effects in tauopathies [26]. Now it is accepted that while NFTs are presumably involved in tau toxicity, they may even result from an endogenous rescuing mechanism to eliminate neurotoxic, soluble oligomeric species from the neuronal environment [27]. Tau oligomers are neurotoxic in a number of preclinical [28,29] and clinical environments [28,30,31]. They are found in the extracellular space [32], and propagate from cell to cell in cellular [33,34] and animal models [35,36].

The aggregation of tau, its intermediate and final stages (including the bioanalytical characterization of small oligomers, protofibrils, fibrils, and NFTs), and the identification of prospective points of intervention for anti-aggregation drugs targeted against tau are dealt with extensively in recent reviews [37,38] and in the biology-oriented companion book [7]. Our limited knowledge about authentic *in vivo* aggregation of tau isoforms in NDDs provides some guidance, but more information (possibly through the identification of selective and effective small molecule inhibitors of tau aggregation) is acutely needed.

Rather than focusing on a single aggregation step and naming it a target for tau anti-aggregating agents, a thorough description of known modulators of tau aggregation, mostly identified by *in vitro* high throughput (HT) assays that recapitulate the tau aggregation process [37,39], is provided here. Compounds are classified depending on their chemotype, rather than according to their effects on the aggregation of tau.

The phenothiazine salt *methylene blue (MB, methylthioninium chloride, MTC, rember™,* **6.1a**, Figure 6.1) is an aromatic dye known since late nineteenth century for its pharmacological activities [40]. Its heat shock protein 70 (Hsp70)-binding ability and its activity as a CHIP co-chaperone protein modulator are briefly mentioned in section 3.2 (see also **3.4**, Figure 3.1).

MB is used as an oral/systemic therapeutic agent against malaria, if-osfamid-induced neurotoxicity in cancer patients, methemoglobinemia, and acute catecholamine-refractory vasoplegia [41]. It prevents urinary tract infections in aged patients, and is used as a staining agent for the visualization of tissues and organs [41]. MB is a first generation tau aggregation inhibitor (TAI) that shows reasonable tolerability and significant clinical efficacy in a Phase II study on more than 300 mild to moderate Alzheimer's disease (AD) patients [42]. The highest oral dose (60 mg/kg, three times per day) slows disease progression at 24 weeks, and stabilizes treated patients after 50 weeks. Molecular imaging (single photon emission computer tomography/SPECT and positron emission tomography/PET) shows that brain areas protected by MB correspond to brain regions mostly affected by AD [43].

The methyl thioninium MB cation (**6.1**) is the ionized-oxidized form in a tri-component redox equilibrium (Figure 6.1). Reduced, ring-neutral leuco-methylene blue species (LMT, **6.3**) may experience hydrogen abstraction to form mesomerism-stabilized radical species **6.2**. Oxidative loss of an electron leads to stable, ionized MB resonance structures bearing the positive charge on the N- and on the S-atom (respectively **6.1** and **6.1'**, Figure 6.1) [40]. The redox equilibrium, and consequently the relative ratio

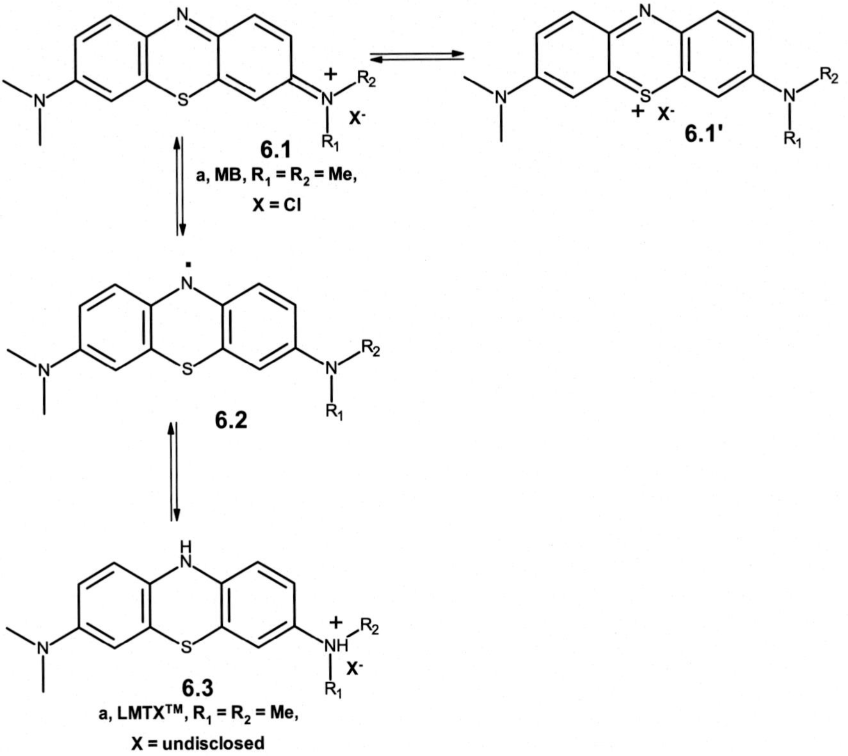

FIGURE 6.1   Methylene blue (MB): redox cycling, chemical structures, **6.1–6.3**.

of MB and LMT species *in vitro* and *in vivo*, is influenced by experimental conditions (e.g., presence of reactive oxygen species (ROS)) and by enzymatic activities (e.g., NADH- and NADPH-dependent, LMT-producing reductases) [44].

Oral absorption of MB is poor, due to its charged structure. It depends on the enzymatic conversion of MB into LMT, which is claimed to happen in the stomach at low pH (unpublished results) [45]. The sub-optimal MB tolerability profile requires its administration with food, but its conversion into LMT is inhibited by food intake [45]. A second generation, reduced leuco-MB TAI (*LMTX*™, *Trx0237*, **6.3a**, Figure 6.1, undisclosed counterion) shows much higher tolerability and better absorption at higher doses. Unpublished preclinical, Phase I and Phase II studies using LMTX determine its optimal dosages and expected clinical outcomes for Phase III studies [46]. Two Phase III studies on AD (NCT01689246, 150 and 250 mg/kg daily, >800 mild to moderate AD patients, enrolling, completion expected by October 2015; NCT01689233, 200 mg/kg daily, 700 mild AD patients, enrolling, completion expected by January 2016) and one Phase III study on behavioral frontotemporal dementia/bvFTD (NCT01689246, 100 mg/kg daily, 180 bvFTD patients, enrolling, completion expected by August 2015) could soon provide an approved, disease-modifying drug against tauopathies [46,47].

MB shows moderate, low μM activity in an *in vitro* aggregation assay involving Glu391-truncated tau [48]. Similar inhibitor potencies for MB are observed in a heparin-dependent *in vitro* aggregation assay relying on the fluorescent compound thioflavin S (ThS) [49]. All three or four microtubule binding repeats (MTBRs) in 3R and 4R tau isoforms are needed for MB to inhibit tau aggregation [50]. Nuclear magnetic resonance (NMR) shows interactions between MB and the ^{306}VQIVYK^{311} sequence, N-terminal Y18 and Y29, and C291 and C322 residues in tau [51]. Two reports suggest that Cys oxidation (either to sulfenic acid [51] or promoting an *intra*-molecular Cys291–Cys322 disulfide bridge formation in 4R tau [52]) determines the anti-aggregation properties of MB. The single Cys322 residue on 3R tau isoforms (lack of Cys291) either causes an *inter*-molecular, fibrillation-permissive Cys322–Cys322 disulfide bond [52], or reduces the impact of MB-induced Cys oxidation to sulfenic acid [51]. In both cases, 3R tau-dependent tauopathies should be less responsive to MB. An unspecific Cys oxidation-related activity of MB on multiple proteins is not observed *in vivo*, possibly due to glutathione-dependent regulation [52], which may minimize the impact of MB-induced Cys oxidation *in vivo*.

MB displays *in vitro* activity on a number of central nervous system (CNS)-related targets that could influence its *in vivo* efficacy/tolerability profile [53]. MB is an inhibitor of soluble guanylate cyclase [54], nitric oxide (NO) synthase [55], cholinesterase [56], glutathione reductase (GSR) [57], and monoamine oxidase A [58] and B [59]. MB inhibits noradrenaline neuronal uptake [60], reduces blood serum tumor necrosis factor α (TNFα) levels [61], and its MB–LMT redox cycling positively influences mitochondrial activity [62]. A recent review from the developers of rember™ and LMTX™ questions the relevance of most of these interactions at physiological concentrations [38]. Unfortunately, the mentioned preclinical studies by the same research group are still unpublished [63].

The intricate network of MB-derived metabolites (Figure 6.1) is mirrored by a complex *in vivo* pharmacokinetic (PK) and pharmacological behavior. MB is metabolically transformed into mono- (**6.1b**, azure B, Figure 6.2) or bis-demethylated compound (**6.1c**, azure A) [51]. Azure B and A enter an MB-like redox cycle, leading respectively to leuco/reduced analogues **6.3b** and **6.3c** as shown for MB/**6.1a** in Figure 6.1. In addition to S-charged resonance structures **6.1b'** and **6.1c'**, azure A and B may assume the neutral resonance structures **6.1b** and **6.1c** inaccessible to MB (Figure 6.2). These neutral structures increase their cell membrane/blood–brain barrier (BBB) penetration when compared to MB. Any species in Figures 6.1 and 6.2 shows a different physicochemical, biological,

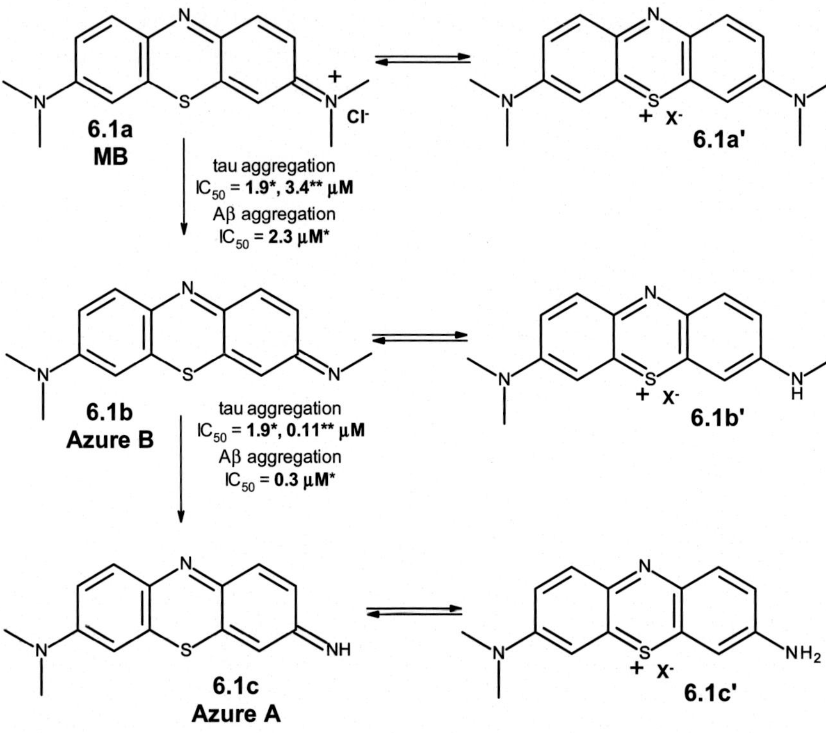

*, ref. 49; **, ref. 48

FIGURE 6.2   Methylene blue (MB), azure A and B: chemical structures **6.1a–c, 6.1a'–c'**.

and pharmacological profile. Table 6.1 contains some published information about MB/**6.1a** and azure B/**6.1b**.

Azure A and B are more potent tau anti-aggregation agents than MB [48,49]. NMR studies highlight interactions of azure A and B with tau regions containing aromatic amino acids (AAs), in addition to previously mentioned MB–tau interactions [51]. Interestingly, further demethylation (azure C—one methyl left, and thionine—two primary amines, no methyls) decreases anti-aggregation potency [49]. Azure B shows similar potencies on acetyl (AChE-) and butyrylcholinesterase (BuChE) [64], and on GSR [57]. It is an even better substrate for GSR than MB [57], hinting towards a more efficient *in vivo* redox cycling for azure B.

A single report describes the *in vivo* administration of azure B in mice models of endotoxic shock and tumor growth. Azure B shows significant efficacy in both models, while MB is marginally active [61]. *In vivo* testing of MB is frequently reported, but azure B is detectable in equal amounts after oral administration of MB in rats [65]. *Post-mortem* analysis of the organs of a septic shock patient dosed with MB shortly before his death

TABLE 6.1 Comparison between MB and Azure B

Compound/Data	MB/6.1a	Azure B/6.1b
Tau aggregation, IC_{50}, μM	3.4^a, 1.9^b	0.11^a, 1.9^b
AchE, IC_{50}, μM	0.21	0.49
BuChE, IC_{50}, μM	0.39	1.99
GSR, IC_{50}, μM	16	5
GSR, substrate efficiency, $M^{-1} s^{-1}$	4800	9200
TNFα suppression, blood serum, %	≈90%	≈50%
Urine distribution, 2.5 mg/kg MB p.o.	≈50%	≈50%
Organ distribution, *post-mortem*, liquid chromatography (LC)/mass spectrometry (MS)/MS	$5–11\%^c$	$71–79\%^d$
In vivo endotoxic shock, mice	inactive	protection
In vivo tumor growth, mice	inactive	arrested^e

^{a} *Ref. 48*
^{b} *Ref. 49*
^{c} *quantified as MB–LMT mixture*
^{d} *missing to 100%: Azure A*
^{e} *only effective in female mice.*

identifies both azure A and B (the latter in large excess compared with MB levels) [66].

MB shows a $t_{1/2}$ ≈10 hours after i.v. (50 mg) administration, and a ≈73% bioavailability following p.o. (500 mg) administration in humans [67]. Its blood concentration reaches up to ≈20 μM, with most MB bound to erythrocytes (nM concentrations of free MB in plasma) [57]. The toxicity of MB is far higher than therapeutically relevant dosages (oral LD_{50} = 1180 and 3500 mg/kg in rats and mice [68]; i.v. LD_{50} = 77 and 42.3 mg/kg in mice and sheep [69]). Gastrointestinal toxicity in humans is observed after MB injection at 7 mg/kg, i.e., a dosage far exceeding the ≈2 mg/kg usually employed in therapy.

MB penetrates the BBB and accumulates in the human brain, with an ≈10 brain/plasma concentration 1 hour after i.v. or p.o. administration [68]. Its brain levels are much higher after intravenous rather than oral administration [70]. The BBB permeability of MB is likely to stem from the neutral, lipophilic LMT form that is produced by GSR and other reductases in erythrocytes and in various tissues [71,72].

MB administration improves cognition, enhances memory, and prevents neuronal damage in animal models [73]. A single dose of MB (1 mg/kg i.v.) enhances memory retention in rats when administered after training, while a much higher dosage (50 mg/kg) shows detrimental effects [74]. Similar observations depend on hormetic dose–response curves for

MB [75]. Low dosages (typically 0.5–1 μM *in vitro*, 2–4 mg/kg i.v. or i.p. *in vivo*) elicit positive biochemical, behavioral, and physiological effects. Conversely, high dosages (typically ≥10 μM *in vitro*, ≥20 mg/kg i.v. or i.p. *in vivo*) negatively influence the same endpoints [75]. A concentration-induced switch of the redox cycle is suggested to determine hormesis. Low MB concentrations act as metabolic energy enhancers/electron donors/ stimulators of mitochondrial pathways. Higher concentrations cause an MB-driven subtraction of electrons from the same pathways that, as a result, become impaired [75].

Positive effects on spatial memory retention in a food search task are observed after i.p. injections of MB (1 mg/kg daily, 5 days) [76]. I.p.-administered MB (1 mg/kg daily, 5 days [77] or 4 mg/kg, single dose [78]) rescues cognitive deficits in a rat model with sodium azide-dependent cytochrome oxidase impairment, supporting its action through enhancement of mitochondrial respiration. MB (4 mg/kg i.p. daily, 5 days) assists the return to normal behavior/extinction of conditioned fear in normal rats [79]. The same dosing schedule compensates for the conditioned fear extinction deficits in congenitally helpless rats [80]. The hormetic effect of MB is confirmed in a study on rats, where a single 4 mg/kg i.p. dose improves object recognition and long-term habituation, while raising the dosage to 10 mg/kg i.p. does not show any memory enhancement [81]. MB reverses cognitive deficits in scopolamine-treated rats after single i.p. administration at 1 and 4 mg/kg [82].

Neuroprotective effects of MB are observed in a therapeutic (50 mg/ kg daily i.v., 6 days) or preventive (50 mg/kg daily i.v., 4 days) treatment schedule on cancer patients treated with iphosphamide, an agent that may cause severe encephalopathy, used against solid tumors [83]. In animal models, MB reverts the neuropathic retinal damage induced by the mitochondrial complex I inhibitor rotenone at 70 μg/kg intravitreal concentrations [84]. MB administration shows neuroprotection in rotenone-induced Parkinsonism and cerebral ischemia [85], rotenone-induced striatal neurodegeneration [86], and methylmalonate toxin-induced neural impairment [87].

In vivo efficacy of MB in cognitive/neurotoxic models is mostly ascribed to the stimulation and protection of mitochondrial activity [62]. MB, though, shows effects on NDD-related disease-modifying targets covered in this book, and described in detail in the biology-oriented companion book [7]. As to the *chaperone machinery* (Chapter 2 here and in [7]), a first study [88] identifies MB as one among five agents causing a significant reduction of tau levels in a screening campaign on 880 clinically tested compounds. The same research group characterizes MB as an inhibitor of the ATPase activity of Hsp70 from a high throughput screening (HTS) campaign on 2800 bioactive compounds [89]. The Hsp70 inhibition-dependent lowering of tau levels is observed at ≈50 μM for MB, possibly by preventing Hsp70-driven refolding/rescuing of tau. MB reduces the levels of tau in general, and of the pathological pS396/pS404 tau isoform in particular [89]. A chemo-genomic study with MB, targeting the tau interactome, identifies Hsp70 and Hsp90 among the proteins with MB-dependent decreased and increased tau binding, respectively [90]. Hsp70 ATPase inhibition should cause the release of Hsp70 from tau. Tau subsequently binds with Hsp90. The Hsp90–tau interaction should drive tau towards proteasomal degradation [90]. MB acts on the stress-induced Hsp72 protein through oxidation of its Cys306 residue, inducing conformational changes and eventually ATPase inhibition [91]. Conversely, structurally similar, constitutive Hsc70 protein lacks Cys306 and is resistant to MB inhibition [91]. As to the *ubiquitin–proteasome system* (*UPS*, Chapter 3 here and in [7]), a single report [92] shows an increase of proteasomal activity in MB-treated triple transgenic (TG) mice that leads to the decrease of soluble $A\beta$ peptide levels. As to *autophagy* (Chapter 4 and 5 here and in [7]), P301L tau-bearing TG mice show MB-dependent induction of autophagy, leading to increased levels of autophagy markers cathepsin D, beclin-1, and light chain 3-phosphatidylethanolamine (LC3-II) [93]. Organotypic slice cultures from the same TG mice, when treated with nM concentrations of MB, show reduced levels of hyperphosphorylated (HP)-tau levels and of insoluble tau aggregates [93]. MB (100 mg/kg daily, 14 days) induces macroautophagy (MA) and increases LC3-II levels in the hippocampus and cortex of mice in a mammalian target of rapamycin (mTOR)-independent manner [94]. Rather, MB induces MA through activation of the 5'-adenosine monophosphate-activated protein kinase (AMPK) pathway [94]. The relevance of these mechanisms in the *in vivo* CNS activity of MB is questionable [38] due to the high brain concentration needed to affect chaperones and autophagy. As such brain levels are observed by some groups, these mechanisms are likely to contribute to neuroprotection and memory enhancement.

The efficacy of MB as a TAI in animal models is confirmed [42]. The developing company claims preclinical efficacy for MB when tested on two proprietary TG mouse models that respectively express a short human tau construct that aggregates into neurotoxic oligomers without aggregating further into tangles, and full length (FL) P301S and G335D-mutated tau [95]. In the former model, a significant reduction in oligomer-positive neurons is obtained with oral and i.v. MB (respectively 15 mg/kg and 5 mg/kg, unspecified duration of the treatment). A 45 mg/kg oral dose of MB elicits cognitive enhancements in 7-month-old mice [95]. Treatment of the FL, double mutant mice with oral MB (10 mg/kg, unspecified duration of the treatment) is claimed to provide behavioral benefits and biochemical evidence of TAI-driven efficacy [95]. A more detailed report on preclinical characterization of MB on these and other preclinical models is surely needed.

Preclinical testing of MB in animal models of tau aggregation by other research groups provides a complex and sometimes controversial picture. MB treatment (100 μM solutions, up to 6 days post-fertilization) of FTD-recapitulating TG zebrafish embryos does not cause a reduction in HP-tau, does not rescue decreased axonal growth and swimming behavior, and does not prevent neurotoxic cell death [96]. Administration of MB (10 mg/kg daily in water, 12 weeks) shows concentration-dependent neuroprotective effects in 3-month-old P301L TG mice [97]. TG mice with brain MB ≥470 μM (LC-MS quantitation) display cognitive improvements correlated with lower soluble tau levels. Conversely, pathology progression does not correlate with the reduction of tangles, or with the count of viable neurons [97]. Six-month-old triple TG mice (M146V/presenilin-1-PSEN1, K670N-M671L Sweden mutation/amyloid precursor protein-APP, P301L/tau) treated with MB (0.025% w/w solution daily, *ad libitum*, 6 months) show MB-driven biochemical effects on Aβ peptides, and positive effects on learning and memory [92]. Conversely, they do not show reduced tau levels in general and HP-tau in particular in the hippocampus [93]. A reduction in HP-tau levels is observed in 3-month-old P301L tau mice after oral gavage (20 mg/kg daily, 5 days per week, 2 weeks) [93]. The reduction in tau insoluble aggregates observed in organotypic slices is not reproduced in TG animals, possibly due to insufficient brain concentration of MB [93]. Oral administration of MB (0.3 to 1 mg/kg daily in drinking water, 5 months) to aged P301L mice (8 to 11 months old) decreases insoluble tau aggregates (TAI role) without variations in total tau levels (no effects on tau expression) [98]. Administration of ≈165 μM MB in drinking water for 1.5 months in 16-month-old P301L mice does not reduce the number of neurons showing extensive tau pathology [99]. Conversely, the levels of soluble tau are significantly reduced. A slow process of tau tangle disaggregation, solubilization, and clearance may result from long-term administration (≥18 months) of MB in TG mice [99]. The observed reversal of tau pathology by stoppage of tau expression for 6 months in 18-month-old tau TG mice [100] supports this hypothesis.

MB shows anti-aggregation activity on amyloidogenic, aggregation-prone CNS proteins. MB shows low μM potency in a heparin-dependent Aβ aggregation *in vitro* assay, similar to its effect on tau [49]. Interestingly, demethylated azure A-C show stronger, sub-μM potency [49,101]. MB de-stabilizes neurotoxic, soluble Aβ oligomers, and promotes their fibrillation [102]. MB binding transforms neurotoxic Aβ oligomers into aggregation-prone species that rapidly add monomeric Aβ peptides to form less toxic fibrils [103]. The preferred addition of monomeric Aβ peptides to growing fibrils reduces the levels of newly formed, soluble Aβ oligomers [102]. It must be mentioned that the reliability of thioflavin T (ThT)-based *in vitro* Aβ fibrillation assays is questionable [104,105]. Structurally similar small molecules may displace ThT from its binding sites on Aβ fibrils rather than truly inhibiting Aβ fibrillation. Using a label-free assay based on matrix assisted laser desorption/ionization (MALDI)-time of flight (TOF) MS, MB is classified as a false positive, without effects on Aβ fibrillation, while azure C shows μM rather than nM potency [104].

Atomic force microscopy (AFM) and transmission electron microscopy (TEM) measurements show that MB addition causes a conformational switch to a more amorphous aggregate structure for growing Aβ oligomers, and for pre-formed Aβ fibrils [106]. Double TG Sweden APP-A246E PSEN1 mice treated either in a preventive (2-month-old mice, 0.80 mg/kg daily in drinking water, 3 months) or in a therapeutic mode (6-month-old mice, 0.80 mg/kg daily in drinking water or 2 mg/kg daily i.p., 3 months) show prevention of the Aβ-driven pathology (preventive) and significant reduction of behavioral and cognitive impairments (therapeutic) [107]. The effects are ascribed to the stimulation of mitochondrial pathways rather than to a TAI behavior, due to the estimated brain concentration reached by MB in both administration schedules [107].

MB shows potent *in vitro* and *in vivo* beneficial effects as an anti-aggregation agent targeted against mutant *huntingtin* (*Htt*) [108]. MB inhibits the aggregation of an N-terminal Htt fragment containing a 53 polyQ repeat (Httex1Q53). MQ slows the aggregation of monomeric, oligomeric, and even fibrillar Httex1Q53 *in vitro* [108]. MB (100 nM) reduces both oligomer and fibril formation in Httex1Q53-transfected rat primary cortical neurons, and increases their mean survival time [108]. Treatment with MB (25 mg in 100 mg of food/chow daily, 4 weeks) of 5-week-old, 115 Q repeat-expressing TG R6/2 mice shows partial effects on body weight loss and behavioral impairments. Htt aggregates are reduced, while oligomeric Htt levels are unchanged [108]. Positive effects of MB treatment are observed also in a *Drosophila* model, in terms of a reduction of aggregate levels and of embryo viability [108]. The lack of neuroprotection, even while MB reduces Htt aggregation in another study on TG *Drosophila* [96], is justified by structural differences in the polyQ Htt constructs.

An NMR study determines a highly ionizable region on native *prion proteins* (*PrPs*) that binds MB with high affinity at neutral pH [109]. The structural instability of the MB binding region may facilitate the switch to aggregation-prone PrP conformations, which is prevented by MB binding. MB reduces the PrP oligomerization rate and suppresses fibril formation at ≥2 MB/PrP molar ratios [109]. MB administration in ScN2a neuroblastoma cells infected with PrPs is toxic at >1 μM, preventing the observation of anti-aggregation beneficial effects at higher MB concentrations [110]. MB, azure A, and azure C do not interfere with the aggregation of *α-synuclein* up to 80 μM concentrations [111].

MB interferes with the aggregation of *TAR DNA-binding protein 43* (*TDP-43*). Two human neuroblastoma SH-SY5Y cell models expressing mutant TDP-43 constructs show an ≈50% reduction in the number of aggregate-containing cells when treated with 50 nM MB ($\geq$100 nM MB concentrations show toxicity on SH-SY5Y cells) [112]. Administration of MB (30 to 60 μM concentrations, dissolved in growing medium since hatching) to C. *elegans* and zebrafish TG models expressing mutated TDP-43 (respectively A315T and G348C) causes an average 90% recovery from mutation-induced biochemical and behavioral defects [113]. Similar effects are observed with the same MB dosage on C. *elegans* and zebrafish TG models expressing mutated *fused in sarcoma* (*FUS*) (respectively S57Δ and R521H). MB treatment of TG C. *elegans* from day 5 post-fertilization reduces its neuroprotective effects (55% vs. 90% behavioral recovery) [113].

Two TG mice models of amyotrophic lateral sclerosis (ALS) are insensitive to MB [114]. Surprisingly, 6-month-old TG mice bearing the G348C TDP-43 mutation, when treated with MB (1 mg/kg i.p. every other day, 6 months) do not show any biochemical or cognitive improvement [114]. Lack of efficacy is observed with higher MB dosages (up to 10 mg/kg i.p. every other day, until the death of mice) for 3-month-old TG mice bearing a neurotoxic G93A-mutated *superoxide dismutase* (*SOD1*) protein [114]. The inactivity of MB against mutated SOD1-dependent ALS-like TG mice is confirmed in another study, where up to 25 mg/kg MB daily is administered [115]. Finally, limited efficacy is observed when MB is locally administered in the spinal cord of G93A SOD1 mice, delaying the disease onset without extending the lifespan of TG mice [116].

Brain concentrations of MB and of its (possibly more potent) metabolites, their moderate potency on a large number of molecular targets (including multiple aggregation-prone proteins), and the effect of pathological environments on the metabolism and distribution of MB determine the putative therapeutic usefulness of MB and of its derivatives. The results of three ongoing Phase III studies with the second generation, more bioavailable/less toxic LMTX^{TM} will influence the continuation and the expansion of research efforts targeted towards tau aggregation inhibitors and disassembling agents.

Clinically tested aggregation inhibitors developed against Aβ *plaques and oligomers*, but often active also on tau, are studied as putative sources of disease-modifying agents against AD [117,118]. Scyllo-inositol/ELND005/ AZD-103 (**6.4**, Figure 6.3) is a polyhydroxylated cyclohexane developed up to Phase II with mixed results [119]. Its development appears to be in stand-by.

The development of homotaurine/tramiprosate/Alzhemed^{TM}(**6.5**), an aminosulfonate, is currently halted after a negative Phase III study [120]. The active principle is available in the nutraceutical agent Vivimind^{TM}. A prodrug of tramiprosate, BLU8499/ALZ-801, has completed Phase I studies [121]. Both compounds are now being progressed [122]. Tramiprosate binds tau on its C-terminal region, promoting its aggregation in a heparin-like mode [123]. It does not affect the interaction between tau

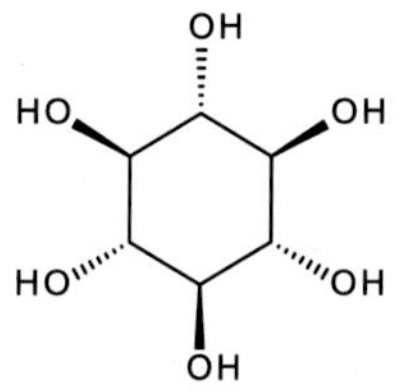

6.4
prevention of Aβ aggregation (5 μM)
at **3 mM** concentration

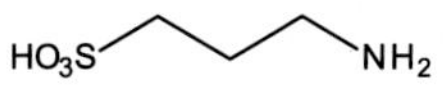

6.5
≈ **22 %** decrease in cell death of neurons
in presence of Aβ (5 μM)
at **100 μM** concentration

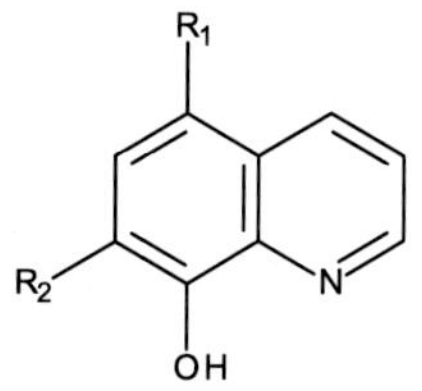

6.6a,b
a, R_1 = Cl, R_2 = I,
≈ **70 %** dissolution, insoluble $A\beta_{42}$, 10 μM,
≈ **2-fold** increase of Cu^{2+} in cells, 5 μM
b, R_1 = R_2 = H,
≈ **40 %** dissolution, insoluble $A\beta_{42}$, 10 μM,
≈ **25-fold** increase of Cu^{2+} in cells, 5 μM

FIGURE 6.3 Clinically tested small molecule inhibitors of Aβ and tau aggregation: chemical structures, **6.4–6.6b**.

and microtubules (MTs), and does not have neurotoxic effects on neuronal cells [123].

The Zn^{2+}- and Cu^{2+}-selective metal chelator hydroxyquinoline clioquinol/PBT-1 (**6.6a**) shows efficacy in Phase II studies, but has limited BBB permeability and displays toxicity at high dosages [124]. Its analogue PBT-2 (**6.6b**, Figure 6.3) is a more BBB-permeable and more tolerated metal chaperone [124]. PBT-2 is currently evaluated in Phase II clinical trials [125,126]. PBT-2 promotes an increase of Ser9-phosphorylated glycogen synthase kinase 3 beta (GSK-3β) by inhibiting the phosphatase calcineurin, and consequently reduces the levels of pathological HP-tau [127]. Exebryl-1/PTI-80 (**6.7**, undisclosed structure) is a BBB-permeable modulator of Aβ and tau pathologies, with anti-aggregation and disaggregation ability [36,128] that is currently undergoing Phase I clinical studies [129].

Naturally occurring polyphenols act against the aggregation of tau [49] and Aβ peptides [130]. Epigallocathechine gallate (EGCG, sunphenon, **6.8a**, Figure 6.4) is currently tested in Phase II/III on early stage AD patients [131,132]. Its Hsp70-targeted activity is briefly mentioned in section 2.3.1 (see also **2.39**, Figure 2.8).

EGCG displays low μM affinity/anti-aggregation potency against tau and Aβ [49]. Biochemical (reduction of Aβ plaques, reduction of HP-tau species) and cognitive benefits (reduction of memory impairment) are observed in Sweden APP TG mice when treated with EGCG (either 20 mg/kg i.p. daily, 12-month-old mice, 60 days [133], or 50 mg/kg in drinking water daily, 8-month-old mice, 6 months [134]). Modulation of β-secretase (BACE) activity and anti-oxidant effects contribute to the observed effects of EGCG [134]. EGCG directs the aggregation of small Aβ oligomers towards less neurotoxic, off-pathway aggregates [135]. It remodels mature Aβ fibrils into disordered, non-toxic aggregates [136]. Binding of EGCG with amyloidogenic proteins is supposed to happen *via* its oxidized **6.8b**

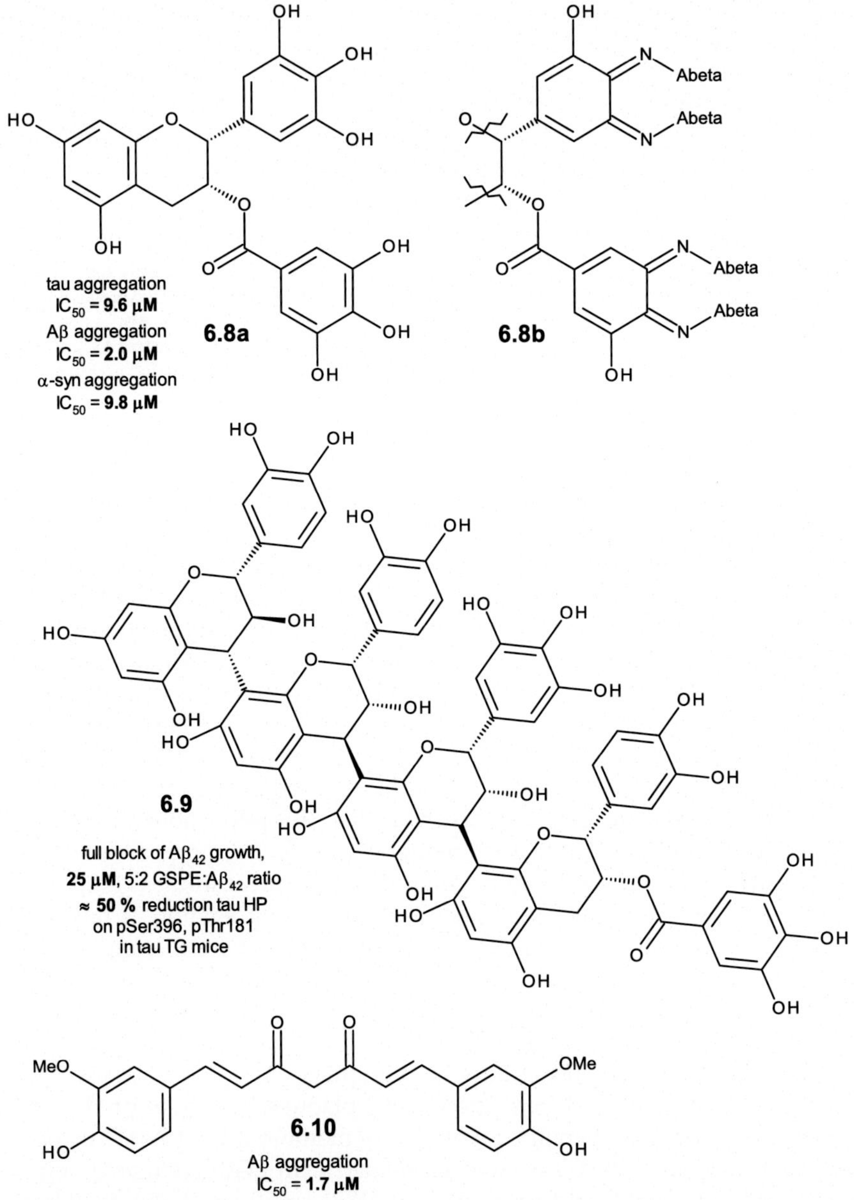

FIGURE 6.4   Small molecule polyphenolic inhibitors of the aggregation of tau and other amyloidogenic proteins: chemical structures, **6.8a–6.10**.

form, which should establish covalent bonds with free NH_2 (represented as a 4:1 Aβ:**6.8b** adduct in Figure 6.4, real stoichiometry unknown) and/or SH groups of amyloidogenic proteins, contributing to neurotoxicity-reducing protein remodeling [137]. The metal chelation ability of EGCG supports its remodeling activity on naturally occurring, neurotoxicity-involved metal-Aβ species [138].

Grape seed polyphenolic extracts (GSPEs, Meganatural™-BP, **6.9**—illustrative structure) are structurally characterized mixtures of catechin-, epicatechin-, epigallocatechin-, and EGCG-containing monomeric, oligomeric, and polymeric species [139]. GSPEs are commercialized antioxidant nutraceuticals with beneficial effects against hypertension [140]. They are evaluated in a Phase II study on AD patients [141], due to an extensive preclinical characterization as tau- and Aβ-directed anti-aggregation agents. *In vivo* administration of GSPEs (200 mg/kg in drinking water daily, 5 months) is safe and reduces Aβ oligomerization, with mitigation of cognitive impairments, in the brains of adult Sweden APP TG mice [139]. *In vitro* and *in vivo* mechanistic studies suggest that their therapeutic effects are mainly due to the inhibition of Aβ oligomerization, and to the reduction of soluble oligomeric Aβ levels [139]. GSPEs work either in preventive (inhibition of Aβ oligomer assembly from Aβ monomers) or in therapeutic cellular models (disassembly of pre-assembled Aβ oligomers) [142]. GSPEs inhibit *in vitro* the aggregation of the tau [306]VQIVYK[311] amyloidogenic sequence, and disassemble pre-formed [306]VQIVYK[311] aggregates [143]. GSPEs bind to the sequences, inhibit their conformational switch and aggregation into oligomeric β-sheet aggregates. Rather, disordered, less neurotoxic GSPE–tau [306]VQIVYK[311] aggregates are formed [144]. GSPEs (200 mg/kg daily, 2 months) administered to 4-month-old V337M-R406W double mutant tau-carrying TG mice in a preventive schedule delay the appearance of insoluble tau aggregates [144]. Their effects are due both to inhibition of tau aggregation and to GSPEs-mediated inhibition of tau hyperphosphorylation by extracellular signal-regulated kinases (ERK1/2) [144]. GSPEs (150 mg/kg daily in drinking water, 6 months) administered to adult 9-month-old P301L mutant TG mice produce a reduction of motor impairment and lower HP-tau levels [145]. GSPEs-treated TG mice show up to 90% reduction of insoluble tau-containing neurons in the spinal cord, and up to 80% decrease of insoluble HP-tau aggregates [145]. Interestingly, GSPEs disaggregate tau paired helical filaments (PHFs) from AD patients [146]. They bind to the fuzzy coat-external and to the core-internal portions of tau PHFs, causing their disorganization and increase in width, followed by gradual disassembly/disintegration [146].

The polyphenolic stilbene curcumin (**6.10**, Figure 6.4) is the main component of the turmeric spice. Curcumin is clinically evaluated in many indications [147,148], although its bioavailability is limited [149]. Curcumin is undergoing at least two Phase II clinical trials as a dietary supplement for mild to moderate AD patients [150]. Completed and published results from a Phase II study (1 and 4 g daily curcumin, oral, 6 months [151]) are inconclusive, as the placebo group did not show neurodegeneration. $A\beta_{40}$ levels are unchanged in AD patients treated with placebo and curcumin, although a non-significant raise of serum $A\beta_{40}$ levels may indicate some fibril disaggregation [151]. Curcumin (2 and 4 g daily curcumin, oral, 48 weeks) does not show clinical or biochemical benefits in another Phase II study [152]. Both studies show low curcumin concentrations in plasma, due to poor bioavailability and to extensive metabolism, which may explain the negative results [152].

The activity of curcumin as a deubiquitinase (DUB) inhibitor is briefly mentioned in section 3.3 (see also **3.19**, Figure 3.4). Its multiple beneficial effects in neurodegeneration are ascribed to anti-oxidant and anti-inflammatory properties [153]. Curcumin inhibits BACE and AChE [154]. Its low μM anti-aggregation potency on Aβ peptides is attributed to pan-inhibition of oligomer formation/elongation/fibril disassembly [155,156], or to an oligomeric species-specific inhibition [101]. The anti-aggregation effects of curcumin are questioned by some research groups [104]. The interference of fluorescent curcumin with thioflavin T and other dyes in aggregation assays *in vitro*, and the different experimental protocols used in published studies, may explain the controversial observations [157].

Bioanalytical studies detect a strong interaction between the [16]KLVFF[20] core Aβ peptide sequence and curcumin, which is suggested to break the β-sheet ordered Aβ oligomers [158]. Solid state NMR shows how curcumin destabilizes an *inter*-molecular, fibril-promoting salt bridge between Asp23 and Lys28 residues on two Aβ peptides [159]. It drives Aβ aggregation towards an off-pathway of less neurotoxic aggregates. Zn^{2+} ions cause similar effects on Aβ peptides [159]. Curcumin reduces neurotoxic Aβ–lipid membrane interactions (see also Chapter 6 of the biology-oriented companion book [7]) and preserves membrane integrity [160]. This effect could be due to Aβ–curcumin interactions, but also to curcumin interactions with the core (lipophilic) and the head groups (hydrophilic) of the lipid membrane [160]. Curcumin treatment of primary cortical rat neurons induces up-regulation of the Bcl2-associated athanogene 2 (BAG-2) co-chaperone (see also section 2.2.2 of the biology-oriented companion book [7]) [161]. BAG-2 up-regulation should down-regulate the levels of HP-tau and of NFTs, as observed in an okadaic acid (OKA)-induced model of tau hyperphosphorylation [161].

Curcumin is effective in neurodegeneration models *in vivo*. Biochemical (plaque numbers) and cognitive deficits (water maze) induced by *intra*-cerebroventricular (i.c.v.) infusion of Aβ in middle-aged wild-type (WT) rats are ameliorated by preventive p.o. [162] and therapeutic i.p. [153] treatment with curcumin. Curcumin restores physiological levels of the *post*-synaptic density protein 95 (PDS-95) marker [162] and of synaptophysin

[153]. Aβ-dependent GSK-3β activation, leading to HP-tau, is inhibited by curcumin [153]. Ten to 12-month-old Sweden APP TG mice treated for 4 to 6 months with curcumin p.o. show a 40–50% reduction of soluble and insoluble Aβ, of plaque burden, and lower levels of oxidized proteins in their brains [163,164]. Lower levels of any neurotoxic Aβ species are observed in the brain of middle-aged and aged Sweden APP TG mice [156]. Interestingly, higher curcumin dosages are reported either to be ineffective in lowering the levels of Aβ species in Sweden APP TG mice [163], or to affect only oligomeric Aβ species/reduce Aβ oligomerization, but not Aβ fibrillation [165]. Double Sweden APP–ΔE9 PSEN1 TG mice treated with curcumin i.v. for a week show reduced plaque burden and diminished dendritic abnormalities [166]. Cognitive deficits shown by 3-month-old double APP–PSEN1 TG mice are rescued by 3 months' p.o. treatment with curcumin [167]. Curcumin reduces the expression of monomeric $Aβ_{40}$ and $Aβ_{42}$, and of presenilin 2 (PSEN2, a component of the γ-secretase complex). It increases the expression of the insulin-degrading enzyme and of neprilysin (IDE and NEP, two Aβ-degrading enzymes) [167]. Cognitive (memory test) and biochemical benefits (reduced c-Jun N-terminal kinase/JNK-dependent tau hyperphosphorylation) are observed after p.o. administration of curcumin (4 months) to 5-month-old triple TG mice (M146V/PSEN1, Sweden/APP, P301L/tau) [168]. The combination of curcumin and omega-3 fatty acid docosahexaenoic acid (DHA) shows synergistic effects on the same TG mice [168].

Curcumin shows anti-aggregation-dependent effects in *C. elegans* [169] and *Drosophila* models [170]. As to the former, curcumin decreases the percentage of paralyzed worms expressing an aggregation-prone Aβ peptide and shows a lifespan-extending tendency [169]. Four Aβ- and one tau-carrying TG *Drosophila* models show varying biochemical (highest for mutant $Aβ_{42}$ Arctic/E22G-expressing flies) and behavioral benefits (highest for tau-expressing flies) [170] when treated with curcumin. Interestingly, Aβ fibrillation seems to be promoted while Aβ (and possibly tau) oligomerization is inhibited, with an overall neuroprotective effect [170].

Curcumin shows efficacy *in vivo* against tau-dependent neurodegeneration. Mice treated with i.c.v. OKA show extensive memory impairment, oxidative-nitrosative stress, impaired energy metabolism, and higher HP-tau levels, leading to PHF and NFT formation [171]. Curcumin (50 mg/kg p.o. daily, 13 days) improves cognitive and locomotor deficits, and attenuates OKA-dependent biochemical abnormalities [171]. Fifteen- to sixteen-month-old mice expressing WT human tau develop severe tau pathology [172]. Treatment with curcumin (500 p.p.m. daily p.o., chow, 4–5 months) attenuates the behavioral deficits of aged human tau-expressing mice. Curcumin selectively decreases the levels of neurotoxic 140 kDa tau dimers, without altering monomeric and insoluble tau levels, and it increases the levels of Hsp90, Hsp70, and Hsc70 chaperones [172]. As a result, dysregulated synaptic excitatory proteins are restored and neurotoxic tau-dependent effects are attenuated. Lower 140 kDa tau levels and higher chaperone-mediated clearance of insoluble tau aggregates contribute to the biochemical effect of curcumin [172].

The multitargeted profile of curcumin against neurodegeneration is confirmed by its activity against other aggregation-prone proteins/diseases. As to *Parkinson's disease* (*PD*), curcumin inhibits the aggregation of α-synuclein, and disassembles pre-formed aggregates *in vitro* in cell-free [173] and in SH-SY5Y cell-based models [174]. Its *in vitro* and *in vivo* efficacy on PD models (due also to positive effects on mitochondrial dysfunctions, on nitrosative and oxidative stress) is thoroughly reviewed [175]. Although curcumin is reported to promote polyQ-dependent neurotoxicity in PC12 cells [176], it reduces mutant polyQ aggregation and disassembles pre-existing polyQ aggregates in a yeast model [177]. Curcumin provides partial behavioral improvements and polyQ-related biochemical benefits in the CAG 140 KI mouse model of *Huntington's disease* (*HD*) [178]. Curcumin selectively binds to, inhibits the aggregation of, and disassembles non-native, disease-related *PrP* isoforms [179]. It protects scrapie-infected neuroblastoma (scNB) [180] and murine neuroblastoma (N2a) cells [181] from the accumulation of neurotoxic PrP aggregates.

The low bioavailability and metabolic instability of curcumin stimulate the search for alternative formulations [182]. Inorganic and polymeric nanoparticles, liposomes, and micelles are popular curcumin nanoformulations [183]. Some among them are targeted against NDDs. A curcumin–pyrazole–polyethylenglycol (PEG) lipid conjugate [184] protrudes from liposomes ($\approx$200 nm mean diameter) and inhibits $A\beta_{42}$ aggregation in a dose-dependent manner, while lipid-conjugated liposomes are inactive [184]. Water-soluble, poly-(lactic-co-glycolic acid) (PLGA) nanoparticles (150–200 nm mean diameter) decorated with a motor neuron-targeting Tet-1 peptide encapsulate curcumin, promote its bioavailability and biological activity [185]. The targeting peptide determines the preferential entry of curcumin-loaded PLGA nanoparticles into GI-1 glioma cells, does not cause cytotoxicity, and fully disassembles amyloid aggregates in 72 hours [185]. Similar curcumin-loaded PLGA nanoparticles (80–120 nm mean diameter) confirm lack of toxicity, and protect human SK-N-SH neuroblastoma cells from H_2O_2-induced elevation of ROS and consumption of glutathione [186]. Curcumin-functionalized, Si-coated, amide/ester-connected Au nanoparticles (**6.11**, Figure 6.5, 10–25 nm hydrodynamic diameter) preferentially inhibit the elongation of $A\beta_{40}$ oligomers, and promote the disaggregation of $A\beta_{40}$ fibrils [187]. $A\beta_{40}$ fibrils disaggregated by Au-supported curcumin are less neurotoxic on mouse neural crest-derived neuro2a cells than either native $A\beta_{40}$ fibrils or soluble curcumin-disassembled $A\beta_{40}$ fibrils [187].

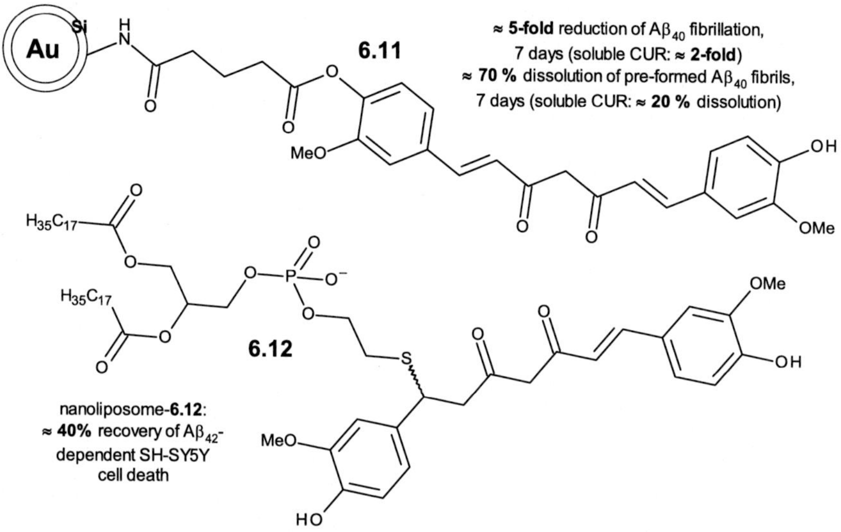

FIGURE 6.5 Curcumin–nanoparticle conjugates as inhibitors of Aβ aggregation: chemical structures, **6.11–6.12**.

The curcumin–phospholipid conjugate **6.12** (Figure 6.5), synthesized by Michael addition, is used to prepare curcumin-decorated nanoliposomes (63–200 nm mean diameter) [188]. Curcumin-loaded nanoliposomes reduce Aβ toxicity in human embryonic kidney (HEK) cells and decrease Aβ levels. Once administered i.c.v. in mice, fluorescent curcumin-decorated nanoliposomes selectively stain amyloid plaques *in vivo* and in *postmortem* samples [188]. WT rats, once injected i.c.v. with Aβ$_{42}$ peptides and treated with curcumin-containing lipid nanoparticles (≈200 nm mean diameter) or with soluble curcumin, are protected from Aβ-dependent neurodegeneration (reduced cognitive impairment, higher synaptophysin levels, lower HP-tau and pro-inflammatory cytokine levels, and higher neurothropic factor levels) [153]. Free curcumin (50 mg/kg daily, 10 days, 3 days after Aβ i.c.v. injection) and curcumin-containing lipid nanoparticles (equivalent to 2.5 mg/kg daily of curcumin, 10 days, 3 days after Aβ i.c.v. injection) are similarly effective, notwithstanding the 20-fold lower nanoparticle dosage [153]. Neuroprotective effects are observed in Aβ-stereotaxically injected rats with PGLA-based nanoparticles at a lower dosage (equivalent to 0.5 mg/kg daily of curcumin, 14 days, 7 days after Aβ stereotaxic injection) [189]. The same free curcumin/weight dose is ineffective [189]. Curcumin encapsulated into ≈60–70 nm PEG–polylactic acid (PLA) nanoparticles is orally administered (equivalent to 23 mg/kg weekly of curcumin, 3 months) to 9-month-old Sweden APP TG mice [190]. An ≈five times higher brain concentration of curcumin is reached with nanoparticle-encapsulated with respect to soluble curcumin.

Partial recovery from biochemical and behavioral impairment of adult Sweden APP TG mice is observed [190]. Nanolipid-encapsulated curcumin (≈150 nm mean diameter) is effective against a 3-nitrophenol-induced model of HD in rats [191]. Curcumin-loaded nanoparticles (equivalent to 40 mg/kg daily of curcumin, gavage, 7 days) restore locomotor deficits, and show amelioration of 3-nitrophenol-induced mitochondrial dysfunctions [191].

Curcumin analogues are another popular area of research [192]. A simple structure–activity relationship (SAR) for curcumin-like compounds to bind to the Aβ peptide, and to act as anti-aggregation agents, is determined (Figure 6.5) [193]. An 8- to 16 Å-wide rigid linker must separate two aromatic portions, and a hydroxylic group on one or both aromatic portions respectively ensures or increases Aβ affinity [193]. The symmetrical, constrained piperidine–bispiperazine curcumin derivative **6.13** [194] (Figure 6.6) shows improved physicochemical properties, bioavailability, and stability. It is a more potent inhibitor of Aβ aggregation, anti-oxidant, and metal-chelating agent than curcumin in cell-free assays [194]. It acts on Aβ peptides by disfavoring the conformational transition to aggregation-prone β-sheets, hindering the formation of both Aβ fibrils and oligomers in an Aβ aggregation assay, and decreasing oxidative stress in SH-SY5Y cells [194].

The amphiphilic, sugar-bearing curcumin derivative **6.14** shows a ≈1000-fold higher solubility in water than curcumin [195]. It displays an impressively low nM anti-aggregation potency *in vitro* on both Aβ peptides and aggregation-prone tau [306]VQIVYK[311] sequences. Unfortunately, its anti-aggregation potency is gradually lost by increasing the concentration of the amphiphilic curcumin, possibly due to micelle formation [195]. The pyrazole-containing compounds **6.15a,b** (Figure 6.6) are heterocycle-bridged curcumin analogues displaying potency as Aβ and tau anti-aggregating agents, and as γ-secretase inhibitors [196]. Compound **6.15a** shows moderate potency, but displays off-target toxicity. N-substituted **6.15b** is a potent tau aggregation and γ-secretase inhibitor [196].

Other polyphenolic/flavonoid structures, including gossypetin, hypericin, and purpurogallin, prevent Aβ and/or tau aggregation [49]. Oleuropein aglycone (**6.16a**, Figure 6.7), the active principle from oleuropein (**6.16b**), the major polyphenol in olive oil, reduces the Aβ burden, prolongs the lifespan, and decreases the population of paralyzed worms in an Aβ-expressing TG C. *elegans* model [197]. Soluble Aβ levels are unaffected, while Aβ oligomerization, fibrillation, and insolubilization are inhibited [197].

WT rats intracranially injected with pre-formed Aβ aggregates in the presence or absence of oleuropein aglycone display a different biochemical scenario after sacrifice (30 days *post*-injection) [198]. Amyloidogenic Aβ-dependent neurotoxicity is observed in neurons from rats exposed to

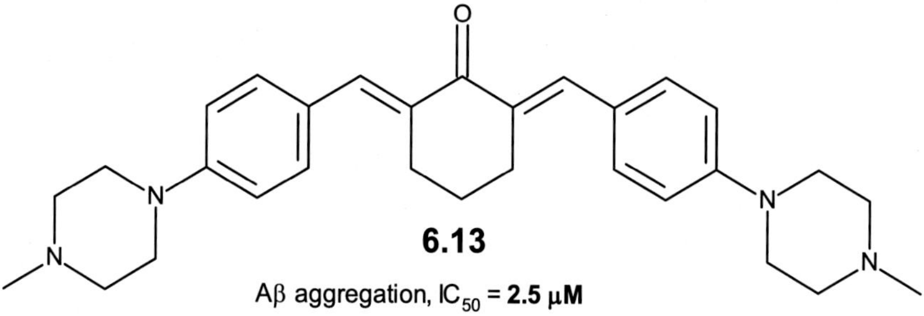

6.13

Aβ aggregation, IC_{50} = 2.5 μM
(curcumin IC_{50} = 12.1 μM in this study)

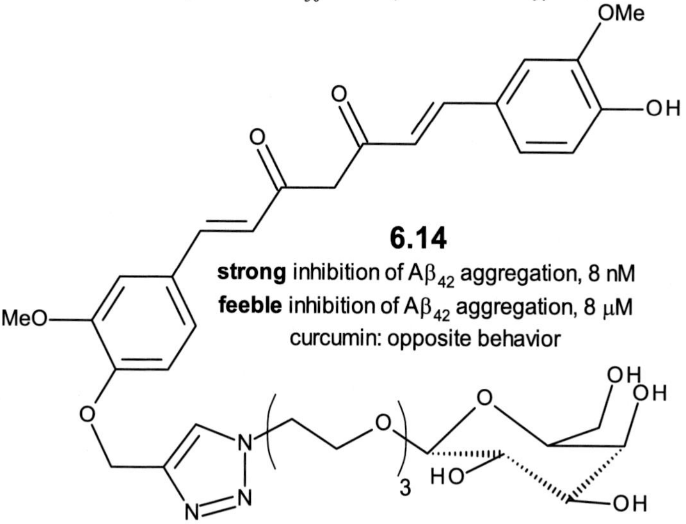

6.14

strong inhibition of $Aβ_{42}$ aggregation, 8 nM
feeble inhibition of $Aβ_{42}$ aggregation, 8 μM
curcumin: opposite behavior

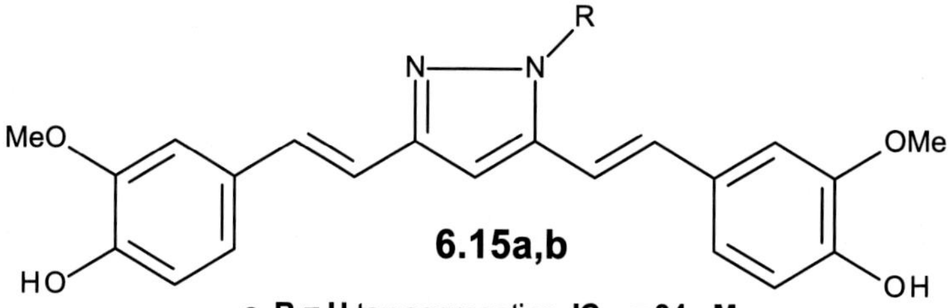

6.15a,b

a, R = H tau aggregation, IC_{50} = 34 μM
tau disassembly, IC_{50} = 17 μM
Aβ affinity, IC_{50} = 480 nM
$Aβ_{42}$ γ-secretase inhibition, IC_{50} = 5.2 μM
b, R = p-NO$_2$Ph tau aggregation, IC_{50} = 1.2 μM
tau disassembly, IC_{50} = 1.1 μM
Aβ affinity, IC_{50} > 1 μM
$Aβ_{42}$ γ-secretase inhibition, IC_{50} = 3.5 μM

FIGURE 6.6 Curcumin analogues as inhibitors of Aβ and tau aggregation: chemical structures, **6.13–6.15b**.

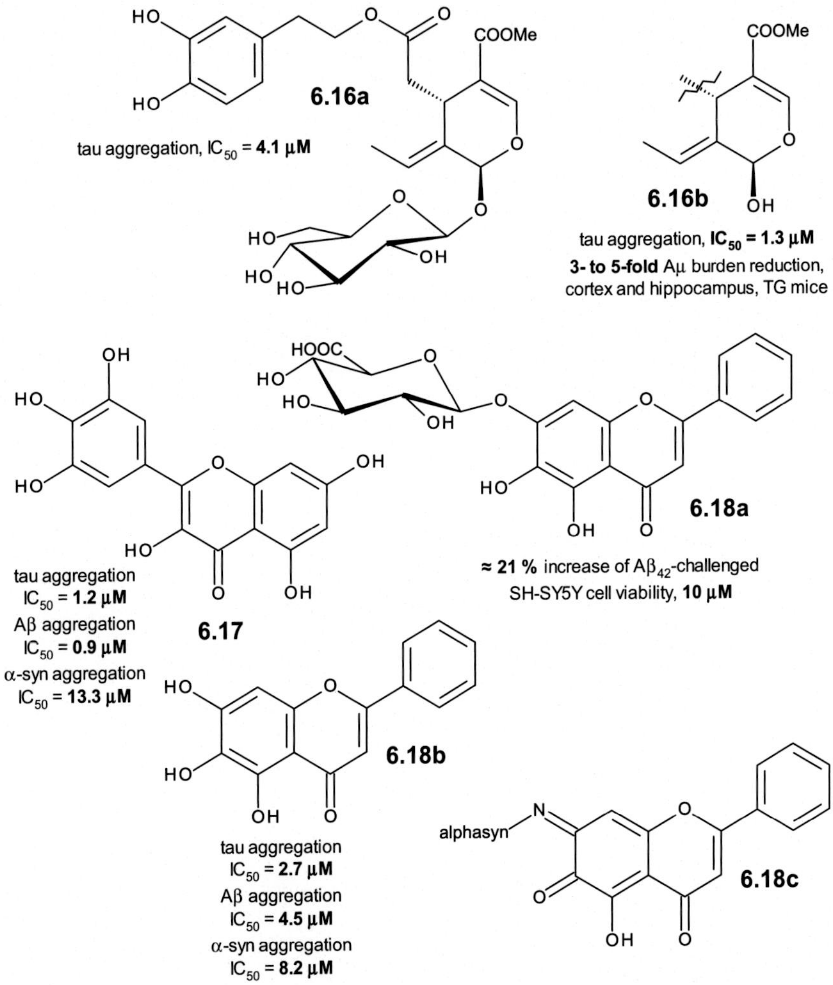

FIGURE 6.7   Small molecule polyphenolic inhibitors of the aggregation of tau and other amyloidogenic proteins: chemical structures, **6.16a–6.18c.**

ordered Aβ fibrils and oligomers (absence of oleuropein aglycone). Disordered Aβ–oleuropein aggregates resulting from pre-incubation with oleuropein aglycone, conversely, are not neurotoxic to neurons [198]. The binding between Aβ peptides and oleuropein is structurally characterized by NMR (hydrophobic interactions with Leu17-Phe20 region, hydrophilic interactions with N-terminus and His13-Lys16 regions) [199]. Three-month-old double Sweden-Indiana/V717F TG mice treated with oleuropein aglycone (50 mg/kg daily with food, 8 weeks) show memory improvements and a lower Aβ burden [200]. The observed aggregates are made by less packed/partially disassembled Aβ plaques, if compared to aggregates from untreated TG mice [200]. Remarkably, oleuropein aglycone induces and sustains autophagy in treated/disease-free TG mice, while untreated/diseased TG mice show impairments in late autophagy steps. Inhibition of the negative autophagy regulator mTOR kinase is postulated [200]. Oleuropein aglycone binds to FL human tau and to mutated P301L tau with low μM affinity [201]. Tau aggregation in the presence of oleuropein aglycone yields a smaller number of aggregates with different morphology and smaller size, compared to untreated samples [201].

The flavonoid myricetin (**6.17**) inhibits *in vitro* tau and Aβ aggregation at low μM concentrations, and interferes with α-synuclein aggregation at higher concentrations [111]. Its activity as an Hsp70 ATPase inhibitor is briefly mentioned in section 2.3.1 (see also **2.40b**, Figure 2.8). Myricetin prevents the formation of Aβ oligomers from monomers, their elongation at low nM concentrations, and it disassembles Aβ fibrils at higher concentrations [130]. Administration of myricetin with food for 9 months to 5-month-old Sweden APP TG mice reduces neurotoxic, oligomeric Aβ species in their brains, but does not reduce the Aβ plaque burden [165]. Surface plasmon resonance (SPR) suggests that myricetin has a stronger affinity for Aβ fibrils and oligomers, rather than for monomers, antagonizing their expansion/fibrillation by binding to the growing ends of oligomers and fibrils [202]. Interestingly, myricetin is also a low μM inhibitor of the ATPase activity of Hsp70 [203], possibly contributing to a reduction of tau levels by promoting UPS-driven degradation of tau.

The water soluble, sugar-containing baicalin (**6.18a**) is converted *in vivo* to the biologically active flavanoid baicalein (**6.18b**) [204]. BBB-permeable baicalein inhibits at low μM concentrations cell-free Aβ aggregation, leading to fewer and smaller Aβ fibrils [205]. It protects SH-SY5Y cells from Aβ-induced cytotoxicity, and increases their viability through dose-dependent reduction of the concentration of Aβ-induced neurotoxic H_2O_2 [205]. Baicalin, as a prodrug of baicalein, is neuroprotective by i.p. administration (10 mg/kg daily, 8 weeks) on 6-month-old Sweden APP TG mice [206]. It rescues cognitive deficits in TG mice by preventing Aβ aggregation and by lowering neurotoxic Aβ levels/stimulating non-amyloidogenic processing of APP [206]. Baicalein inhibits aggregation of α-synuclein by covalent reaction in its oxidized form with a Lys residue, and formation of a non-fibrillogenic, baicalein–α-syn Schiff base oligomer (**6.18c**, Figure 6.7) [207]. Baicalein disaggregates pre-formed α-synuclein fibrils by cleavage at internal sites, rather than by disassembly of terminal α-synuclein monomers [207]. Baicalein shows similar effects in SH-SY5Y and HeLa cells, alleviating either α-syn- or Aβ-induced neurotoxicity [208]. Baicalein and other polyphenols prevent membrane permeabilization by α-syn- and Aβ oligomers in cell-free liposome-based and mitochondrial membrane-based assays [209]. Baicalein also shows low μM inhibition of tau aggregation [111].

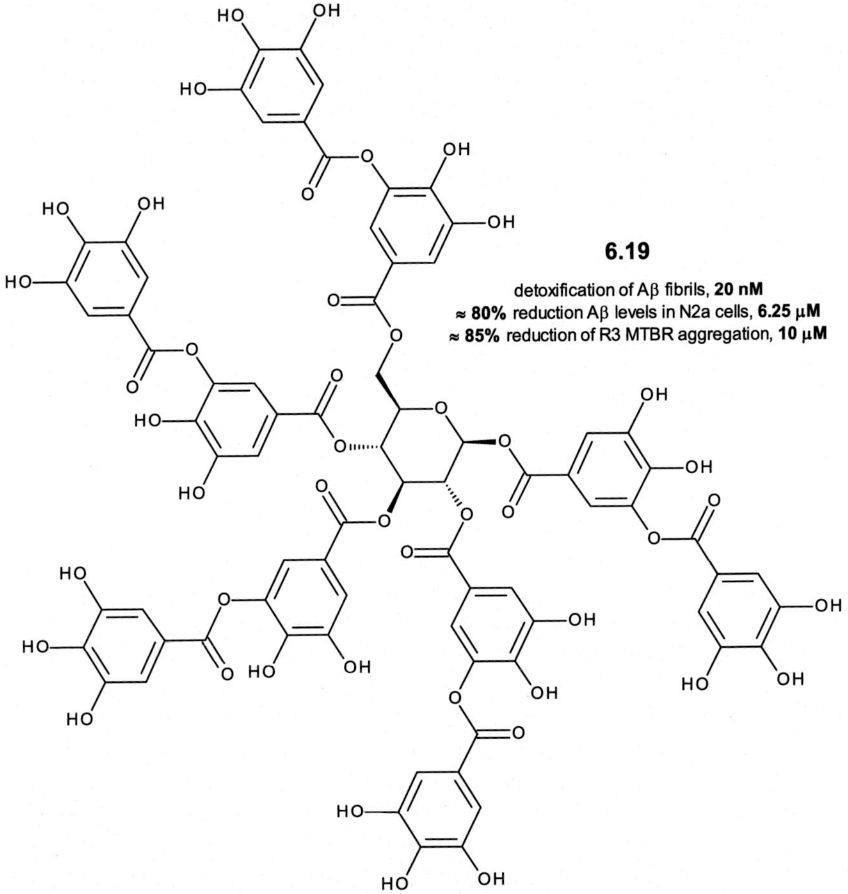

6.19

detoxification of Aβ fibrils, **20 nM**
≈ **80%** reduction Aβ levels in N2a cells, **6.25 μM**
≈ **85%** reduction of R3 MTBR aggregation, **10 μM**

FIGURE 6.8 Small molecule polyphenolic inhibitors of the aggregation of tau and other amyloidogenic proteins: chemical structure, tannic acid, **6.19**.

The polymeric, anti-oxidant tannin **6.19** (tannic acid, Figure 6.8) is a potent/low nM inhibitor of Aβ oligomerization, elongation and fibrillation, and an equipotent promoter of Aβ fibril destabilization [210].

It disaggregates Aβ fibrils and promotes their reassembly into nontoxic, low molecular weight species behaving as disaggregated Aβ [211]. Tannic acid (30 mg/kg daily, gavage, 6 months) supplied to 6-month-old double TG mice (Sweden-APP, ΔE9-PSEN1) prevents the insurgence of behavioral impairments [212]. In addition to reducing Aβ aggregation, it lowers Aβ levels *via* BACE inhibition and it reduces neuroinflammation [212]. Tannic acid inhibits *in vitro* heparin-promoted aggregation of the tau three repeat/R3 MTBR isoforms at low μM concentrations [213]. It acts by dose-dependently preventing the aggregation-required β-sheet conformational shift of R3 tau, and it forces tannic acid-bound R3 tau to assume a hairpin-like conformation. Tannic acid acts as an aggregation inhibitor also on FL human tau, although at higher μM concentrations [213].

Anthraquinones with tau and Aβ anti-aggregating properties include PHF016, daunorubicine, and adriamycine [49,111,214]. Exifone (**6.20**, Figure 6.9) inhibits at low μM concentration the aggregation of tau, Aβ [49],

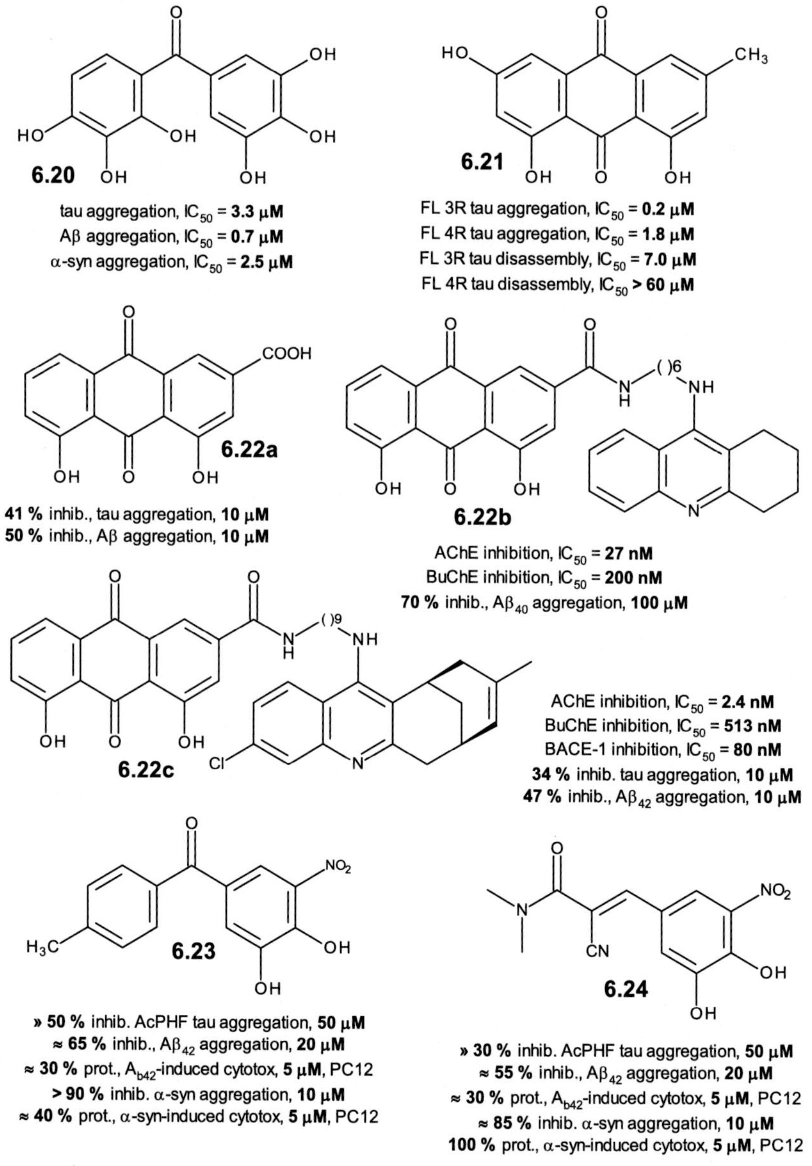

6.20

tau aggregation, IC_{50} = 3.3 μM
Aβ aggregation, IC_{50} = 0.7 μM
α-syn aggregation, IC_{50} = 2.5 μM

6.21

FL 3R tau aggregation, IC_{50} = 0.2 μM
FL 4R tau aggregation, IC_{50} = 1.8 μM
FL 3R tau disassembly, IC_{50} = 7.0 μM
FL 4R tau disassembly, IC_{50} > 60 μM

6.22a

41 % inhib., tau aggregation, **10** μM
50 % inhib., Aβ aggregation, **10** μM

6.22b

AChE inhibition, IC_{50} = **27 nM**
BuChE inhibition, IC_{50} = **200 nM**
70 % inhib., Aβ_{40} aggregation, 100 μM

6.22c

AChE inhibition, IC_{50} = **2.4 nM**
BuChE inhibition, IC_{50} = **513 nM**
BACE-1 inhibition, IC_{50} = **80 nM**
34 % inhib. tau aggregation, 10 μM
47 % inhib., Aβ_{42} aggregation, 10 μM

6.23

» **50 %** inhib. AcPHF tau aggregation, **50** μM
≈ **65 %** inhib., Aβ_{42} aggregation, **20** μM
≈ **30 %** prot., A$_{b42}$-induced cytotox, **5** μM, PC12
> **90 %** inhib. α-syn aggregation, **10** μM
≈ **40 %** prot., α-syn-induced cytotox, **5** μM, PC12

6.24

» **30 %** inhib. AcPHF tau aggregation, **50** μM
≈ **55 %** inhib., Aβ_{42} aggregation, **20** μM
≈ **30 %** prot., A$_{b42}$-induced cytotox, **5** μM, PC12
≈ **85 %** inhib. α-syn aggregation, **10** μM
100 % prot., α-syn-induced cytotox, **5** μM, PC12

FIGURE 6.9 Small molecule anthraquinone inhibitors of the aggregation of tau and other amyloidogenic proteins: chemical structures, **6.20–6.24**.

and α-synuclein [111]. It is suggested that exifone and similar compounds covalently bind to oligomeric tau through their adjacent phenolic hydroxyls and prevent their further aggregation [49].

Emodin (6.21) inhibits the aggregation of R3 and R4 MTBRs in tau, and their disassembly [214]. Its inhibition of casein kinase 2 (CK2), a regulatory kinase for Hsp90 chaperones, is briefly mentioned in section 2.4.3 (see also 2.71a, Figure 2.13). The anti-aggregating effect of emodin is largely retained on FL 3R and 4R human tau isoforms, while its disassembling potency is reduced on FL 3R tau and almost lost on FL 4R tau [214]. Emodin antagonizes the toxicity of $A\beta_{25-35}$ peptides in cortical neurons [215].

Rhein (6.22a) reduces the levels of Aβ peptides and of ROS in a senescence-accelerated mouse model (SAMP8, 4-month-old, treated with 25 and 50 mg/kg daily in drinking water, 6 months) [216]. Rhein and emodin are poorly active against two *in vitro* Aβ aggregation assays [217]. Amide conjugates of rhein with tacrine (6.22b [218]) and huprine Y (6.22c [219]) are endowed with multi-targeted activity against AD. The conjugates display cholinesterase (low nM–AChE, mid-high nM–BuChE) inhibition, and moderate anti-aggregation activity on Aβ peptides (mid μM concentration). Compound 6.22b shows an Fe^{2+}- and Cu^{2+}-targeted chelating ability [218]. Compound 6.22c inhibits BACE (medium nM concentration), is highly BBB-permeable, and inhibits tau aggregation at higher concentrations [219]. Hippocampal slices from 2-month-old mice treated with $A\beta_{42}$ are protected from synaptic failure by compound 6.22c. I.p. treatment with 6.22c (2 mg/kg daily, three times per week, 4 weeks) of 7-month-old/young and 11-month-old/aged double TG mice (Sweden-APP, ΔE9-PSEN1) decreases the concentration of neuro- and synaptotoxic Aβ hexamers and dodecamers [219].

The *nitrocatechol moiety* in marketed catechol O-methyl transferase inhibitors tolcapone and entacapone (respectively 6.23 and 6.24, Figure 6.9) contribute to their pan-anti-aggregating profile [220]. The compounds inhibit cell-free seeding, oligomerization, and fibrillation of α-synuclein and $A\beta_{42}$, and protect PC-12 cells from α-synuclein- and $A\beta_{42}$-induced cytotoxicity [220]. They prevent the aggregation of the acetyl-[306]VQIVYK[311]-amide tau hexapeptide in cell-free systems [221].

Memoquin (6.25, Figure 6.10) is a multi-targeted, neuroprotective quinone acting as an AChE, BuChe, and BACE-1 inhibitor [222]. It inhibits Aβ aggregation indirectly by preventing the Aβ peptide assembly-accelerating AChE–Aβ interaction [222], and directly by binding to ordered Aβ β-sheets [223]. Memoquin prevents scopolamine-induced memory impairments (10 and 15 mg/kg p.o., 20 minutes before scopolamine treatment) [224]. It prevents cognitive impairments caused by i.c.v. $A\beta_{42}$ injection (15 mg/kg p.o. daily, 6 days after i.c.v. $A\beta_{42}$ injection) [224].

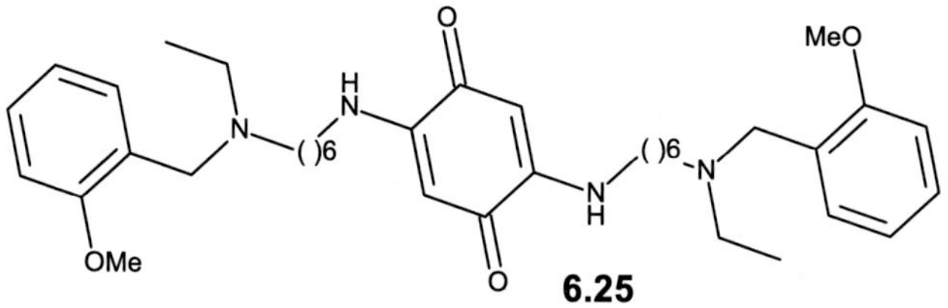

6.25

AChE inhibition, IC_{50} = **1.55 nM**
BuChE inhibition, IC_{50} = **144 nM**
BACE-1 inhibition, IC_{50} ≈ **1 μM**
≈ **67 %** inhib., $A\beta_{42}$ aggregation, **10 μM**

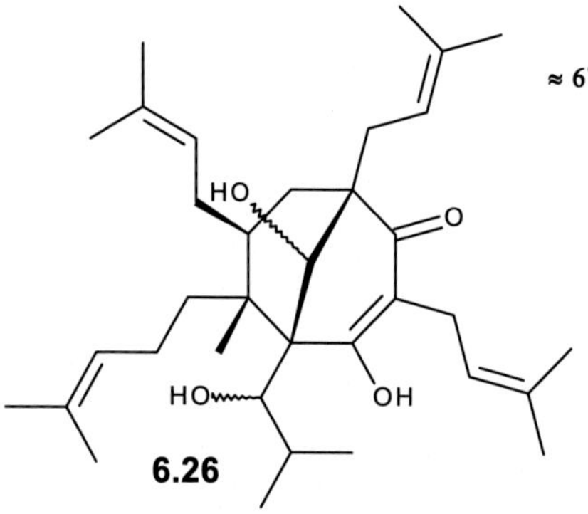

6.26

≈ **30 %** reduction $A\beta$, total burden
and oligomers, in Swe-PS1 TG mice
≈ **80 %** reduction, PHF-1 positive cells
close to $A\beta$ plaques, in Swe-PS1 TG mice

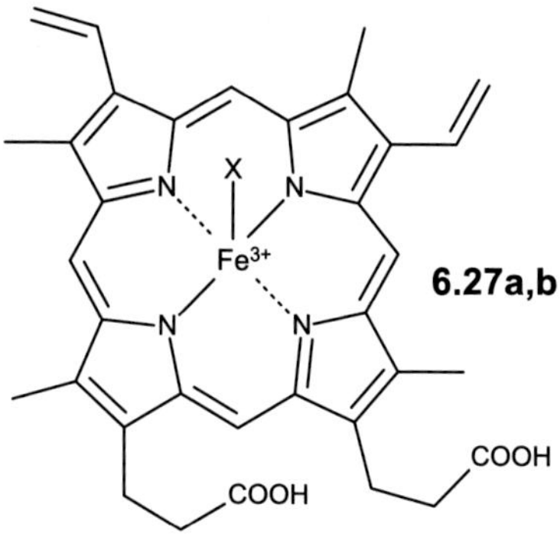

6.27a,b

a, X = Cl
tau aggregation
IC_{50} = **8.2 μM**
$A\beta$ aggregation
IC_{50} = **0.1 μM**

b, X = OH (bovine)
tau aggregation
IC_{50} = **10.4 μM**
$A\beta$ aggregation
IC_{50} = **0.2 μM**
α-syn aggregation
IC_{50} = **15 μM**

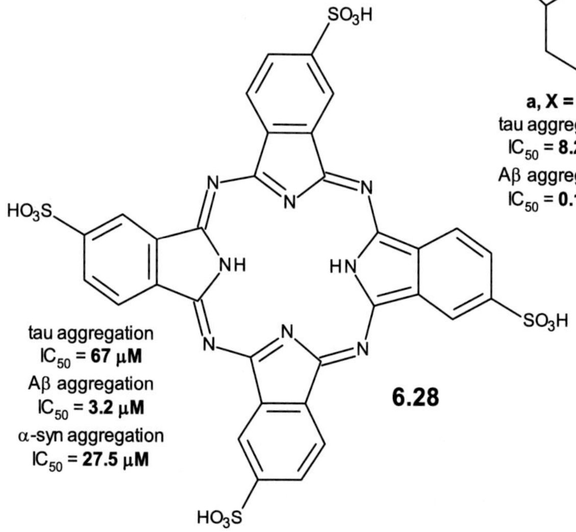

tau aggregation
IC_{50} = **67 μM**
$A\beta$ aggregation
IC_{50} = **3.2 μM**
α-syn aggregation
IC_{50} = **27.5 μM**

6.28

FIGURE 6.10 Small molecule inhibitors of the aggregation of tau and other amyloido-genic proteins: chemical structures, **6.25–6.28**.

The acylphloroglucinol *tetrahydrohyperforin* (THP, IDN5706, **6.26**) is a stable, orally bioavailable and neuroprotective semi-synthetic derivative of naturally occurring hyperforin [225]. As its parent compound, THP alleviates the consequences of Aβ aggregate-dependent pathologies by, *inter alia*, disrupting the AChE–Aβ interaction and disassembling pre-formed Aβ aggregates [226]. It alleviates Aβ-induced memory decline, decreases the levels of larger size Aβ deposits, and prevents astroglyosis and oxidative damage triggered by high Aβ deposit levels in 12-month-old double Sweden-APP, ΔE9-PSEN1 TG mice (2 mg/kg p.o. daily, 1 month) [227]. A different treatment schedule (4 mg/kg i.p. three times per week, 10 weeks) on 5-month-old Sweden-ΔE9TG mice causes a reduction in synaptotoxic Aβ oligomers and in HP-tau species, and restores the physiological levels of synaptic proteins [228].

Porphyrins are anti-aggregating agents against α-synuclein [111], Aβ peptides [229], and tau [49]. Hemin and hematin (respectively **6.27a,b**) bind preferentially to human Aβ peptides [230]. They contribute to the inhibition of Aβ aggregation and to the dismantling of Aβ fibrils in neurons in basal conditions [231]. The porphyrin hemin ring interacts with Phe residues (Phe19 in particular) and impairs the Phe–Phe *inter*-strand, Aβ fibrillation-promoting interactions. The hemin iron center binds to His residues in the Aβ N-terminus, regulates the overall heme–Aβ binding affinity and the disassembly of pre-existing Aβ aggregates [231]. Structurally related phthalocyanine tetrasulfonate (PcTS, **6.28**, Figure 6.10) prevents aggregation of α-synuclein [232], Aβ peptides [233], and prion proteins [234]. It prevents the aggregation of FL 4R human tau (low μM concentration), and to a lower extent of the 3R isoform [235]. NMR and electron paramagnetic resonance (EPR) show that PcTS specifically interacts with five aromatic Phe and Tyr residues outside of the N-terminus of tau. PcTS promotes the formation of soluble, off-pathway, disordered tau aggregates containing between 7 and 24 tau monomers [235]. Conversely, it prevents the formation of neurotoxic, globular, β-sheet/ordered tau oligomers with an apparent trimeric size. PcTS prevents the formation of PHF-1-like tau aggregates in an inducible model of aggregation-prone, mutated tau-expressing neuroblastoma N2a cells [235].

Cyanine dyes are small molecules binding to aggregation-prone proteins. In particular, *benzothiazolium-based* thioflavin S (**6.29a,b**, ThS, Figure 6.11, see also **2.45a,b**, Figure 2.9 and section 2.3.2—inhibitor of Hsp70-BAG-1 complexes, and **3.1a,b**, Figure 3.1—CHIP modulator) and thioflavin T (**6.30a**, ThT) are the most popular histology dyes used to set up *in vitro* aggregation assays [157].

Neutral benzothiazoles such as the Pittsburgh B compound (PiB, **6.30b** [236]) are as active and more bioavailable than their charged counterparts. Acyclic benzothiazolium–benzothiazole cyanines include N744

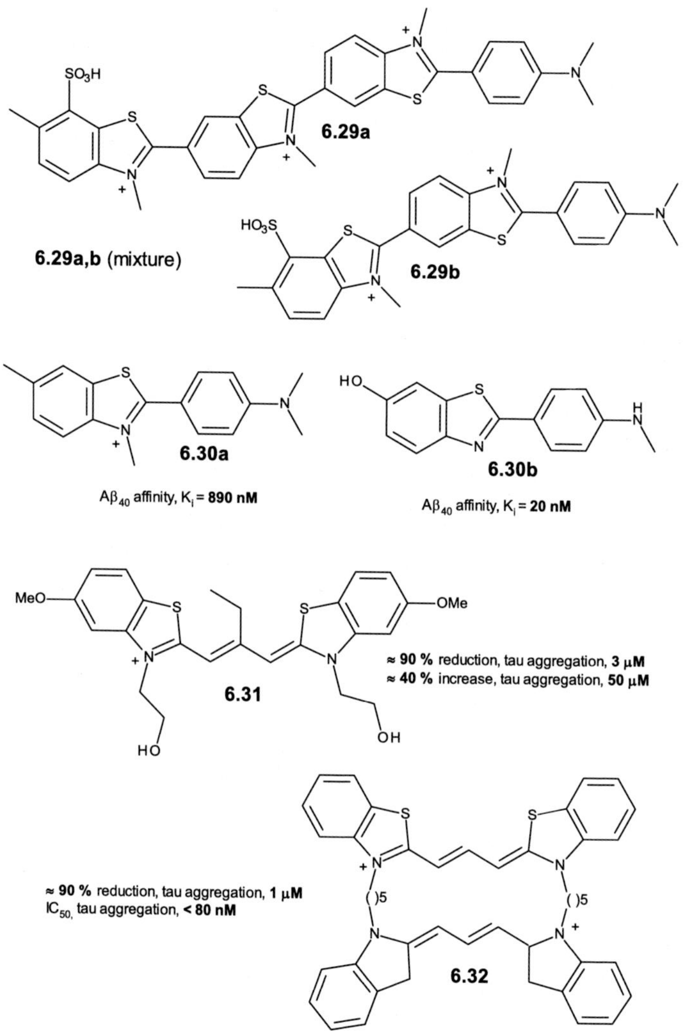

FIGURE 6.11 Small molecule benzothiazole-based inhibitors of the aggregation of tau and other amyloidogenic proteins: chemical structures, **6.29a–6.32**.

(**6.31**), a hit identified in an HTS campaign targeted against tau aggregation inhibitors [237]. N744 is a sub-μM aggregation inhibitor acting on 3R and 4R tau isoforms, on nucleation and elongation of tau oligomers and fibrils, and on tau disassembly by endwise tau residue disaggregation [238,239]. It shows an ≈10-fold selectivity for tau aggregation *vs.* Aβ and α-synuclein aggregation [238]. Interestingly, the anti-aggregation activity of N744 is rapidly reverted at concentrations of N744 >4 μM, with an increase of tau aggregation at higher concentration [240]. N744 itself aggregates at high concentrations, and N744 monomers and oligomers compete with tau (thus freeing it for tau aggregation) for further N744 binding. Pathology-related PTMs of tau, such as hyperphosphorylation, further decrease the aptitude of N744 as a tau anti-aggregation agent [240]. A similar biphasic behavior is shown by other cyanines [241,242] in cellular models of tau aggregation characterized by mutant and WT tau isoforms [243]. Multivalent, macrocyclic benzothiazolium–benzothiazole cyanines connected through their N atoms (e.g., **6.32**, Figure 6.11) are stronger tau aggregation inhibitors due to multivalency [244,245]. The most potent macrocycles adopt an "open"/multi-presenting conformation, while "closed" macrocycles are less active. Multivalent macrocyclic cyanines retain the class-specific biphasic behavior on tau aggregation (inhibition–low concentration/stimulation–high concentration) [245].

Water soluble, BBB-permeable *benzothiazole anilines* (*BTAs*) such as **6.33** (BTA-EG$_4$, Figure 6.12) possess strong affinity/inhibition potency against Aβ aggregation [246]. Their amphiphilic nature ensures coating of Aβ peptides, preventing their neurotoxic interactions with proteins (e.g., peroxidase catalase, amyloid-binding alcohol dehydrogenase ABAD) and regulating Aβ–lipid membrane contacts [247].

BTA-EG$_4$ (15 and 30 mg/kg i.p. daily, 2 weeks) reduces Aβ levels, promotes dendritic spine density, and improves cognitive performances in WT mice [248]. It increases the cell surface levels of APP with a Ras-dependent signaling mechanism, and promotes the non-toxic α-secretase processing of APP [248]. Triple TG mice (M146V/PSEN1, Sweden/APP, P301L/tau) treated at 2–3, 6–10, and 12–13 months of age with BTA-EG$_4$ (30 mg/kg i.p. daily, 2 weeks) show cognitive enhancement and synaptic density improvement following an age-dependent pattern [249]. Preliminary results indicate a potential usefulness for BTA-EG$_4$ in early AD pathology, and limited beneficial effects on tau pathology [249]. Neutral benzothiazoles bearing a metal chelating moiety, such as **6.34**, take advantage of a PBT-1-like (see **6.6**, Figure 6.3) metal chelation/anti-aggregation effect implanted onto a ThT-like, Aβ-binding chemiotype [250]. The compound is a strong binder of Aβ$_{42}$, and does increase its binding affinity in the presence of Cu^{2+} and Zn^{2+}. It leads to the disappearance (Cu^{2+}) or to the reduction (Zn^{2+}) of Aβ fibrils, to the disaggregation of pre-formed Aβ aggregates in the presence of both metal ions, and to the increase of neurotoxic Aβ$_{42}$ oligomers [250]. The last observation explains the neurotoxicity of **6.34** on tau-inducible N2A cells, and warns against indiscriminate aggregate dissolution of anti-fibril/plaque/ tangle by anti-aggregating agents. Benzothiazole-based cyanines are

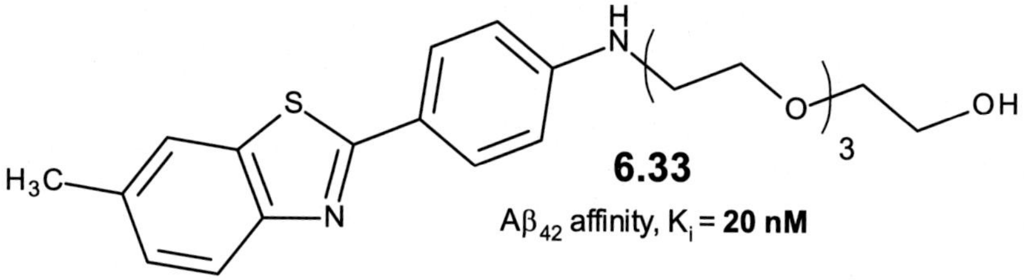

6.33

$A\beta_{42}$ affinity, $K_i = 20$ nM

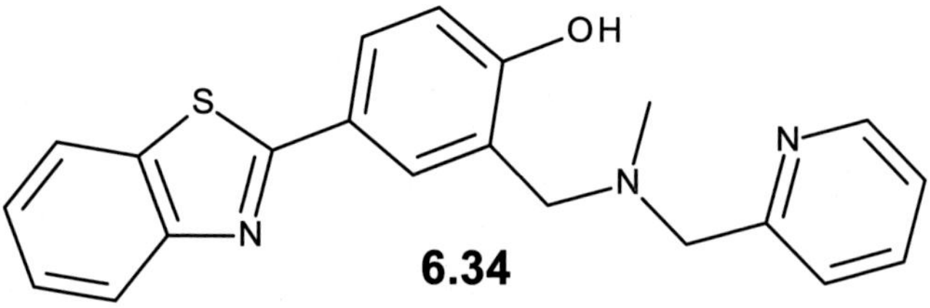

6.34

$A\beta_{42}$ affinity, $K_i = 36$ nM

$A\beta_{42}$-Zn^{2+} affinity $K_i = 275$ nM

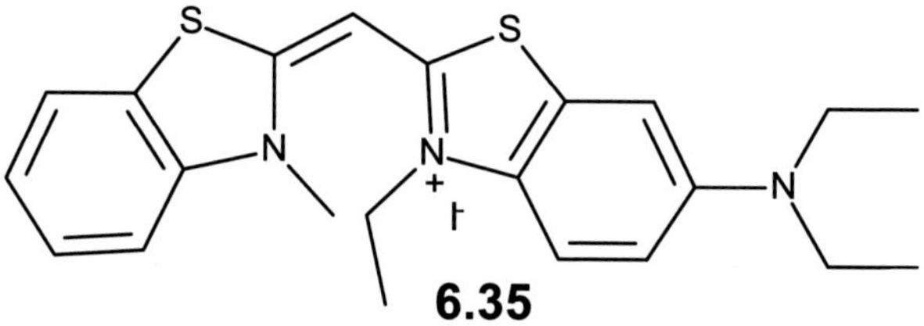

6.35

α-syn fibril height: $\approx$ **6.9** Angstroem (β-sheet)

α-syn fibril height with **6.33**: $\approx$ **8.0** Angstroem (insertion by **6.33**)

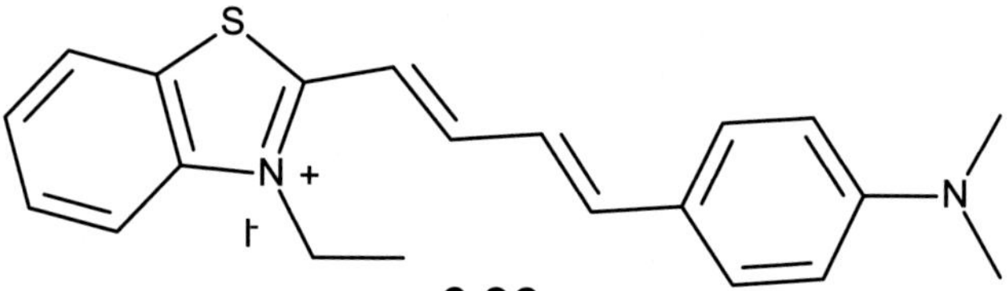

6.36

FIGURE 6.12 Small molecule benzothiazole-based inhibitors of the aggregation of tau and other amyloidogenic proteins: chemical structures, **6.33–6.36**.

α-synuclein-selective anti-aggregation compounds (T-284, **6.35** [251]), and tau-selective imaging agents (i.e., PBB-5, **6.36**, Figure 6.12 [252,253]). A recent review [254] covers the main features of benzothiazole-based diagnostics and therapeutics targeted against AD.

Rhodanines (**6.37a–c**, Figure 6.13) identified through an HTS campaign inhibit tau aggregation, and promote disaggregation of tau oligomers/fibrils [214].

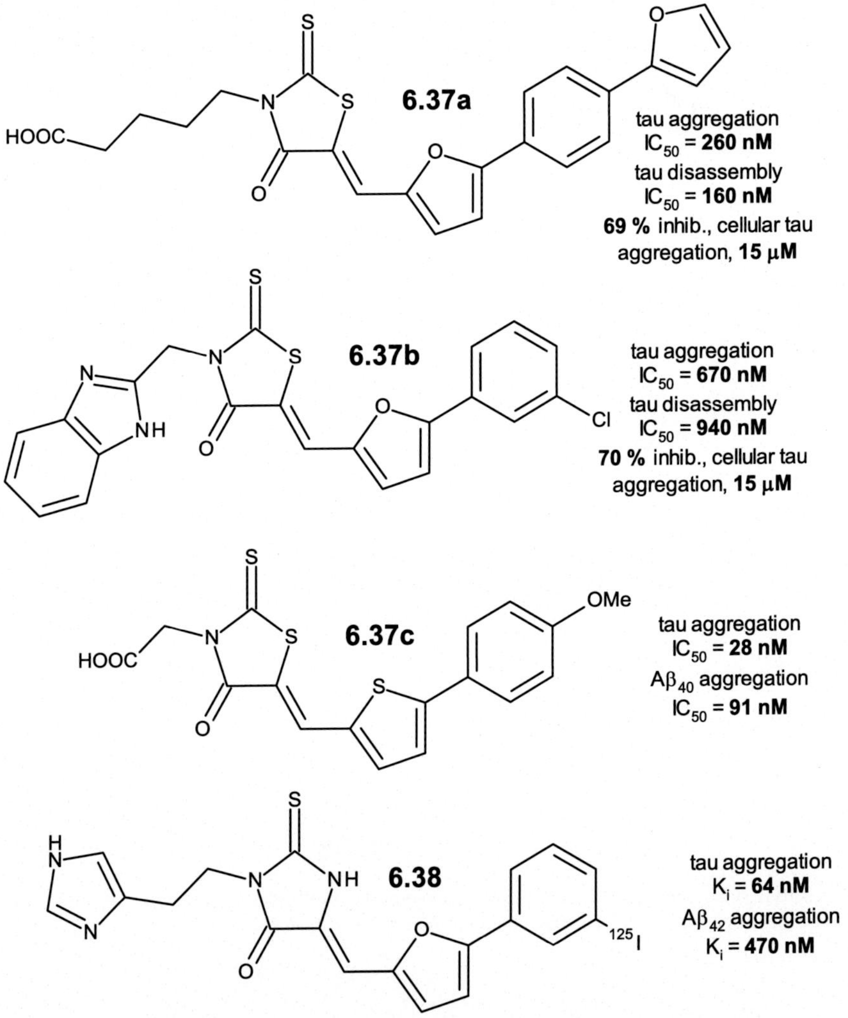
6.37a
HOOC
tau aggregation
IC50 = 260 nM
tau disassembly
IC50 = 160 nM
69 % inhib., cellular tau aggregation, 15 μM
6.37b
Cl
tau aggregation
IC50 = 670 nM
tau disassembly
IC50 = 940 nM
70 % inhib., cellular tau aggregation, 15 μM
6.37c
HOOC
OMe
tau aggregation
IC50 = 28 nM
Aβ40 aggregation
IC50 = 91 nM
6.38
125I
tau aggregation
Ki = 64 nM
Aβ42 aggregation
Ki = 470 nM

FIGURE 6.13 Small molecule rhodanine- and thiohydantoine-based inhibitors of the aggregation of tau and other amyloidogenic proteins: chemical structures, **6.37a–6.38**.

A small library of rhodanines determines an *in vitro* SAR with respect to tau aggregation and disassembly, and identifies cell-active rhodanines [255]. Increased ligand polarizability, through a larger conjugated π-electron network, augments the effect of rhodanines on tau assembly [243]. Rhodanines **6.37a,b**, respectively, represent the most potent cell-free binder/tau aggregation inhibitor (with poor permeability due to the free COOH group), and the most active tau aggregation inhibitor in tau-inducible neuroblastoma N2a cells [255]. Hippocampal organotypic slice cultures from TG mice expressing the mutant Δ280K 4R MTBR human tau fragment show co-aggregation of mouse and human tau, reduction of dendritic spine density, *intra*-cellular Ca^{++} dysregulation, and caspase 3-promoted neurotoxicity [256]. Compound **6.37b** partially rescues the pathological phenotype in hippocampal organotypic slice cultures [256]. Closely related, ^{125}I-labeled thiohydantoin **6.38** (Figure 6.13) is an NFT-specific radiolabeled ligand with high cellular uptake, good BBB penetration, and fast washout in mice after i.v. administration [257]. Fluorescent rhodanine-based imaging agents such as **6.37c** are NFT-selective staining agents in *post-mortem* samples from the brains of AD patients [258]. They are non-toxic either in HepG2 cells and in zebrafish embryos at high concentrations. Interestingly, their preference for NFT *vs.* Aβ plaque staining is not predicted by *in vitro* aggregation assays, where **6.37c** is almost equipotent (inhibition at low nM concentration) on tau and Aβ [258].

Phenylthiazolyl hydrazides (*PTHs*) result from a pharmacophore-targeted computational HTS screen aiming for tau aggregation inhibitors [259]. A structural optimization project from few commercially available hits leads to compound **6.39** (Figure 6.14), selected from ≈50 synthesized PTH analogues [259]. It is a potent, non-toxic PTH derivative capable of rescuing mutant tau-derived neurotoxicity in inducible N2a cells [260]. Compound **6.39** does not have any effect on Aβ aggregation [261]. *N'-benzylidene-N-benzoyl hydrazides* such as **6.40a** (BSc3504) are a tau-targeted, anti-aggregation/pro-disassembly chemiotype [262]. Amphiphilic, pyrogallol-containing, bulky hydrazides such as **6.40a** are potent tau binders, with high selectivity *vs.* Aβ aggregation. Conversely, smaller, non-hydroxylated compounds such as **6.40b** (BSc3297) preferentially interact with Aβ plaques *vs.* NFT tangles in *post-mortem* AD brain tissues [262].

N-aryl/heteroaryl amines such as **6.41** are low μM inhibitors of tau aggregation, and promoters of pre-formed PHF disaggregation [263]. Amine **6.41** reduces up to 70% the levels of PHF-bearing N2a cells, and cleans pre-formed PHFs from up to 50% of the same cells [263,264]. The adrenergic receptor antagonist carvedilol (**6.42**, Figure 6.14) is built on a carbazole scaffold that mimicks a constrained N-bisaryl amine. Carvedilol prevents the transition of Aβ monomers/oligomers from disordered to ordered/α-helix and β-sheet conformations [265], and reduces the levels of neurotoxic Aβ oligomers in cell-free and cellular assays [266]. It reduces

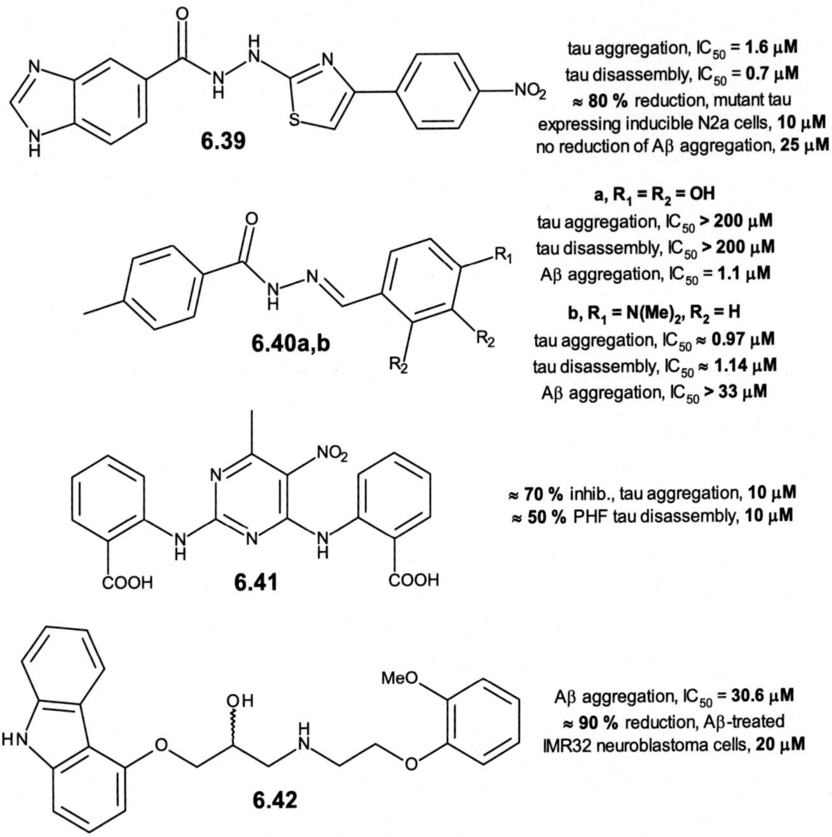

tau aggregation, IC_{50} = 1.6 µM
tau disassembly, IC_{50} = 0.7 µM
≈ 80 % reduction, mutant tau
expressing inducible N2a cells, 10 µM
no reduction of Aβ aggregation, 25 µM

6.39

a, R_1 = R_2 = OH
tau aggregation, IC_{50} > 200 µM
tau disassembly, IC_{50} > 200 µM
Aβ aggregation, IC_{50} = 1.1 µM

b, R_1 = $N(Me)_2$, R_2 = H
tau aggregation, IC_{50} ≈ 0.97 µM
tau disassembly, IC_{50} ≈ 1.14 µM
Aβ aggregation, IC_{50} > 33 µM

6.40a,b

≈ 70 % inhib., tau aggregation, 10 µM
≈ 50 % PHF tau disassembly, 10 µM

6.41

Aβ aggregation, IC_{50} = 30.6 µM
≈ 90 % reduction, Aβ-treated
IMR32 neuroblastoma cells, 20 µM

6.42

FIGURE 6.14   Small molecule hydrazide- and amine-based inhibitors of the aggregation of tau and other amyloidogenic proteins: chemical structures, **6.39–6.42**.

the number and length of Aβ fibrils, increasing the levels of disordered aggregates. Carvedilol lowers Aβ levels in APP Sweden TG mice [267]. Two-month-old double APP Sweden-Indiana TG mice treated with carvedilol (1.5 mg/kg daily in drinking water, 5 months) show reduced Aβ oligomerization and fibrillation, strengthened basal synaptic transmission, and attenuated Aβ-dependent cognitive deterioration [265]. Carvedilol is evaluated in a clinical trial for its effects on AD patients [268].

Quinoxalines such as **6.43** (Figure 6.15) are hits from an HTS campaign, carried out on ≈51K diverse compounds, that show low µM potency as inhibitors of the aggregation of P301L K18, an aggregation-prone, mutated tau fragment containing four MTBRs [269]. A preliminary quinoxaline SAR, deducible from inactive analogues included in the HTS collection, identifies 2,3-(difuran-2-yl)quinoxalines as novel chemotypes against tau aggregation [269].

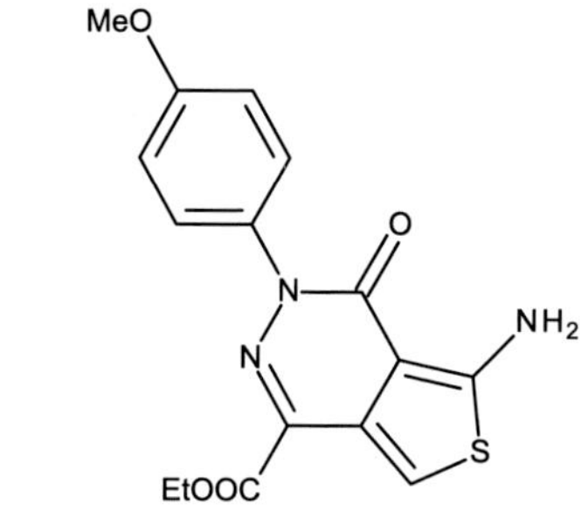

6.44a

≈ **86 %** inhib., tau aggregation, **80 μM**
≈ **28 %** inhib., Aβ$_{42}$ aggregation, **80 μM**
≈ **50 %** increase, soluble P301L K18
after centrifugation, **100 μM**

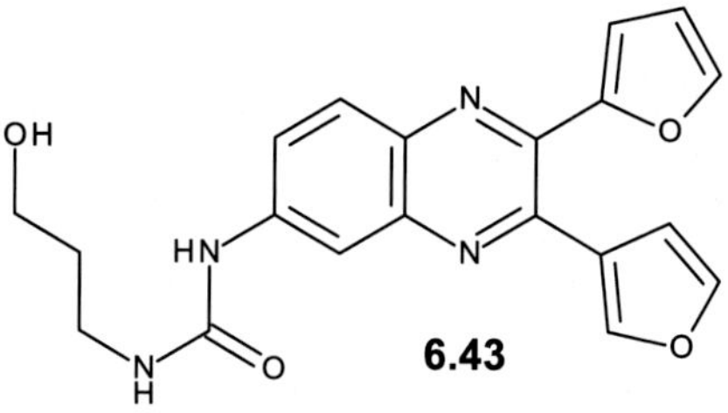

6.43

tau aggregation, IC$_{50}$ = **2.4 μM**

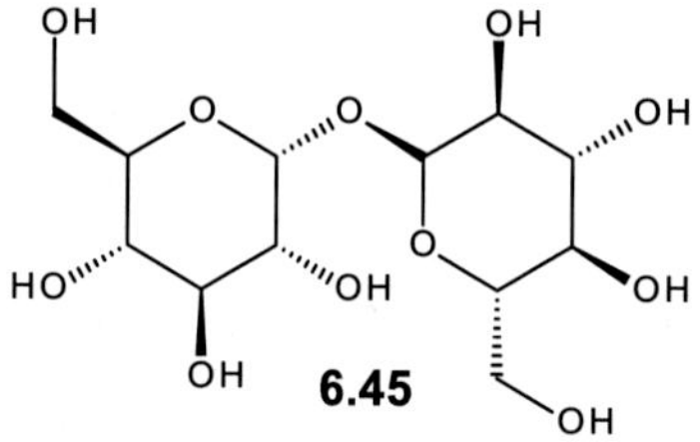

6.45

≈ **100 %** inhib., Aβ$_{40}$ aggregation, **100 mM**
≈ **50 %** inhib., Aβ$_{42}$ aggregation, **100 mM**
≈ **60 %** inhib., 4R MTBR Δ280K aggregation, **50 mM**
≈ **30 %** reduction, polyQ-positive N2a cells, **50 μM**

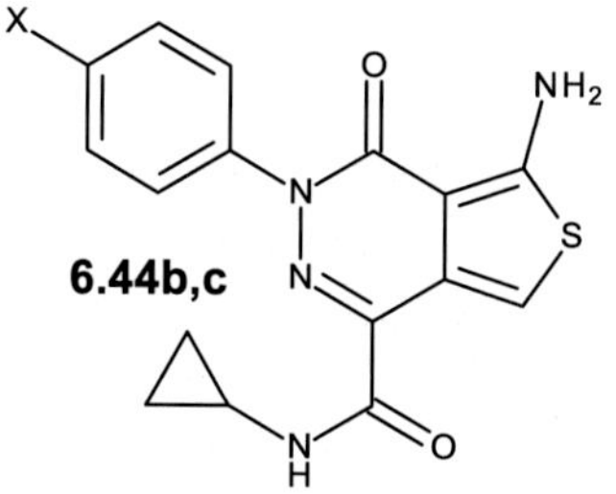

6.44b,c

b, X = Cl
P301L K18 aggregation, IC$_{50}$ = **8.1 μM**
≈ **58 %** increase, soluble FL tau after
centrifugation, **50 μM**
brain/plasma ratio: ≈ **1.6 (1.1 %** unbound)
water solubility: ≈ **32 μM**

c, X = F
P301L K18 aggregation, IC$_{50}$ = **2.5 μM**
≈ **78 %** increase, soluble FL tau after
centrifugation, **50 μM**
brain/plasma ratio: ≈ **1.09 (4.3 %** unbound)
water solubility: > **200 μM**

FIGURE 6.15 Small molecule inhibitors of the aggregation of tau and other amyloidogenic proteins: chemical structures, **6.43–6.45**.

An HTS campaign on P301L K18 tau, using an ≈291K collection from the National Institute of Health (NIH) and other public institutions, identifies *aminothienopyridazines (ATPZs)* as selective, potent inhibitors of tau aggregation [270]. Two ATPZ hits, and 21 analogues prepared in a preliminary SAR effort, show varying potency on tau aggregation and selectivity *vs.* Aβ aggregation. The best ATPZ ester (MLS000062428, **6.44a**) prevents the fibrillation of P301L K18 tau, and increases the levels of monomeric and oligomeric seeding-incompetent tau species [270]. ATPZ amides potently inhibit FL human tau aggregation, show drug-like properties in terms of lipophilicity and bioavailability in mouse brains, and are not cytotoxic up to 100 μM [271]. Amide **6.44b** shows a favorable PK profile after p.o. and i.v. administration (respectively 5 and 2 mg/kg), high brain/plasma ratios, and good metabolic stability. Amide **6.44b** shows efficacy at high concentrations *in vivo* (*C. elegans*, 100 μM, improved locomotion [272]), but its use is limited by low water solubility and low unbound brain fraction (respectively ≈30% and ≈1%) [271]. Profiling of an array of ≈40 ATPZs identifies amide **6.44c** as a potent, soluble, bioavailable, and brain-penetrant inhibitor of tau aggregation [273]. It is tolerated after a chronic oral administration schedule (50 mg/kg daily, 1 month) without signs of gross/organ toxicity. The brain concentration of **6.44c** does not show accumulation, while it could provide beneficial effects on tau-dependent pathologies in diseased animals [273]. ATPZs promote an *intra*-molecular Cys291–Cys322 disulfide bridge formation, as described for the phenothiazine MB, that may determine their anti-aggregation properties and could limit their efficacy on 3R tau-dependent tauopathies due to the single Cys322 residue on 3R tau isoforms [52]. Unspecific, Cys oxidation-related toxicity of MB and ATPZ is not reported *in vivo*, but further investigations would be desirable.

Trehalose (**6.45**, Figure 6.15) is a disaccharidic chemical chaperone exerting a wealth of neuroprotective effects in animal models of proteinopathies. Its effects are often attributed to mTOR-independent stimulation of autophagy (see Chapter 4), which enhances the clearance of α-synuclein and mutant huntingtin [274], prion proteins [275], SOD1 [276], and tau [277]. Trehalose, though, regulates the aggregation of amyloidogenic proteins. It inhibits oligomerization and fibrillation of $A\beta_{40}$, reducing its toxicity on SH-SY5Y cells [278]. Conversely, it inhibits only fibrillation of $A\beta_{42}$ and consequently it does not reduce its oligomer-dependent cellular toxicity [278]. Trehalose impairs nucleation and elongation of $A\beta_{40}$ by preventing its conformational switch to β-sheet aggregation-prone forms [279]. It forms a hydrophilic layer around $A\beta_{40}$ monomers that weakens *inter*-monomeric, aggregation-prone hydrophobic interactions [279]. Trehalose interacts with the head groups of phospholipids in lipid membranes, and its presence (100 mM) causes the rapid insertion of Aβ monomers in the membrane [280]. Trehalose stabilizes an α-helix conformation of Aβ peptides, and reduces the lag time needed for their deep insertion into the membrane. Membrane-inserted Aβ species are more disordered/less toxic than their counterparts in trehalose-free medium, and may contribute to an overall neuroprotective effect (lower membrane disruption in Aβ peptide-challenged neuronal cells treated with mM trehalose) [280]. The direct interaction between trehalose and polyQ proteins [281], prion proteins [282], and α-synuclein [283] is reported in the literature, to substantiate a mixed autophagy inducer/anti-aggregation profile for trehalose in NDDs. Trehalose is a weak, direct inhibitor of tau aggregation, acting on tau oligomerization and fibrillation [284]. Trehalose is a holder and a solubilizer for unfolded polyQ proteins in a yeast model, providing a longer time window to refolding chaperones such as Hsp104 to rescue such proteins [285]. This mechanism may act in conjunction with autophagy stimulation in other NDDs.

6.3 HSP110-DRIVEN DISAGGREGATION

Members of the *HSP100 protein family* act in cooperation with co-chaperones to disaggregate protein aggregates [286]. The HSP100 protein family member Hsp104 [287], in combination with the Hsp70 complex [288] and small chaperones Hsp26 and Hsp42 [289], shows significant amyloid-disaggregating activity on polyQ and α-synuclein fibers [289].

Metazoan homologues of Hsp104 are unknown, possibly because their amyloid-dismantling disaggregase activity could produce incompletely disassembled, neurotoxic small oligomers (a cure worse than the disease) [290]. Metazoan *constitutive Hsp110 family*, coupled with the Hsp70–Hsp40 complex and small chaperone HspB5, is capable of α-synuclein disassembly and renaturation [289], although with a lower amyloid-disaggregating activity than yeast Hsp104 complexes [291]. A slow disaggregase activity promoted by Hsp110 may still modulate *in vivo* the slow progression of human NDDs.

The physiological and pathological relevance of Hsp110–Hsp70–Hsp40–HspB5-driven disassembly of NDD-relevant amyloid aggregates, and the effect of Hsp110 modulation by biologicals and small molecule modulators on NDDs, remains to be determined. Expression of Hsp110, either alone or with Hsp70 and Hsp40 members, rescues *in vivo* neurotoxicity in *Drosophila* [292] and squid [293] models. Yeast Hsp104 synergizes with the mammalian chaperone machinery to provide amyloid-disaggregating capacity *in vitro* [291,294]. Expressed Hsp104 reduces polyQ levels and prolonges the lifespan of HD-like TG mice [295], and is neuroprotective in a *Drosophila* model of spinocerebellar ataxia type 3 (SCA3) [296]. Neuroprotection is observed by expression of Hsp104 in a rat model of PD [297]. Recombinant, potentiated disaggregases obtained from rational mutagenesis of yeast Hsp104 show stronger unfoldase, translocase, and ATPase activity [298]. They dissolve pre-formed α-synuclein, TDP-43, and FUS aggregates, and rescue their induced toxicity in yeast cells. Potentiated disaggregases prevent neurodegeneration in a *C. elegans* model of PD [298]. Further efforts on artificial, potentiated networks containing Hsp110/Hsp104 family members together with sHsp, Hsp70, and/or Hsp40 congeners could be useful as disease-modifying treatments for NDDs [299].

Small molecule activators of Hsp110, or of any of its yeast and bacterial homologues, are unknown. Although enzyme inhibitors are more easily conceived and rationally optimized, protein activation is now becoming more accessible for medicinal chemists [300]. Small molecule activators of Hsp110, and maybe their combination with known small molecule modulators of sHsps and Hsp70 (either as co-administered compounds, or as hybrid dual action compounds), represent a meaningful goal for future research efforts.

6.4 RECAP

This chapter deals with small molecule modulators of neuropathological alterations related to the aggregation of tau, and to the development of tauopathies. Sixty compounds/scaffolds acting on tau as disease-modifying, anti-aggregating, and/or as disassembly agents are reported in Figures 6.1 to 6.15, and are briefly characterized in Table 6.2. The chemical core of each scaffold/compound is structurally defined; its target—tau alone, tau, and other amyloidogenic proteins—is mentioned; the developing laboratory (either public or private) is listed; and the development status—according to publicly available information—is finally provided.

TABLE 6.2   Compounds **6.1a–6.45** Chemical Class, Mechanism, Developing Organization, Development Status

Number	Chemical cpd./class	Target	Organization	Dev. status
6.1a	Methylene blue (MB)	Aβ, tau, huntingtin, TDP-43, prion	TauRX	Ph II
6.1b–d	Azure A–C	Aβ, tau, huntingtin, TDP-43, prion	TauRX	PE
6.3a	LMTX™	Tau, TDP-43, α-synuclein	TauRX	Ph III
6.4	Scyllo-inositol, ELND005, AZD-103	Aβ	Transition Therapeutics— Elan	Ph II
6.5	Alzhemed™, Vivimind™, tramiprosate	Aβ, tau	Neurochem	Ph III; Nu
6.6a,b	Clioquinol (PBT-1), PBT-2	Aβ, tau	Prana	Ph II
6.7	Exebryl-1	Aβ, tau	ProteoTech	Ph I
6.8a,b	EGCG, sunphenon™	Aβ, tau	Taiyo	Ph II, Nu
6.9	GSPE, Meganatural™	Aβ, tau	Polyphenolics	Ph II, Nu
6.10	Curcumin	Aβ, tau	John Douglas French Foundation	Ph II
6.11	Curcumin–Au nanoparticles	Aβ, tau	Natl. Brain Res. Centre, Manesar, India	DD
6.12	Curcumin– phospholipid conjugates	Aβ, tau	Pitiè Salpetriere, Paris	DD

TABLE 6.2 Compounds **6.1a–6.45** Chemical Class, Mechanism, Developing
Organization, Development Status *(cont.)*

Number	Chemical cpd./class	Target	Organization	Dev. status
6.13	Amine-containing curcumin analogues	Aβ, tau	Sun-Yat Sen Univ., China	DD
6.14	Sugar–curcumin conjugates	Aβ, tau	University of New York	LO
6.15	Curcumin-derived heterocycles	Aβ, tau	Max Planck, Hamburg	DD
6.16	Oleuropein	Aβ, tau	University of Florence, Italy	LO, Nu
6.17	Myricetin	Aβ, tau, α-synuclein	RIKEN	LO
6.18a,b	Baicalin (6.18a), baicalein (6.18b)	Aβ, α-synuclein	Tongji University, China	LO, Nu
6.19	Tannic acid	Aβ, tau	Tongji University, China	DD
6.20	Exifon	Aβ, tau	Tokyo University	DD
6.21	Emodin	Aβ, tau	Max Planck, Hamburg	DD
6.22a–c	Rheins	Aβ, tau	University of Barcelona	LO
6.23	Tolcapone	Aβ, tau, α-synuclein	Swiss Inst. Technology, Lausanne, CH	DD
6.24	Entacapone	Aβ, tau, α-synuclein	Swiss Inst. Technology, Lausanne, CH	DD
6.25	Memoquin	Aβ	Univ. of Bologna, Italy	PE
6.26	Tetrahydroperforin	Aβ, tau	Univ. Santiago, Chile	LO
6.27a,b	Hemin (6.27a), hematin (6.27b)	Aβ, tau, α-synuclein	Tokyo Institute of Psychiatry	DD
6.28	Phthalocyanine tetrasulfonate	Aβ, tau, α-synuclein, prion	Max Planck Inst., Goettingen and Bonn	DD
6.29a,b	Thioflavin S	Aβ, tau	Cancer Research, UK	LO
6.30a	Thioflavin T	Aβ, tau	University of Pittsburgh	PE
6.30b	Pittsburgh compound B	Aβ, tau	University of Pittsburgh	PE
6.31	Benzothiazolium cyanines	Aβ, tau	Ohio State University	PE

(Continued)

TABLE 6.2   Compounds **6.1a–6.45** Chemical Class, Mechanism, Developing Organization, Development Status *(cont.)*

Number	Chemical cpd./class	Target	Organization	Dev. status
6.32	Multivalent benzothiazole–benzothiazolium cyanines	Aβ, tau	Ohio State University	DD
6.33	BTA-EG₄	Aβ, tau	Georgetown University, Washington	LO
6.34	Benzothiazole–metal chelating hybrids	Aβ	Washington Univ., Missouri	DD
6.35	T-284	α-synuclein	Natl. Acad. Sci., Ukraine	DD
6.36	PBB-5	Tau	Mol. Imaging Center, Chiba, Japan	LO
6.37a–c	Rhodanins	Tau	Max Planck, Hamburg	LO
6.38	Thiohydantoins	Tau	Kyoto University	LO
6.39	Benzothiazolyl hydrazides	Tau	Max Planck, Hamburg	LO
6.40a,b	Pyrogalloyl phenylhydrazides	Aβ, tau	Max Planck, Hamburg	DD
6.41	N-Phenylamines	Tau	Max Planck, Hamburg	DD
6.42	Carvedilol	Aβ	Mount Sinai School of Medicine	MKTD
6.43	2,3-(Difuran-2-yl) quinoxalines	Tau	University of Pennsylvania	DD
6.44a–c	Aminothienopyrid-azines (ATPZs)	Tau	University of Pennsylvania	PE
6.45	Trehalose	Tau	Max Planck, Hamburg	LO

Not progressed, NP; early discovery, DD; lead optimization, LO; preclinical evaluation, PE; nutraceutical, Nu; clinical Phase I–II–III, Ph I–Ph III; marketed, MKTD.

References

[1] Kelly, J. W. The alternative conformations of amyloidogenic proteins and their multi-step assembly pathways. *Curr. Opin. Struct. Biol.* **1998**, *8*, 101–106.
[2] Chiti, F.; Dobson, C. M. Protein misfolding, functional amyloid, and human disease. *Annu. Rev. Biochem.* **2006**, *75*, 333–366.
[3] Sunde, M.; Blake, C. The structure of amyloid fibrils by electron microscopy and X-ray diffraction. *Adv. Protein Chem.* **1997**, *50*, 123–159.
[4] Sarkar, N.; Dubey, V. K. Exploring critical determinants of protein amyloidogenesis: a review. *J. Pept. Sci.* **2013**, *19*, 529–536.

[5] McClellan, A. J.; Tam, S.; Kaganovich, D.; Frydman, J. Protein quality control: chaperones culling corrupt conformations. *Nature Cell Biol.* **2005**, *7*, 736–741.

[6] Kaganovich, D.; Kopito, R.; Frydman, J. Misfolded proteins partition between two distinct quality control compartments. *Nature* **2008**, *454*, 1088–1095.

[7] Seneci, P. *Molecular targets in protein misfolding and neurodegenerative disease.* Academic Press, **2014**, 278 pages.

[8] Uversky, V. N. Mysterious oligomerization of the amyloidogenic proteins. *FEBS J.* **2010**, *277*, 2940–2953.

[9] Relini, A.; Marano, N.; Gliozzi, A. Misfolding of amyloidogenic proteins and their interactions with membranes. *Biomolecules* **2014**, *4*, 20–55.

[10] David, D. C. Aging and the aggregating proteome. *Front. Genet.* **2012**, *3*, 347.

[11] Pekar, A. H.; Frank, B. H. Conformation of proinsulin: a comparison of insulin and proinsulin self-association at neutral pH. *Biochemistry* **1972**, *11*, 4013–4016.

[12] Alford, J. R.; Kendrick, B. S.; Carpenter, J. F.; Randolph, T. W. High concentration formulations of recombinant human interleukin-1 receptor antagonist: II. Aggregation kinetics. *J. Pharm. Sci.* **2007**, *97*, 3005–3021.

[13] Invernizzi, G.; Papaleo, E.; Sabate, R.; Ventura, S. Protein aggregation: mechanisms and functional consequences. *Int. J. Biochem. Cell Biol.* **2012**, *44*, 1541–1554.

[14] Pedersen, J. T.; Heegaard, N. H. H. Analysis of protein aggregation in neurodegenerative disease. *Anal. Chem.* **2013**, *85*, 4215–4227.

[15] Sabate, R.; Gallardo, M.; Estelrich, J. An autocatalytic reaction as a model for the kinetics of the aggregation of beta-amyloid. *Biopolymers* **2003**, *71*, 190–195.

[16] Cohen, S. I. A.; Vendruscolo, M.; Welland, M. E.; Dobson, C. M.; Terentjev, E. M.; Knowles, T. P. J. Nucleated polymerization with secondary pathways. I. Time evolution of the principal moments. *J. Chem. Phys.* **2011**, *135*, 065105.

[17] Andersen, C. B.; Yagi, H.; Manno, M.; Martorana, V.; Ban, T.; Christiansen, G., et al. Branching in amyloid fibril growth. *Biophys. J.* **2009**, *96*, 1529–1536.

[18] Lorenzo, A.; Yankner, B. A. Beta-amyloid neurotoxicity requires fibril formation and is inhibited by Congo Red. *Proc. Natl. Acad. Sci. U.S.A.* **1994**, *91*, 12243–12247.

[19] Novitskaya, V.; Bocharova, O. V.; Bronstein, I.; Baskakov, I. V. Amyloid fibrils of mammalian prion protein are highly toxic to cultured cells and primary neurons. *J. Biol. Chem.* **2006**, *281*, 13828–13836.

[20] Ferreira, S. T.; Vieira, M. N.; De Felice, F. G. Soluble protein oligomers as emerging toxins in Alzheimer's and other amyloid diseases. *IUBMB Life* **2007**, *59*, 332–345.

[21] Bucciantini, M.; Rigacci, S.; Stefani, M. Amyloid aggregation: role of biological membranes and the aggregate–membrane system. *J. Phys. Chem. Lett.* **2014**, *5*, 517–527.

[22] Burke, K. A.; Yates, E. A.; Legleiter, J. Biophysical insights into how surfaces, including lipid membranes, modulate protein aggregation related to neurodegeneration. *Front. Neurol.* **2013**, *4*, 17.

[23] Brody, D. L.; Holtzman, D. M. Active and passive immunotherapy for neurodegenerative disorders. *Annu. Rev. Neurosci.* **2008**, *31*, 175–193.

[24] Ross, C. A.; Poirier, M. A. Protein aggregation and neurodegenerative disease. *Nat. Med.* **2004**, *10*, S10–S17.

[25] Lotz, G. P.; Legleiter, J. The role of amyloidogenic protein oligomerization in neurodegenerative disease. *J. Mol. Med.* **2013**, *91*, 653–664.

[26] Arriagada, P. V.; Growdon, J. H.; Hedley- Whyte, E. T.; Hyman, B. T. Neurofibrillary tangles but not senile plaques parallel duration and severity of Alzheimer's disease. *Neurology* **1992**, *42*, 631–639.

[27] Spires-Jones, T. L.; Kopeikina, K. J.; Koffie, R. M.; de Calignon, A.; Hyman, B. T. Are tangles as toxic as they look? *J. Mol. Neurosci.* **2011**, *45*, 438–444.

[28] Berger, Z.; Roder, H.; Hanna, A.; Carlson, A.; Rangachari, V.; Yue, M., et al. Accumulation of pathological tau species and memory loss in a conditional model of tauopathy. *J. Neurosci.* **2007**, *27*, 3650–3662.

[29] Wittmann, C. W.; Wszolek, M. F.; Shulman, J. M.; Salvaterra, P. M.; Lewis, J.; Hutton, M.; Feany, M. B. Tauopathy in drosophila: neurodegeneration without neurofibrillary tangles. *Science* **2001**, *293*, 711–714.

[30] Patterson, K. R.; Remmers, C.; Fu, Y.; Brooker, S.; Kanaan, N. M.; Vana, L., et al. Characterization of prefibrillar tau oligomers in vitro and in Alzheimer disease. *J. Biol. Chem.* **2011**, *286*, 23063–23076.

[31] Maeda, S.; Sahara, N.; Saito, Y.; Murayama, S.; Ikai, A.; Takashima, A. Increased levels of granular tau oligomers: an early sign of brain aging and Alzheimer's disease. *Neurosci. Res.* **2006**, *54*, 197–201.

[32] Lasagna-Reeves, C. A.; Castillo-Carranza, D. L.; Sengupta, U.; Sarmiento, J.; Troncoso, J.; Jackson, G. R.; Kayed, R. Identification of oligomers at early stages of tau aggregation in Alzheimer's disease. *FASEB J.* **2012**, *26*, 1946–1959.

[33] Wu, J. W.; Herman, M.; Liu, L.; Simoes, S.; Acker, C. M.; Figueroa, H., et al. Small misfolded Tau species are internalized via bulk endocytosis and anterogradely and retrogradely transported in neurons. *J. Biol. Chem.* **2013**, *288*, 1856–1870.

[34] Michel, C. H.; Kumar, S.; Pinotsi, D.; Tunnacliffe, A.; St. George-Hyslop, P.; Mandelkow, E., et al. Extracellular monomeric tau protein is sufficient to initiate the spread of tau protein pathology. *J. Biol. Chem.* **2014**, *289*, 956–967.

[35] de Calignon, A.; Polydoro, M.; Suárez-Calvet, M.; William, C.; Adamowicz, D. H.; Kopeikina, K. J., et al. Propagation of tau pathology in a model of early Alzheimer's disease. *Neuron* **2012**, *73*, 685–697.

[36] Clavaguera, F.; Akatsu, H.; Fraser, G.; Crowther, R. A.; Frank, S.; Hench, J., et al. Brain homogenates from human tauopathies induce tau inclusions in mouse brain. *Proc. Natl. Acad. Sci. U.S.A.* **2013**, *110*, 9535–9540.

[37] Bulic, B.; Pickhardt, M.; Mandelkow, E. Progress and developments in tau aggregation inhibitors for Alzheimer disease. *J. Med. Chem.* **2013**, *56*, 4135–4155.

[38] Wischik, C. M.; Harrington, C. R.; Storey, J. M. D. Tau-aggregation inhibitor therapy for Alzheimer's disease. *Biochem. Pharmacol.* **2014**, *88*, 529–539.

[39] Bulic, B.; Pickhardt, M.; Schmidt, B.; Mandelkow, E. -M.; Waldmann, H.; Mandelkow, E. Development of tau aggregation inhibitors for Alzheimer's Disease. *Angew. Chem. Int. Ed.* **2009**, *48*, 1740–1752.

[40] Ohlow, M. J.; Moosmann, B. Phenothiazine: the seven lives of pharmacology's first lead structure. *Drug Discov. Today* **2011**, *16*, 119–131.

[41] Schirmer, R. H.; Adler, H.; Pickhardt, M.; Mandelkow, E. Lest we forget you—methylene blue.... *Neurobiol. Aging* **2011**, *32*, 2325e7–2325e16.

[42] Wischik, C. M.; Bentham, P.; Wischik, D. J.; Seng, K. M. Tau aggregation inhibitor (TAI) therapy with rember™ arrests disease progression in mild and moderate Alzheimer's disease over 50 weeks. *Alzheimer's Dementia* **2008**, *4*, T167.

[43] http://taurx.com/clinical-data.html.

[44] Atamna, H.; Nguyen, A.; Schultz, C.; Boyle, K.; Newberry, J.; Kato, H.; Ames, B. N. Methylene blue delays cellular senescence and enhances key mitochondrial biochemical pathways. *FASEB J.* **2008**, *22*, 703–712.

[45] http://taurx.com/1st-and-2nd-gen-tais.html.

[46] http://clinicaltrials.gov/ct2/results?term=TRx0237&Search=Search.

[47] Wischik, C. M. TauRx global Phase 3 trial in Alzheimer's disease with tau aggregation inhibitor LMTX. *Neurobiol. Aging* **2014**, *35*, S26.

[48] Wischik, C. M.; Edwards, P. C.; Lai, R. Y.; Roth, M.; Harrington, C. R. Selective inhibition of Alzheimer disease-like tau aggregation by phenothiazines. *Proc. Natl. Acad. Sci. U.S.A.* **1996**, *93*, 11213–11218.

[49] Taniguchi, S.; Suzuki, N.; Masuda, M.; Hisanaga, S.; Iwatsubo, T.; Goedert, M.; Hasegawa, M. Inhibition of heparin-induced tau filament formation by phenothiazines, polyphenols, and porphyrins. *J. Biol. Chem.* **2005**, *280*, 7614–7623.

[50] Hattori, M.; Sugino, E.; Minoura, K.; In, Y.; Sumida, M.; Taniguchi, T., et al. Different inhibitory response of cyanidin and methylene blue for filament formation of tau microtubule-binding domain. *Biochem. Biophys. Res. Commun.* **2008**, *374*, 158–163.

[51] Akoury, E.; Pickhardt, M.; Gajda, M.; Biernat, J.; Mandelkow, E.; Zweckstetter, M. Mechanistic basis of phenothiazine-driven inhibition of tau aggregation. *Angew. Chem. Int. Ed.* **2013**, *52*, 3511–3515.

[52] Crowe, A.; James, M. J.; Lee, V. M. -Y.; Smith, A. B., III.; Trojanowski, J. Q.; Ballatore, C.; Brunden, K. R. Aminothienopyridazines and methylene blue affect tau fibrillization via cysteine oxidation. *J. Biol. Chem.* **2013**, *288*, 11024–11037.

[53] Oz, M.; Lorke, D. E.; Hasan, M.; Petroianu, G. A. Cellular and molecular actions of methylene blue in the nervous system. *Med. Res. Rev.* **2011**, *31*, 93–117.

[54] Masaki, E.; Kondo, I. Methylene blue, a soluble guanylyl cyclase inhibitor, reduces the sevoflurane minimum alveolar anesthetic concentration and decreases the brain cyclic guanosine monophosphate content in rats. *Anesth. Analg.* **1999**, *89*, 484–489.

[55] Mayer, B.; Brunner, F.; Schmidt, K. Inhibition of nitric oxide synthesis by methylene blue. *Biochem. Pharmacol.* **1993**, *45*, 367–374.

[56] Pfaffendorf, M.; Bruning, T. A.; Batnik, H. D.; van Zwieten, P. A. The interaction between methylene blue and the cholinergic system. *Br. J. Pharmacol.* **1997**, *122*, 95–98.

[57] Buchholz, K.; Schirmer, R. H.; Eubel, J. K.; Akoachere, M. B.; Dandekar, T.; Becker, K.; Gromer, S. Interactions of methylene blue with human disulfide reductases and their orthologues from *Plasmodium falciparum*. *Antimicrob. Agents Chemother.* **2008**, *52*, 183–191.

[58] Ramsay, R. R.; Dunford, C.; Gillman, P. K. Methylene blue and serotonin toxicity: inhibition of monoamine oxidase A (MAO A) confirms a theoretical prediction. *Br. J. Pharmacol.* **2007**, *152*, 946–951.

[59] Harvey, B. H.; Duvenhage, I.; Viljoen, F.; Scheepers, N.; Malan, S. F.; Wegener, G., et al. Role of monoamine oxidase, nitric oxide synthase and regional brain monoamines in the antidepressant-like effects of methylene blue and selected structural analogues. *Biochem. Pharmacol.* **2010**, *80*, 1580–1591.

[60] Chies, A. B.; Custódio, R. C.; de Souza, G. L.; Correa, F. M. A.; Pereira, O. C. M. Pharmacological evidence that methylene blue inhibits noradrenaline neuronal uptake in the vas deferens. *Pol. J. Pharmacol.* **2003**, *55*, 573–579.

[61] Culo, F.; Sabolovic, D.; Somogyi, L.; Marusic, M.; Berbiguier, N.; Galey, L. Antitumoral and antiinflammatory effects of biological stains. *Agents Actions* **1991**, *34*, 424–428.

[62] Atamna, H.; Kumar, R. Protective role of methylene blue in Alzheimer's disease via mitochondria and cytochrome *c* oxidase. *J. Alzheimer's Dis.* **2010**, *20*, S439–S452.

[63] http://taurx.com/ad-development-history.html.

[64] Petzer, A.; Harvey, B. H.; Petzer, J. P. The interactions of azure B, a metabolite of methylene blue, with acetylcholinesterase and butyrylcholinesterase. *Toxicol. Applied Pharmacol.* **2014**, *274*, 488–493.

[65] Gaudette, N. F.; Lodge, J. W. Determination of methylene blue and leucomethylene blue in male and female Fischer 344 rat urine and B6C3F1 mouse urine. *J. Anal. Toxicol.* **2005**, *29*, 28–33.

[66] Warth, A.; Goeppert, B.; Bopp, C.; Schirmacher, P.; Flechtenmacher, C.; Burhenne, J. Turquoise to dark green organs at autopsy. *Virchows Arch.* **2009**, *454*, 341–344.

[67] Walter-Sack, I.; Rengelshausen, J.; Oberwittler, H.; Burhenne, J.; Müller, O.; Meissner, P.; Mikus, G. High absolute bioavailability of methylene blue given as an aqueous oral formulation. *Eur. J. Clin. Pharmacol.* **2009**, *65*, 179–189.

[68] Oz, M.; Lorke, D. E.; Petroianu, G. A. Methylene blue and Alzheimer's disease. *Biochem. Pharmacol.* **2009**, *78*, 927–932.

[69] Peter, C.; Hongwan, D.; Kupfer, A.; Lauterburg, B. H. Pharmacokinetics and organ distribution of intravenous and oral methylene blue. *Eur. J. Clin. Pharmacol.* **2000**, *56*, 247–250.

[70] Burrows, G. E. Methylene blue: effects and disposition in sheep. *J. Vet. Pharmacol. Ther.* **1984**, *7*, 225–231.

[71] DiSanto, A. R.; Wagner, J. G. Pharmacokinetics of highly ionized drugs. II. Methylene blue-absorption, metabolism, and excretion in man and dog after oral administration. *J. Pharm. Sci.* **1972**, *61*, 1086–1090.

[72] Clifton, I. I. J.; Leikin, J. B. Methylene blue. *Am. J. Ther.* **2003**, *10*, 289–291.

[73] Rojas, J. C.; Bruchey, A. K.; Gonzalez-Lima, F. Neurometabolic mechanisms for memory enhancement and neuroprotection of methylene blue. *Progr. Neurobiol.* **2012**, *96*, 32–45.

[74] Martinez, J., Jr.; Jensen, R. A.; Vasquez, B.; McGuiness, T.; McGaugh, J. L. Methylene blue alters retention of inhibitory avoidance responses. *Physiol. Psychol.* **1978**, *6*, 387–390.

[75] Bruchey, A. K.; Gonzalez-Lima, F. Behavioral, physiological and biochemical hormetic responses to the autoxidizable dye methylene blue. *Am. J. Pharm. Toxicol.* **2008**, *3*, 72–79.

[76] Callaway, N. L.; Riha, P. D.; Bruchey, A. K.; Munshi, Z.; Gonzalez-Lima, F. Methylene blue improves brain oxidative metabolism and memory retention in rats. *Pharmacol. Biochem. Behav.* **2004**, *77*, 175–181.

[77] Callaway, N. L.; Riha, P. D.; Wrubel, K. M.; McCollum, D.; Gonzalez-Lima, F. Methylene blue restores spatial memory retention impaired by an inhibitor of cytochrome oxidase in rats. *Neurosci. Lett.* **2002**, *332*, 83–86.

[78] Riha, P. D.; Rojas, J. C.; Gonzalez-Lima, F. Beneficial network effects of methylene blue in an amnestic model. *Neuroimage* **2011**, *54*, 2623–2634.

[79] Gonzalez-Lima, F.; Bruchey, A. K. Extinction memory improvement by the metabolic enhancer methylene blue. *Learn. Mem.* **2004**, *11*, 633–640.

[80] Wrubel, K. M.; Barrett, D.; Shumake, J.; Johnson, S. E.; Gonzalez-Lima, F. Methylene blue facilitates the extinction of fear in an animal model of susceptibility to learned helplessness. *Neurobiol. Learn. Mem.* **2007**, *87*, 209–217.

[81] Riha, P. D.; Bruchey, A. K.; Echevarria, D. J.; Gonzalez-Lima, F. Memory facilitation by methylene blue: dose-dependent effect on behavior and brain oxygen consumption. *Eur. J. Pharmacol.* **2005**, *511*, 151–158.

[82] Deiana, S.; Harrington, C. R.; Wischik, C. M.; Riedel, G. Methylthioninium chloride reverses cognitive deficits induced by scopolamine: comparison with rivastigmine. *Psychopharmacology* **2009**, *202*, 53–65.

[83] Pelgrims, J.; De Vos, F.; Van den Brande, J.; Schrijvers, D.; Prove, A.; Vermorken, J. B. Methylene blue in the treatment and prevention of ifosfamide-induced encephalopathy: report of 12 cases and a review of the literature. *Br. J. Cancer* **2000**, *82*, 291–294.

[84] Rojas, J. C.; John, J. M.; Lee, J.; Gonzalez-Lima, F. Methylene blue provides behavioral and metabolic neuroprotection against optic neuropathy. *Neurotox. Res.* **2009**, *15*, 260–273.

[85] Wen, Y.; Li, W.; Poteet, E. C.; Xie, L.; Tan, C.; Yan, L. J., et al. Alternative mitochondrial electron transfer as a novel strategy for neuroprotection. *J. Biol. Chem.* **2011**, *286*, 16504–16515.

[86] Rojas, J. C.; Simola, N.; Kermath, B. A.; Kane, J. R.; Schallert, T.; Gonzalez-Lima, F. Striatal neuroprotection with methylene blue. *Neuroscience* **2009**, *163*, 877–889.

[87] Furian, A. F.; Fighera, M. R.; Oliveira, M. S.; Ferreira, A. P.; Fiorenza, N. G.; de Carvalho Myskiw, J., et al. Methylene blue prevents methylmalonate-induced seizures and oxidative damage in rat striatum. *Neurochem. Int.* **2007**, *50*, 164–171.

[88] Dickey, C. A.; Ash, P.; Klosak, N.; Lee, W. C.; Petrucelli, L.; Hutton, M.; Eckman, C. B. Pharmacologic reductions of total tau levels; implications for the role of microtubule dynamics in regulating tau expression. *Mol. Neurobiol.* **2006**, *1*, 6.

[89] Jinwal, U. K.; Miyata, Y.; Koren, J., III.; Jones, J. R.; Trotter, J. H.; Chang, L., et al. Chemical manipulation of Hsp70 ATPase activity regulates tau stability. *J. Neurosci.* **2009**, *29*, 12079–12088.

[90] Thompson, A. D.; Scaglione, K. M.; Prensner, J.; Gillies, A. T.; Chinnaiyan, A.; Paulson, H. L., et al. Analysis of the tau-associated proteome reveals that exchange of Hsp70 for Hsp90 is involved in tau degradation. *ACS Chem. Biol.* **2012**, *7*, 1677–1686.

[91] Miyata, Y.; Rauch, J. N.; Jinwal, U. K.; Thompson, A. D.; Srinivasan, S.; Dickey, C. A.; Gestwicki, J. E. Cysteine reactivity distinguishes redox sensing by the heat-inducible and constitutive forms of heat shock protein 70. *Chem. Biol.* **2012**, *19*, 1391–1399.

[92] Medina, D. X.; Caccamo, A.; Oddo, S. Methylene blue reduces abeta levels and rescues early cognitive deficit by increasing proteasome activity. *Brain Pathol.* **2011**, *21*, 140–149.

[93] Congdon, E. E.; Wu, J. W.; Myeku, N.; Figueroa, Y. H.; Herman, M.; Marinec, P. S., et al. Methylthioninium chloride (methylene blue) induces autophagy and attenuates tauopathy in vitro and in vivo. *Autophagy* **2012**, *8*, 609–622.

[94] Xie, L.; Li, W.; Winters, A.; Yuan, F.; Jin, K.; Yang, S. Methylene blue induces macroautophagy through 5′-adenosine monophosphate-activated protein kinase pathway to protect neurons from serum deprivation. *Front. Cell. Neurosci.* **2013**, *7*, 56.

[95] Wischik, C. M.; Wischik, D. J.; Storey, J. M. D.; Harrington, C. R. Rationale for Tau-aggregation inhibitor therapy in Alzheimer's disease and other tauopathies. *RSC Drug Discovery Series* **2010**, *2*, 210–232.

[96] van Bebber, F.; Paquet, D.; Hruscha, A.; Schmid, B.; Haass, C. Methylene blue fails to inhibit Tau and polyglutamine protein dependent toxicity in zebrafish. *Neurobiol. Dis.* **2010**, *39*, 265–271.

[97] O'Leary, J.; Li, Q.; Marinec, P.; Blair, L.; Congdon, E.; Johnson, A. G., et al. Phenothiazine mediated rescue of cognition in tau transgenic mice requires neuroprotection and reduced soluble tau burden. *Mol. Neurodegener.* **2010**, *5*, 45.

[98] Hosokawa, M.; Arai, T.; Masuda-Suzukake, M.; Nonaka, T.; Yamashita, M.; Akiyama, H.; Hasegawa, M. Methylene blue reduced abnormal tau accumulation in P301L tau transgenic mice. *PLoS ONE* **2012**, *7*, e52389.

[99] Spires-Jones, T. L.; Friedman, T.; Pitstick, R.; Polydoro, M.; Roe, A.; Carlson, G. A.; Hyman, B. T. Methylene blue does not reverse existing neurofibrillary tangle pathology in the rTg4510 mouse model of tauopathy. *Neurosci. Lett.* **2014**, *562*, 63–68.

[100] Polydoro, M.; de Calignon, A.; Suarez-Calvet, M.; Sanchez, L.; Kay, K. R.; Nicholls, S. B., et al. Reversal of neurofibrillary tangles and tau-associated phenotype in the rTg-TauEC model of early Alzheimer's disease. *J. Neurosci.* **2013**, *33*, 13300–13311.

[101] Necula, M.; Kayed, R.; Milton, S.; Glabe, C. G. Small molecule inhibitors distinguish between amyloid β oligomerization and fibrillization pathways. *J. Biol. Chem.* **2007**, *282*, 10311–10324.

[102] Necula, M.; Breydo, L.; Milton, S.; Kayed, R.; van der Veer, W. E.; Tone, P.; Glabe, C. G. Methylene blue inhibits amyloid Aβ oligomerization by promoting fibrillization. *Biochemistry* **2007**, *46*, 8850–8860.

[103] Ladiwala, A. R. A.; Dordick, J. S.; Tessier, P. M. Aromatic small molecules remodel toxic soluble oligomers of amyloid β through three independent pathways. *J. Biol. Chem.* **2011**, *286*, 3209–3218.

[104] Zovo, K.; Helk, E.; Karafin, A.; Tougu, V.; Palumaa, P. Label-free high-throughput screening assay for inhibitors of Alzheimer's amyloid-b peptide aggregation based on MALDI MS. *Analyt. Chem.* **2010**, *82*, 8558–8565.

[105] Noormägi, A.; Primar, K.; Tõugu, V.; Palumaa, P. Interference of low-molecular substances with the thioflavin-T fluorescence assay of amyloid fibrils. *J. Pept. Sci.* **2012**, *18*, 59–64.

[106] Irwin, J. A.; Wong, H. E.; Kwon, I. Different fates of Alzheimer's disease amyloid-β fibrils remodeled by biocompatible small molecules. *Biomacromolecules* **2013**, *14*, 264–274.

[107] Paban, V.; Manrique, C.; Filali, M.; Maunoir-Regimbal, S.; Fauvelle, F.; Alescio-Lautier, B. Therapeutic and preventive effects of methylene blue on Alzheimer's disease pathology in a transgenic mouse model. *Neuropharmacology* **2014**, *76*, 68–79.

[108] Mitchell Sontag, E.; Lotz, G. P.; Agrawal, N.; Tran, A.; Aron, R.; Yang, G., et al. Methylene blue modulates huntingtin aggregation intermediates and is protective in Huntington's disease models. *J. Neurosci.* **2012**, *32*, 11109–11119.

[109] Cavaliere, P.; Torrent, J.; Prigent, S.; Granata, V.; Pauwels, K.; Pastore, A., et al. Binding of methylene blue to a surface cleft inhibits the oligomerization and fibrillization of prion protein. *Biochim. Biophys. Acta* **1832**, *2013*, 20–28.

[110] Korth, C.; May, B. C.; Cohen, F. E.; Prusiner, S. B. Acridine and phenothiazine derivatives as pharmacotherapeutics for prion disease. *Proc. Natl. Acad. Sci. U.S.A.* **2001**, *98*, 9836–9841.

[111] Masuda, M.; Suzuki, N.; Taniguchi, S.; Oikawa, T.; Nonaka, T.; Iwatsubo, T., et al. Small molecule inhibitors of α-synuclein filament assembly. *Biochemistry* **2006**, *45*, 6085–6094.

[112] Yamashita, M.; Nonaka, T.; Arai, T.; Kametani, F.; Buchman, V. L.; Ninkina, N., et al. Methylene blue and dimebon inhibit aggregation of TDP-43 in cellular models. *FEBS Lett.* **2009**, *583*, 2419–2424.

[113] Vaccaro, A.; Patten, S. A.; Ciura, S.; Maios, C.; Therrien, M.; Drapeau, P., et al. Methylene blue protects against TDP-43 and FUS neuronal toxicity in *C. elegans* and *D. rerio*. *PLoS ONE* **2012**, *7*, e42117.

[114] Audet, J. N.; Soucy, G.; Julien, J. -P. Methylene blue administration fails to confer neuroprotection in two amyotrophic lateral sclerosis mouse models. *Neuroscience* **2012**, *209*, 136–143.

[115] Dibaj, P.; Zschuntzsch, J.; Steffens, H.; Scheffel, J.; Goricke, B.; Weishaupt, J. H., et al. Influence of methylene blue on microglia-induced inflammation and motor neuron degeneration in the SOD1G93A model for ALS. *PLoS ONE* **2012**, *7*, e43963.

[116] Lougheed, R.; Turnbull, J. Lack of effect of methylene blue in the SOD1 G93A mouse model of amyotrophic lateral sclerosis. *PLoS ONE* **2011**, *6*, e23141.

[117] Herrmann, N.; Chau, S. A.; Kircanski, I.; Lanctot, K. L. Current and emerging drug treatment options for Alzheimer's disease. *Drugs* **2011**, *71*, 2031–2065.

[118] Reitz, C. Alzheimer's disease and the amyloid cascade hypothesis: a critical review. *Int. J. Alzheimer's Dis.* **2012**, 369808.

[119] Salloway, S.; Sperling, R.; Keren, R.; Porsteinsson, A. P.; van Dyck, C. H.; Tariot, P. N., et al. A phase 2 randomized trial of ELND005, scyllo-inositol, in mild to moderate Alzheimer disease. *Neurology* **2011**, *77*, 1253–1262.

[120] Aisen, P. S.; Gauthier, S.; Ferris, S. H.; Saumier, D.; Haine, D.; Garceau, D., et al. Tramiprosate in mild-to-moderate Alzheimer's disease—a randomized, double-blind, placebo-controlled, multi-centre study (the Alphase Study). *Arch. Med. Sci.* **2011**, *7*, 102–111.

[121] Mullane, K.; Williams, M. Alzheimer's therapeutics: continued clinical failures question the validity of the amyloid hypothesis—but what lies beyond? *Biochem. Pharmacol.* **2013**, *85*, 289–305.

[122] http://www.bellushealth.com/English/news/news-releases/News-Release-Details/2013/BELLUS-Health-Out-Licenses-VIVIMIND-BLU8499-and-Its-Analogs/default.aspx.

[123] Santa-Maria, I.; Hernández, F.; Del Rio, J.; Moreno, F. J.; Avila, J. Tramiprosate, a drug of potential interest for the treatment of Alzheimer's disease, promotes an abnormal aggregation of tau. *Mol. Neurodeg.* **2007**, *2*, 17.

[124] Adlard, P. A.; Cherny, R. A.; Finkelstein, D. I.; Gautier, E.; Robb, E.; Cortes, M., et al. Rapid restoration of cognition in Alzheimer's transgenic mice with 8-hydroxy quinoline analogs is associated with decreased interstitial Abeta. *Neuron* **2008**, *59*, 43–55.

[125] Adlard, P. A.; Bush, A. I. Metal chaperones: a holistic approach to the treatment of Alzheimer's disease. *Fr. Psychiatry* **2012**, *3*, 15.

[126] Lannfelt, L.; Blennow, K.; Zetterberg, H.; Batsman, S.; Ames, D.; Harrison, J., et al. Safety, efficacy, and biomarker findings of PBT2 in targeting Abeta as a modifying therapy for Alzheimer's disease: a phase IIa, double-blind, randomised, placebo-controlled trial. *Lancet Neurol.* **2008**, *7*, 779–786.

[127] Crouch, P. J.; Savva, M. S.; Hung, L. W.; Donnelly, P. S.; Mot, A. I.; Parker, S. J., et al. The Alzheimer's therapeutic PBT2 promotes amyloid-beta degradation and GSK3 phosphorylation via a metal chaperone activity. *J. Neurochem.* **2011**, *119*, 220–230.

[128] Snow, A. D.; Cummings, J.; Lake, T.; Hu, Q.; Esposito, L.; Cam, J., et al. Exebryl-1: a novel small molecule currently in human clinical trials as a disease-modifying drug for the treatment of Alzheimer's disease. *Alzheimer's Dementia* **2009**, *5*, P418.

[129] http://www.proteotech.com/pipeline/pipeline_overview.html.

[130] Ono, K.; Yoshiike, Y.; Takashima, A.; Hasegawa, K.; Naiki, H.; Yamada, M. Potent antiamyloidogenic and fibril destabilizing effects of polyphenols in vitro: implications for the prevention and therapeutics of Alzheimer's disease. *J. Neurochem.* **2003**, *87*, 172–181.

[131] http://clinicaltrials.gov/ct2/show/NCT00951834?term=egcg&cond=%22Alzheimer+ Disease%22&rank=1&submit_fld_opt=. Charite University, Berlin, Germany. Sunphenon EGCg (Epigallocatechin-Gallate) in the early stage of Alzheimer's disease (SUN-AK).

[132] http://www.sunphenon.com/about-green-tea/what-green-tea/.

[133] Rezai-Zadeh, K.; Shytle, D.; Sun, N.; Mori, T.; Hou, H.; Jeanniton, D., et al. Green tea epigallocatechin-3-gallate (EGCG) modulates amyloid precursor protein cleavage and reduces cerebral amyloidosis in Alzheimer transgenic mice. *J. Neurosci.* **2005**, *25*, 8807–8814.

[134] Rezai-Zadeh, K.; Arendash, G. W.; Hou, H.; Fernandez, F.; Jensen, M.; Runfeldt, M., et al. Green tea epigallocatechin-3-gallate (EGCG) reduces β-amyloid mediated cognitive impairment and modulates tau pathology in Alzheimer transgenic mice. *Brain Res.* **2008**, *1214*, 177–187.

[135] Ehrnhoefer, D. E.; Bieschke, J.; Boeddrich, A.; Herbst, M.; Masino, L.; Lurz, R., et al. EGCG redirects amyloidogenic polypeptides into unstructured, off-pathway oligomers. *Nat. Struct. Mol. Biol.* **2008**, *15*, 558–566.

[136] Bieschke, J.; Russ, J.; Friedrich, R. P.; Ehrnhoefer, D. E.; Wobst, H.; Neugebauer, K.; Wanker, E. E. EGCG remodels mature alpha-synuclein and amyloid-beta fibrils and reduces cellular toxicity. *Proc. Natl. Acad. Sci. U.S.A.* **2010**, *107*, 7710–7715.

[137] Palhano, F. L.; Lee, J.; Grimster, N. P.; Kelly, J. W. Toward the molecular mechanism(s) by which EGCG treatment remodels mature amyloid fibrils. *J. Am. Chem. Soc.* **2013**, *135*, 7503–7510.

[138] Hyunga, S. -J.; DeTomaa, A. S.; Brendera, J. R.; Leec, S.; Vivekanandana, S.; Kochia, A., et al. Insights into antiamyloidogenic properties of the green tea extract (−)-epigallocatechin-3-gallate toward metal-associated amyloid-β species. *Proc. Natl. Acad. Sci. U.S.A.* **2013**, *110*, 3743–3748.

[139] Wang, J.; Ho, L.; Zhao, W.; Ono, K.; Rosensweig, C.; Chen, L., et al. Grape-derived polyphenolics prevent Abeta oligomerization and attenuate cognitive deterioration in a mouse model of Alzheimer's disease. *J. Neurosci.* **2008**, *28*, 6388–6392.

[140] http://www.polyphenolics.com/products/meganatural-bp/.

[141] http://clinicaltrials.gov/ct2/show/NCT02033941?term=grape+seed&cond=%22Alz heimer+Disease%22&rank=1.

[142] Ono, K.; Condron, M. M.; Ho, L.; Wang, J.; Zhao, W.; Pasinetti, G. M.; Teplow, D. B. Effects of grape seed derived polyphenols on amyloid beta protein self assembly and cytotoxicity. *J. Biol. Chem.* **2008**, *283*, 32176–32187.

[143] Ho, L.; Yemul, S.; Wang, J.; Pasinetti, G. M. Grape seed polyphenolic extract as a potential novel therapeutic agent in tauopathies. *J. Alzheimer's Dis.* **2009**, *16*, 433–439.

[144] Wang, J.; Santa-Maria, I.; Ho, L.; Ksiezak-Reding, H.; Ono, K.; Teplow, D. B.; Pasinetti, G. M. Grape derived polyphenols attenuate tau neuropathology in a mouse model of Alzheimer's disease. *J. Alzheimer's Dis.* **2010**, *22*, 653–661.

[145] Santa-Maria, I.; Diaz-Ruiz, C.; Ksiezak-Reding, H.; Chen, A.; Ho, L.; Wang, J.; Pasinetti, G. M. GSPE interferes with tau aggregation in vivo: implication for treating tauopathy. *Neurobiol. Aging* **2012**, *33*, 2072–2081.

[146] Ksiezak-Reding, H.; Ho, L.; Santa-Maria, I.; Diaz-Ruiz, C.; Wang, J.; Pasinetti, G. M. Ultrastructural alterations of Alzheimer's disease paired helical filaments by grape seed-derived polyphenols. *Neurobiol. Aging* **2012**, *33*, 1427–1439.

[147] Gupta, S. C.; Patchva, S.; Koh, W.; Aggarwal, B. B. Discovery of curcumin, a component of golden spice, and its miraculous biological activities. *Clin. Exp. Pharm. Physiol.* **2012**, *39*, 283–299.

[148] Gupta, S. C.; Patchva, S.; Aggarwal, B. B. Therapeutic roles of curcumin: lessons learned from clinical trials. *AAPS J.* **2012**, *15*, 195–218.

[149] Anand, P.; Kunnumakkara, A. B.; Newman, R. A.; Aggarwal, B. B. Bioavailability of curcumin: problems and promises. *Mol. Pharm.* **2007**, *4*, 807–818.

[150] Potter, P. E. Curcumin: a natural substance with potential efficacy in Alzheimer's disease. *J. Exper. Pharmacol.* **2013**, *5*, 23–31.

[151] Baum, L.; Lam, C. W.; Cheung, S. K.; Kwok, T.; Lui, V.; Tsoh, J., et al. Six-month randomized, placebo-controlled, double-blind, pilot clinical trial of curcumin in patients with Alzheimer disease. *J. Clin. Psychopharmacol.* **2008**, *28*, 110–113.

[152] Ringman, J. M.; Frautschy, S. A.; Teng, E.; Begum, A. N.; Bardens, J.; Beigi, M., et al. Oral curcumin for Alzheimer's disease: tolerability and efficacy in a 24-week randomized, double blind, placebo-controlled study. *Alzheimer's Res. Ther.* **2012**, *4*, 43.

[153] Hoppe, J. B.; Coradini, K.; Frozza, R. L.; Oliveira, C. M.; Meneghetti, A. B.; Bernardi, A., et al. Free and nanoencapsulated curcumin suppress β-amyloid-induced cognitive impairments in rats: involvement of BDNF and Akt/GSK-3β signaling pathway. *Neurobiol. Learning Memory* **2013**, *106*, 134–144.

[154] Hamaguchi, T.; Ono, K.; Yamada, M. Curcumin and Alzheimer's disease. *CNS Neurosci. Therap.* **2010**, *16*, 285–297.

[155] Ono, K.; Hasegawa, K.; Naiki, H.; Yamada, M. Curcumin has potent antiamyloidogenic effects for Alzhemier's β-amyloid fibrils in vitro. *J. Neurosci. Res.* **2004**, *75*, 742–750.

[156] Yang, F.; Lim, G. P.; Begum, A. N.; Ubeda, O. J.; Simmons, M. R.; Ambegaokar, S. S., et al. Curcumin inhibits formation of amyloid β oligomers and fibrils, binds plaques, and reduces amyloid in vivo. *J. Biol. Chem.* **2005**, *280*, 5892–5901.

[157] Jameson, L. P.; Smith, N. W.; Dzyuba, S. V. Dye-binding assays for evaluation of the effects of small molecule inhibitors on amyloid (Aβ) self-assembly. *ACS Chem. Neurosci.* **2012**, *3*, 807–819.

[158] Kumaraswamy, P.; Sethuraman, S.; Krishnan, U. M. Mechanistic insights of curcumin interactions with the core-recognition motif of β-amyloid peptide. *J. Agric. Food Chem.* **2013**, *61*, 3278–3285.

[159] Singh Mithu, V.; Sarkar, B.; Bhowmik, D.; Kant Das, A.; Chandrakesan, M.; Maiti, S.; Madhu, P. K. Curcumin alters the salt bridge-containing turn region in amyloid β (1-42) aggregates. *J. Biol. Chem.* **2014**, *289*, 11122–11131.

[160] Thapa, A.; Vernon, B. C.; De la Pena, K.; Soliz, G.; Moreno, H. A.; Lopez, G. P.; Chi, E. Y. Membrane-mediated neuroprotection by curcumin from Amyloid-β-peptide-induced toxicity. *Langmuir* **2013**, *29*, 11713–11723.

[161] Patil, S. P.; Tran, N.; Geekiyanage, H.; Liu, L.; Chan, C. Curcumin-induced upregulation of the antitau cochaperone BAG2 in primary rat cortical neurons. *Neurosci. Lett.* **2013**, *554*, 121–125.

[162] Frautschy, S. A.; Hu, W.; Kim, P.; Miller, S. A.; Chu, T.; Harris-White, M. E.; Cole, G. M. Phenolic antiinflammatory antioxidant reversal of Abeta-induced cognitive deficits and neuropathology. *Neurobiol. Aging* **2001**, *22*, 993–1005.

[163] Lim, G. P.; Chu, T.; Yang, F.; Beech, W.; Frautschy, S. A.; Cole, G. M. The curry spice curcumin reduces oxidative damage and amyloid pathology in an Alzheimer transgenic mouse. *J. Neurosci.* **2001**, *21*, 8370–8377.

[164] Begum, A. N.; Jones, M. R.; Lim, G. P.; Morihara, T.; Kim, P.; Heath, D. D., et al. Curcumin structure-function, bioavailability, and efficacy in models of neuroinflammation and Alzheimer's disease. *J. Pharmacol. Exp. Ther.* **2008**, *326*, 196–208.

[165] Hamaguchi, T.; Ono, K.; Murase, A.; Yamada, M. Phenolic compounds prevent Alzheimer's pathology through different effects on the Amyloid-β aggregation pathway. *Am. J. Pathol.* **2009**, *175*, 2557–2565.

[166] Garcia-Alloza, M.; Borrelli, L. A.; Rozkalne, A.; Hyman, B. T.; Bacskai, B. J. Curcumin labels amyloid pathology in vivo, disrupts existing plaques, and partially restores distorted neurites in an Alzheimer mouse model. *J. Neurochem.* **2007**, *102*, 1095–1104.

[167] Wang, P.; Su, C.; Li, R.; Wang, H.; Ren, Y.; Sun, H., et al. Mechanisms and effects of curcumin on spatial learning and memory improvement in APPswe/PS1dE9 mice. *J. Neurosci. Res.* **2014**, *92*, 218–231.

[168] Ma, Q. -L.; Yang, F.; Rosario, E. R.; Ubeda, O. J.; Beech, W.; Gant, D. J., et al. β-Amyloid oligomers induce phosphorylation of tau and inactivation of insulin receptor substrate via c-Jun N-terminal kinase signaling: suppression by omega-3 fatty acids and curcumin. *J. Neurosci.* **2009**, *29*, 9078–9089.

[169] Alavez, S.; Vantipalli, M. C.; Zucker, D. J.; Klang, I. M.; Lithgow, G. J. Amyloid-binding compounds maintain protein homeostasis during ageing and extend lifespan. *Nature* **2011**, *472*, 226–229.

[170] Caesar, I.; Jonson, M.; Nilsson, K. P. R.; Thor, S.; Hammarstrom, P. Curcumin promotes A-beta fibrillation and reduces neurotoxicity in transgenic Drosophila. *PLoS ONE* **2012**, *7*, e31424.

[171] Rajasekar, N.; Dwivedi, S.; Tota, S. K.; Kamat, P. K.; Hanif, K.; Nath, C.; Shukla, R. Neuroprotective effect of curcumin on okadaic acid induced memory impairment in mice. *Eur. J. Pharmacol.* **2013**, *715*, 381–394.

[172] Ma, Q. L.; Zuo, X.; Yang, F.; Ubeda, O. J.; Gant, D. J.; Alaverdyan, M., et al. Curcumin suppresses soluble tau dimers and corrects molecular chaperone, synaptic, and behavioral deficits in aged human tau transgenic mice. *J. Biol. Chem.* **2013**, *288*, 4056–4065.

[173] Herva, M. E.; Zibaee, S.; Fraser, G.; Barker, R. A.; Goedert, M.; Spillantini, M. G. Antiamyloid compounds inhibit a-synuclein aggregation induced by protein misfolding cyclic amplification (PMCA). *J. Biol. Chem.* **2014**, *289*, 11897–11905.

[174] Wang, M. S.; Boddapati, S.; Emadi, S.; Sierks, M. R. Curcumin reduces alpha-synuclein induced cytotoxicity in Parkinson's disease cell model. *BMC Neurosci.* **2010**, *11*, 57.

[175] Mythri, R. B.; Srinivas Bharath, M. M. Curcumin: a potential neuroprotective agent in Parkinson's disease. *Curr. Pharm. Des.* **2012**, *18*, 91–99.

[176] Dikshit, P.; Goswami, A.; Mishra, A.; Nukina, N.; Jana, N. R. Curcumin enhances the polyglutamine-expanded truncated N- terminal huntingtin-induced cell death by promoting proteasomal malfunction. *Biochem. Biophys. Res. Commun.* **2006**, *342*, 1323–1328.

[177] Verma, M.; Sharma, A.; Naidu, S.; Kumar Bhadra, A.; Kukreti, R.; Taneja, V. Curcumin prevents formation of polyglutamine aggregates by inhibiting Vps36, a component of the ESCRT-II complex. *PLoS ONE* **2013**, *7*, e42923.

[178] Hickey, M. A.; Zhu, C.; Medvedeva, V.; Lerner, R. P.; Patassini, S.; Franich, N. R., et al. Improvement of neuropathology and transcriptional deficits in CAG 140 knockin mice supports a beneficial effect of dietary curcumin in Huntington's disease. *Mol. Neurodegener.* **2012**, *7*, 12.

[179] Hafner-Bratkovic, I.; Gaspersic, J.; Smid, L. M.; Bresjanac, M.; Jerala, R. Curcumin binds to the alpha-helical intermediate and to the amyloid form of prion protein—a new mechanism for the inhibition of PrP(Sc) accumulation. *J. Neurochem.* **2008**, *104*, 1553–1564.

[180] Caughey, B.; Raymond, L. D.; Raymond, G. J.; Maxson, L.; Silveira, J.; Baron, G. S. Inhibition of protease-resistant prion protein accumulation *in vitro* by curcumin. *J. Virol.* **2003**, *77*, 5499–5502.

[181] Lin, C. -F.; Yu, K. -H.; Jheng, C. -P.; Chung, R.; Lee, C. -I. Curcumin reduces amyloid fibrillation of prion protein and decreases reactive oxidative stress. *Pathogens* **2013**, *2*, 506–519.

[182] Anand, P.; Kunnumakkara, A. B.; Newman, R. A.; Aggarwal, B. B. Bioavailability of curcumin: problems and promises. *Mol. Pharm.* **2007**, *4*, 807–818.

[183] Yallapu, M. M.; Jaggi, M.; Chauhan, S. C. Curcumin nanoformulations: a future nanomedicine for cancer. *Drug Discov. Today* **2012**, *17*, 71–80.

[184] Taylor, M.; Moore, S.; Mourtas, S.; Niarakis, A.; Re, F.; Zona, C., et al. Effect of curcumin-associated and lipid ligand-functionalized nanoliposomes on aggregation of the Alzheimer's Aβ peptide. *Nanomedicine* **2011**, *7*, 541–550.

[185] Mathew, A.; Fukuda, T.; Nagaoka, Y.; Hasumura, T.; Morimoto, H.; Yoshida, Y., et al. Curcumin loaded-PLGA nanoparticles conjugated with Tet-1 peptide for potential use in Alzheimer's disease. *PLoS ONE* **2012**, *7*, e32616.

[186] Doggui, S.; Sahni, J. K.; Arseneault, M.; Dao, L.; Ramassamy, C. Neuronal uptake and neuroprotective effect of curcumin-loaded PLGA nanoparticles on the human SK-N-SH cell line. *J. Alzheimer's Dis.* **2012**, *30*, 377–392.

[187] Palmal, S.; Maity, A. R.; Singh, B. K.; Basu, S.; Jana, N. R.; Jana, N. R. Inhibition of amyloid fibril growth and dissolution of amyloid fibrils by curcumin–gold nanoparticles. *Chem. Eur. J.* **2014**, *20*, 6184–6191.

[188] Lazar, A. N.; Mourtas, S.; Youssef, I.; Parizot, C.; Dauphin, A.; Delatour, B., et al. Curcumin-conjugated nanoliposomes with high affinity for Aβ deposits: possible applications to Alzheimer disease. *Nanomedicine* **2013**, *9*, 712–721.

[189] Tiwari, S. K.; Agarwal, S.; Seth, B.; Yadav, A.; Nair, S.; Bhatnagar, P., et al. Curcumin-loaded nanoparticles potently induce adult neurogenesis and reverse cognitive deficits in Alzheimer's disease model via canonical Wnt/β-catenin pathway. *ACS Nano* **2014**, *8*, 76–103.

[190] Cheng, K. K.; Yeung, C. F.; Ho, S. W.; Chow, S. F.; Chow, A. H. L.; Baum, L. Highly stabilized curcumin nanoparticles tested in an in vitro blood–brain barrier model and in Alzheimer's disease Tg 324-336.2576 mice. *AAPS J.* **2012**, *15*, 324–336.

[191] Sandhir, R.; Yadav, A.; Mehrotra, A.; Sunkaria, A.; Singh, A.; Sharma, S. Curcumin nanoparticles attenuate neurochemical and neurobehavioral deficits in experimental model of Huntington's disease. *Neuromol. Med.* **2014**, *16*, 106–118.

[192] Bairwa, K.; Grover, J.; Kania, M.; Jachak, S. M. Recent developments in chemistry and biology of curcumin analogues. *RSC Adv.* **2014**, *4*, 13946–13978.

[193] Reinke, A. A.; Gestwicki, J. E. Structure–activity relationships of amyloid beta-aggregation inhibitors based on curcumin: influence of linker length and flexibility. *Chem. Biol. Drug Des.* **2007**, *70*, 206–215.

[194] Chen, S. -Y.; Chen, Y.; Li, Y. -P.; Chen, S. -H.; Tan, J. -H.; Ou, T. -M., et al. Design, synthesis, and biological evaluation of curcumin analogues as multifunctional agents for the treatment of Alzheimer's disease. *Bioorg. Med. Chem.* **2011**, *19*, 5596–5604.

[195] Dolai, S.; Shi, W.; Corbo, C.; Sun, C.; Averick, S.; Obeysekera, D., et al. "Clicked" sugar–curcumin conjugate: modulator of amyloid-β and tau peptide aggregation at ultralow concentrations. *ACS Chem. Neurosci.* **2011**, *2*, 694–699.

[196] Narlawar, R.; Pickhardt, M.; Leuchtenberger, S.; Baumann, K.; Krause, S.; Dyrks, T., et al. Curcumin-derived pyrazoles and isoxazoles: Swiss army knives or blunt tools for Alzheimer's disease? *ChemMedChem* **2008**, *3*, 165–172.

[197] Diomede, L.; Rigacci, S.; Romeo, M.; Stefani, M.; Salmona, M. Oleuropein aglycone protects transgenic *C. elegans* strains expressing Aβ42 by reducing plaque load and motor deficit. *PLoS ONE* **2013**, *8*, e58893.

[198] Luccarini, I.; Ed Dami, T.; Grossi, C.; Rigacci, S.; Stefani, M.; Casamenti, F. Oleuropein aglycone counteracts Aβ42 toxicity in the rat brain. *Neurosci. Lett.* **2014**, *558*, 67–72.

[199] Galanakis, P. A.; Bazoti, F. N.; Bergquist, J.; Markides, K.; Spyroulias, G. A.; Tsarbopoulos, A. Study of the interaction between the amyloid beta peptide (1-40) and antioxidant compounds by nuclear magnetic resonance spectroscopy. *Biopolymers* **2011**, *96*, 316–327.

[200] Grossi, C.; Rigacci, S.; Ambrosini, S.; Ed Dami, T.; Luccarini, I.; Traini, C., et al. The polyphenol oleuropein aglycone protects TgCRND8 mice against Aβ plaque pathology. *PLoS ONE* **2013**, *8*, e71702.

[201] Daccache, A.; Lion, C.; Sibille, N.; Gerard, M.; Slomianny, C.; Lippens, G.; Cotelle, P. Oleuropein and derivatives from olives as Tau aggregation inhibitors. *Neurochem. Int.* **2011**, *58*, 700–707.

[202] Hirohata, M.; Hasegawa, K.; Tsutsumi-Yasuhara, S.; Ohhashi, Y.; Ookoshi, T.; Ono, K., et al. The antiamyloidogenic effect is exerted against Alzheimer's α-amyloid fibrils *in vitro* by preferential and reversible binding of flavonoids to the amyloid fibril structure. *Biochemistry* **2007**, *46*, 1888–1899.

[203] Jinwal, U. K.; Miyata, Y.; Koren, J., III.; Jones, J. R.; Trotter, J. H.; ChangF L., et al. Chemical manipulation of Hsp70 ATPase activity regulates Tau stability. *J. Neurosci.* **2009**, *29*, 12079–12088.

[204] Tarrago, T.; Kichik, N.; Claasen, B.; Prades, R.; Teixido, M.; Giralt, E. Baicalin, a prodrug able to reach the CNS, is a prolyl oligopeptidase inhibitor. *Bioorg. Med. Chem.* **2008**, *16*, 7516–7524.

[205] Yin, F.; Liu, J.; Ji, X.; Wang, J.; Zidichouski, J.; Zhang, J. Baicalin prevents the production of hydrogen peroxide and oxidative stress induced by Aβ aggregation in SH-SY5Y cells. *Neurosci. Lett.* **2011**, *492*, 76–79.

[206] Zhang, S. Q.; Obregon, D.; Ehrhart, J.; Deng, J.; Tian, J.; Hou, H., et al. Baicalein reduces β-amyloid and promotes nonamyloidogenic amyloid precursor protein processing in an Alzheimer's disease transgenic mouse model. *J. Neurosci. Res.* **2013**, *91*, 1239–1246.

[207] Zhu, M.; Rajamani, S.; Kaylor, J.; Han, S.; Zhou, F.; Fink, A. L. The flavonoid baicalein inhibits fibrillation of α-synuclein and disaggregates existing fibrils. *J. Biol. Chem.* **2004**, *279*, 26846–26857.

[208] Lu, J. -H.; Ardah, M. T.; Durairajan, S. S. K.; Liu, L. -F.; Xie, L. -X.; Fong, W. -F. D., et al. Baicalein inhibits formation of a-synuclein oligomers within living cells and prevents Aβ peptide fibrillation and oligomerisation. *ChemBioChem* **2011**, *12*, 615–624.

[209] Camilleri, A.; Zarb, C.; Caruana, M.; Ostermeier, U.; Ghio, S.; Högen, T., et al. Mitochondrial membrane permeabilisation by amyloid aggregates and protection by polyphenols. *Biochim. Biophys. Acta* **1828**, *2013*, 2532–2543.

[210] Ono, K.; Hasegawa, K.; Naiki, H.; Yamada, M. Antiamyloidogenic activity of tannic acid and its activity to destabilize Alzheimer's b-amyloid fibrils in vitro. *Biochim. Biophys. Acta* **1690**, *2004*, 193–202.

[211] Ladiwala, A. R. A.; Dordick, J. S.; Tessier, P. M. Aromatic small molecules remodel toxic soluble oligomers of amyloid b through three independent pathways. *J. Biol. Chem.* **2011**, *286*, 3209–3218.

[212] Mori, T.; Rezai-Zadeh, K.; Koyama, N.; Arendash, G. W.; Yamaguchi, H.; Kakuda, N., et al. Tannic acid is a natural β-secretase inhibitor that prevents cognitive impairment and mitigates Alzheimer-like pathology in transgenic mice. *J. Biol. Chem.* **2012**, *287*, 6912–6927.

[213] Yao, J.; Gao, X.; Sun, W.; Yao, T.; Shi, S.; Ji, L. Molecular hairpin: a possible model for inhibition of tau aggregation by tannic acid. *Biochemistry* **2013**, *52*, 1893–1902.

[214] Pickhardt, M.; Gazova, Z.; von Bergen, M.; Khlistunova, I.; Wang, Y.; Hascher, A., et al. Anthraquinones inhibit tau aggregation and dissolve Alzheimer's paired helical filaments in vitro and in cells. *J. Biol. Chem.* **2005**, *280*, 3628–3635.

[215] Liu, T.; Jin, H.; Sun, Q. -R.; Xu, J. -H.; Hu, H. -T. Neuroprotective effects of emodin in rat cortical neurons against (-amyloid-induced neurotoxicity. *Br. Res.* **2010**, *1347*, 149–160.

[216] Liu, J.; Hu, G.; Xu, R.; Qiao, Y.; Wu, H. -P.; Ding, X., et al. Rhein lysinate decreases the generation of β-amyloid in the brain tissues of Alzheimer's disease model mice by inhibiting inflammatory response and oxidative stress. *J. Asian Nat. Prod. Res.* **2013**, *15*, 756–763.

[217] Guo, J. -P.; Yu, S.; McGeer, P. L. Simple in vitro assays to identify amyloid-β aggregation blockers for Alzheimer's disease therapy. *J. Alzheimer's Dis.* **2010**, *19*, 1359–1370.

[218] Li, S. -Y.; Jiang, N.; Xie, S. -S.; Wang, K. D. G.; Wang, X. -B.; Kong, L. -Y. Design, synthesis and evaluation of novel tacrine–rhein hybrids as multifunctional agents for the treatment of Alzheimer's disease. *Org. Biomol. Chem.* **2014**, *12*, 801–814.

[219] Viayna, E.; Sola, I.; Bartolini, M.; De Simone, A.; Tapia-Rojas, C.; Serrano, F. G., et al. Synthesis and multitarget biological profiling of a novel family of rhein derivatives as disease-modifying anti-Alzheimer agents. *J. Med. Chem.* **2014**, *57*, 2549–2567.

[220] Di Giovanni, S.; Eleuteri, S.; Paleologou, K. E.; Yin, G.; Zweckstetter, M.; Carrupt, P. A.; Lashuel, H. A. Entacapone and tolcapone, two catechol-o-methyltransferase inhibitors, block fibril formation of α-synuclein and β-amyloid and protect against amyloid-induced toxicity. *J. Biol. Chem.* **2010**, *285*, 14941–14954.

[221] Mohamed, T.; Hoang, T.; Jelokhani-Niaraki, M.; Rao, P. P. N. Tau-derived-hexapeptide 306VQIVYK311 aggregation inhibitors: nitrocatechol moiety as a pharmacophore in drug design. *ACS Chem. Neurosci.* **2013**, *4*, 1559–1570.

[222] Cavalli, A.; Bolognesi, M. L.; Capsoni, S.; Andrisano, V.; Bartolini, M.; Margotti, E., et al. A small molecule targeting the multifactorial nature of Alzheimer's disease. *Angew. Chem. Int. Ed.* **2007**, *46*, 3689–3692.

[223] Bartolini, M.; Bertucci, C.; Bolognesi, M. L.; Cavalli, A.; Melchiorre, C.; Andrisano, V. Insight into the kinetic of amyloid beta (1-42) peptide self-aggregation: elucidation of inhibitors' mechanism of action. *ChemBioChem* **2007**, *8*, 2152–2161.

[224] Capurro, V.; Busquet, P.; Lopes, J. P.; Bertorelli, R.; Tarozzo, G.; Bolognesi, M. L., et al. Pharmacological characterization of memoquin, a multi-target compound for the treatment of Alzheimer's disease. *PLoS ONE* **2013**, *8*, e56870.

[225] Dinamarca, M. C.; Cerpa, W.; Garrido, J.; Hancke, J. L.; Inestrosa, N. C. Hyperforin prevents β-amyloid neurotoxicity and spatialmemoryimpairments by disaggregation of Alzheimer's amyloid-β-deposits. *Mol. Psychiatry* **2006**, *11*, 1032–1048.

[226] Carvajal, F. J.; Inestrosa, N. C. Interactions of AChE with Aβ aggregates in Alzheimer's brain: therapeutic relevance of IDN 5706. *Front. Mol. Neurosci.* **2011**, *4*, 19.

[227] Cerpa, W.; Hancke, J. L.; Morazzoni, P.; Bombardelli, E.; Riva, A.; Marin, P. P.; Inestrosa, N. C. The hyperforin derivative IDN5706 occludes spatial memory impairments and neuropathological changes in a double transgenic Alzheimer's mouse model. *Curr. Alzheimer Res.* **2010**, *7*, 126–133.

[228] Inestrosa, N. C.; Tapia-Rojas, C.; Griffith, T. N.; Carvajal, F. J.; Benito, M. J.; Rivera-Dictter, A., et al. Tetrahydrohyperforin prevents cognitive deficit, Aβ deposition, tau phosphorylation and synaptotoxicity in the APPswe/PSEN1dE9 model of Alzheimer's disease: a possible effect on APP processing. *Transl. Psychiatry* **2011**, *1*, e20.

[229] Khaengkhan, P.; Nishikaze, Y.; Niidome, T.; Kanaori, K.; Tajima, K.; Ichida, M., et al. Identification of an antiamyloidogenic substance from mulberry leaves. *NeuroReport* **2009**, *20*, 1214–1218.

[230] Atamna, H.; Frey, W. H., 2nd.; Ko, N. Human and rodent amyloid-beta peptides differentially bind heme: relevance to the human susceptibility to Alzheimer's disease. *Arch. Biochem. Biophys.* **2009**, *487*, 59–65.

[231] Yuan, C.; Gao, Z. Aβ interacts with both the iron center and the porphyrin ring of heme: mechanism of heme's action on Aβ aggregation and disaggregation. *Chem. Res. Toxicol.* **2013**, *26*, 262–269.

[232] Lamberto, G. R.; Torres-Monserrat, V.; Bertoncini, C. W.; Salvatella, X.; Zweckstetter, M.; Griesinger, C.; Fernandez, C. O. Toward the discovery of effective polycyclic inhibitors of α-synuclein amyloid assembly. *J. Biol. Chem.* **2011**, *286*, 32036–32044.

[233] Park, J. -W.; Ahn, J. S.; Lee, J. -H.; Bhak, G.; Jung, S.; Paik, S. R. Amyloid fibrillar meshwork formation of iron-induced oligomeric species of Aβ40 with phthalocyanine tetrasulfonate and its toxic consequences. *ChemBioChem* **2008**, *9*, 2602–2605.

[234] Dee, D. R.; Gupta, A. N.; Anikovskiy, M.; Sosova, I.; Grandi, E.; Rivera, L., et al. Phthalocyanine tetrasulfonates bind to multiple sites on natively-folded prion protein. *Biochim. Biophys. Acta* **1824**, *2012*, 826–832.

[235] Akoury, E.; Gajda, M.; Pickhardt, M.; Biernat, J.; Soraya, P.; Griesinger, C., et al. Inhibition of tau filament formation by conformational modulation. *J. Am. Chem. Soc.* **2013**, *135*, 2853–2862.

[236] Klunk, W. E.; Lopresti, B. J.; Ikonomovic, M. D.; Lefterov, I. M.; Koldamova, R. P.; Abrahamson, E. E., et al. Binding of the positron emission tomography tracer Pittsburgh

compound-B reflects the amount of amyloid-beta in Alzheimer's disease brain but not in transgenic mouse brain. *J. Neurosci.* **2005**, *25*, 10598–10606.

[237] Wilson, D. M.; Binder, L. I. Free fatty acids stimulate the polymerization of tau and amyloid beta peptides. In vitro evidence for a common effector of pathogenesis in Alzheimer's disease. *Am. J. Pathol.* **1997**, *150*, 2181–2195.

[238] Chirita, C. N.; Necula, M.; Kuret, J. Ligand-dependent inhibition and reversal of tau filament formation. *Biochemistry* **2004**, *43*, 2879–2887.

[239] Necula, M.; Chirita, C. N.; Kuret, J. Cyanine dye n744 inhibits tau fibrillization by blocking filament extension: implications for the treatment of tauopathic neurodegenerative diseases. *Biochemistry* **2005**, *44*, 10227–10237.

[240] Congdon, E. E.; Necula, M.; Blackstone, R. D.; Kuret, J. Potency of a tau fibrillization inhibitor is influenced by its aggregation state. *Arch. Biochem. Biophys.* **2007**, *465*, 127–135.

[241] Congdon, E. E.; Figueroa, Y. H.; Wang, L.; Toneva, G.; Chang, E.; Kuret, J., et al. Inhibition of tau polymerization with a cyanine dye in two distinct model systems. *J. Biol. Chem.* **2009**, *284*, 20830–20839.

[242] Chang, E.; Congdon, E. E.; Honson, N. S.; Duff, K. E.; Kuret, J. Structure-activity relationship of cyanine tau aggregation inhibitors. *J. Med. Chem.* **2009**, *52*, 3539–3547.

[243] Schafer, K. N.; Cisek, K.; Huseby, C. J.; Chang, E.; Kuret, J. Structural determinants of tau aggregation inhibitor potency. *J. Biol. Chem.* **2013**, *288*, 32599–32611.

[244] Honson, N. S.; Jensen, J. R.; Darby, M. V.; Kuret, J. Potent inhibition of tau fibrillization with a multivalent ligand. *Biochem. Biophys. Res. Commun.* **2007**, *363*, 229–234.

[245] Schafer, K. N.; Murale, D. P.; Kim, K.; Cisek, K.; Kuret, J.; Churchill, D. G. Structure–activity relationship of cyclic thiacarbocyanine tau aggregation inhibitors. *Bioorg. Med. Chem. Lett.* **2011**, *21*, 3273–3276.

[246] Capule, C. C.; Yang, J. Enzyme-linked immunosorbent assay-based method to quantify the association of small moleculeswith aggregated amyloid peptides. *Anal. Chem.* **2012**, *84*, 1786–1791.

[247] Inbar, P.; Li, C. Q.; Takayama, S. A.; Bautista, M. R.; Yang, J. Oligo(ethylene glycol) derivatives of thioflavin T as inhibitors of protein-amyloid interactions. *Chembiochem* **2006**, *7*, 1563–1566.

[248] Megill, A.; Lee, T.; Dibattista, A. M.; Song, J. M.; Spitzer, M. H.; Rubinshtein, M., et al. A tetra(ethylene glycol) derivative of benzothiazole aniline enhances Ras-mediated spinogenesis. *J. Neurosci.* **2013**, *33*, 9306–9318.

[249] Song, J. M.; DiBattista, A. M.; Sung, Y. M.; Ahn, J. M.; Turner, R. S.; Yang, J., et al. A tetra(ethylene glycol) derivative of benzothiazole aniline ameliorates dendritic spine density and cognitive function in a mouse model of Alzheimer's disease. *Exp. Neurol.* **2014**, *252*, 105–113.

[250] Sharma, A. K.; Pavlova, S. T.; Kim, J.; Finkelstein, D.; Hawco, N. J.; Rath, N. P., et al. Bifunctional compounds for controlling metal-mediated aggregation of the $A\beta_{42}$ peptide. *J. Am. Chem. Soc.* **2012**, *134*, 6625–6636.

[251] Volkova, K. D.; Kovalska, V. B.; Losytskyy, M. Y.; Veldhuis, G.; Segers-Nolten, G. M. J.; Tolmachev, O. I., et al. Studies of interaction between cyanine dye T-284 and fibrillar alpha-synuclein. *J. Fluoresc.* **2010**, *20*, 1267–1274.

[252] Maruyama, M.; Shimada, H.; Suhara, T.; Shinotoh, H.; Ji, B.; Maeda, J., et al. Imaging of tau pathology in a tauopathy mouse model and in Alzheimer patients compared to normal controls. *Neuron* **2013**, *79*, 1094–1108.

[253] Gu, J.; Anumala, U. R.; Heyny-von Haussen, R.; Holzer, J.; Goetschy-Meyer, V.; Mall, G., et al. Design, synthesis and biological evaluation of trimethine cyanine dyes as fluorescent probes for the detection of tau fibrils in Alzheimer's disease brain and olfactory epithelium. *ChemMedChem* **2013**, *8*, 891–897.

[254] Noel, S.; Cadet, S.; Gras, E.; Hureau, C. The benzazole scaffold: a SWAT to combat Alzheimer's disease. *Chem. Soc. Rev.* **2013**, *42*, 7747–7762.

[255] Bulic, B.; Pickhardt, M.; Khlistunova, I.; Biernat, J.; Mandelkow, E. M.; Mandelkow, E.; Waldmann, H. Rhodanine-based tau aggregation inhibitors in cell models of tauopathy. *Angew. Chem. Int. Ed. Engl.* **2007**, *46*, 9215–9219.

[256] Messing, L.; Decker, J. M.; Joseph, M.; Mandelkow, E.; Mandelkow, E. -M. Cascade of tau toxicity in inducible hippocampal brain slices and prevention by aggregation inhibitors. *Neurobiol. Aging* **2013**, *34*, 1343–1354.

[257] Ono, M.; Hayashi, S.; Matsumura, K.; Kimura, H.; Okamoto, Y.; Ihara, M., et al. Rhodanine and thiohydantoin derivatives for detecting tau pathology in Alzheimer's brains. *ACS Chem. Neurosci.* **2011**, *2*, 269–275.

[258] Anumala, U. R.; Gu, J.; Lo Monte, F.; Kramer, T.; Heyny-von Haußen, R.; Hölzer, J., et al. Fluorescent rhodanine-3-acetic acids visualize neurofibrillary tangles in Alzheimer's disease brains. *Bioorg. Med. Chem.* **2013**, *21*, 5139–5144.

[259] Larbig, G.; Pickhardt, M.; Lloyd, D. G.; Schmidt, B.; Mandelkow, E. Screening for inhibitors of tau protein aggregation into Alzheimer paired helical filaments: A ligand based approach results in successful scaffold hopping. *Curr. Alzheimer Res.* **2007**, *4*, 315–323.

[260] Pickhardt, M.; Larbig, G.; Khlistunova, I.; Coksezen, A.; Meyer, B.; Mandelkow, E. M., et al. Phenylthiazolylhydrazide and its derivatives are potent inhibitors of tau aggregation and toxicity in vitro and in cells. *Biochemistry* **2007**, *46*, 10016–10023.

[261] Maheshwari, M.; Roberts, J. K.; DeSutter, B.; Duong, K. T.; Tingling, J.; Fawver, J. N., et al. Hydralazine modifies Aβ fibril formation and prevents modification by lipids in vitro. *Biochemistry* **2010**, *49*, 10371–10380.

[262] Taghavi, A.; Nasir, S.; Pickhardt, M.; Heyny-von Haussen, R.; Mall, G.; Mandelkow, E., et al. N'-Benzylidene-benzohydrazides as novel and selective tau-PHF ligands. *J. Alzheimer's Dis.* **2011**, *27*, 835–843.

[263] Pickhardt, M.; Biernat, J.; Khlistunova, I.; Wang, Y. P.; Gazova, Z.; Mandelkow, E. M.; Mandelkow, E. N-Phenylamine derivatives as aggregation inhibitors in cell models of tauopathy. *Curr. Alzheimer Res.* **2007**, *4*, 397–402.

[264] Khlistunova, I.; Biernat, J.; Wang, Y.; Pickhardt, M.; von Bergen, M.; Gazova, Z., et al. Inducible expression of tau repeat domain in cell models of tauopathy. *J. Biol. Chem.* **2006**, *281*, 1205–1214.

[265] Wang, J.; Ono, K.; Dickstein, D. L.; Arrieta-Cruz, I.; Zhao, W.; Qian, X., et al. Carvedilol as a potential novel agent for the treatment of Alzheimer's disease. *Neurobiol. Aging* **2011**, *32*, 2321.e1–2321.e12.

[266] Howlett, D. R.; George, A. R.; Owen, D. E.; Ward, R. V.; Markwell, R. E. Common structural features determine the effectiveness of carvedilol, daunomycin and rolitetracycline as inhibitors of Alzheimer beta-amyloid fibril formation. *Biochem. J.* **1999**, *343*, 419–423.

[267] Wang, J.; Zhao, Z.; Lin, E.; Zhao, W.; Qian, X.; Freire, D., et al. Unintended effects of cardiovascular drugs on the pathogenesis of Alzheimer's disease. *PLoS ONE* **2013**, *8*, e65232.

[268] http://clinicaltrials.gov/ct2/show/NCT01354444?term=carvedilol&cond=%22Alzheimer+Disease%22&rank=1.

[269] Crowe, A.; Ballatore, C.; Hyde, E.; Trojanowski, J. Q.; Lee, V. M. Y. High throughput screening for small molecule inhibitors of heparin-induced tau fibril formation. *Biochem. Biophys. Res. Commun.* **2007**, *358*, 1–6.

[270] Crowe, A.; Huang, W.; Ballatore, C.; Johnson, R. L.; Hogan, A. M.; Huang, R., et al. The identification of aminothienopyridazine inhibitors of Tau assembly by quantitative high-throughput screening. *Biochemistry* **2009**, *48*, 7732–7745.

[271] Ballatore, C.; Brunden, K. R.; Piscitelli, F.; James, M. J.; Crowe, A.; Yao, Y., et al. Discovery of brain-penetrant, orally bioavailable aminothienopyridazine inhibitors of Tau aggregation. *J. Med. Chem.* **2010**, *53*, 3739–3747.

[272] Fatouros, C.; Jeelani Pir, G.; Biernat, J.; Padmanabhan Koushika, S.; Mandelkow, E.; Mandelkow, E.-M., et al. Inhibition of tau aggregation in a novel *Caenorhabditis elegans* model of tauopathy mitigates proteotoxicity. *Hum. Mol. Gen.* **2012**, *21*, 3587–3603.

[273] Ballatore, C.; Crowe, A.; Piscitelli, F.; James, M.; Lou, K.; Rossidivito, G., et al. Aminothienopyridazine inhibitors of Tau aggregation. Evaluation of structure-activity relationship leads to selection of candidates with desirable in vivo properties. *Bioorg. Med. Chem.* **2012**, *20*, 4451–4461.

[274] Sarkar, S.; Davies, J. E.; Huang, Z.; Tunnacliffe, A.; Rubinsztein, D. C. Trehalose, a novel mTOR-independent autophagy enhancer, accelerates the clearance of mutant huntingtin and alpha-synuclein. *J. Biol. Chem.* **2007**, *282*, 5641–5652.

[275] Aguib, Y.; Heiseke, A.; Gilch, S.; Riemer, C.; Baier, M.; Schätzl, H. M.; Ertmer, A. Autophagy induction by trehalose counteracts cellular prion infection. *Autophagy* **2009**, *5*, 361–369.

[276] Gomes, C.; Escrevente, C.; Costa, J. Mutant superoxide dismutase 1 overexpression in NSC-34 cells: effect of trehalose on aggregation, TDP-43 localization and levels of co-expressed glycoproteins. *Neurosci. Lett.* **2010**, *475*, 145–149.

[277] Rodríguez-Navarro, J. A.; Rodríguez, L.; Casarejos, M. J.; Solano, R. M.; Gómez, A.; Perucho, J., et al. Mutant superoxide dismutase 1 overexpression in NSC-34 cells: effect of trehalose on aggregation, TDP-43 localization and levels of co-expressed glycoproteins. *Neurobiol. Dis.* **2010**, *39*, 423–438.

[278] Liu, R.; Barkhordarian, H.; Emadi, S.; Park, C. B.; Sierks, M. R. Trehalose differentially inhibits aggregation and neurotoxicity of beta-amyloid 40 and 42. *Neurobiol. Dis.* **2005**, *20*, 74–81.

[279] Liu, F. -F.; Ji, L.; Dong, X. -Y.; Sun, Y. Molecular insight into the inhibition effect of trehalose on the nucleation and elongation of amyloid b-peptide oligomers. *J. Phys. Chem. B.* **2009**, *113*, 11320–11329.

[280] Izmitli, A.; Schebor, C.; McGovern, M. P.; Reddy, A. S.; Abbott, N. L.; de Pablo, J. J. Effect of trehalose on the interaction of Alzheimer's A(-peptide and anionic lipid monolayers. *Biochim. Biophys. Acta* 1808, **2011**, 26–33.

[281] Tanaka, M.; Machida, Y.; Niu, S.; Ikeda, T.; Jana, N. R.; Doi, H., et al. Trehalose alleviates polyglutamine- mediated pathology in a mouse model of Hunton disease. *Nat. Med.* **2004**, *10*, 148–154.

[282] Beranger, F.; Crozet, C.; Goldsborough, A.; Lehmann, S. Trehalose impairs aggregation of PrPSc molecules and protects prion-infected cells against oxidative damage. *Biochem. Biophys. Res. Commun.* **2008**, *374*, 44–48.

[283] Yu, W. B.; Jiang, T.; Lan, D. M.; Lu, J. H.; Yue, Z. Y.; Wang, J.; Zhou, P. Trehalose inhibits fibrillation of A53T mutant alpha-synuclein and disaggregates existing fibrils. *Arch. Biochem. Biophys.* **2012**, *523*, 144–150.

[284] Krüger, U.; Wang, Y.; Kumar, S.; Mandelkow, E. -M. Autophagic degradation of tau in primary neurons and its enhancement by trehalose. *Neurobiol. Aging* **2012**, *33*, 2291–2305.

[285] Chaudhary, R. K.; Kardani, J.; Singh, K.; Banerjee, R.; Roy, I. Deciphering the roles of trehalose and Hsp104 in the inhibition of aggregation of mutant huntingtin in a yeast model of Huntington's disease. *Neuromol. Med.* **2014**, *16*, 280–291.

[286] Doyle, S. M.; Genest, O.; Wickner, S. Protein rescue from aggregates by powerful molecular chaperone machines. *Nat. Rev. Mol. Cell Biol.* **2013**, *14*, 617–629.

[287] DeSantis, M. E.; Leung, E. H.; Sweeny, E. A.; Jackrel, M. E.; Cushman-Nick, M.; Neuhaus-Follini, A., et al. Operational plasticity enables Hsp104 to disaggregate diverse amyloid and nonamyloid clients. *Cell* **2012**, *151*, 778–793.

[288] Glover, J. R.; Lindquist, S. Hsp104, Hsp70, and Hsp40: a novel chaperone system that rescues previously aggregated proteins. *Cell* **1998**, *94*, 73–82.

[289] Duennwald, M. L.; Echeverria, A.; Shorter, J. Small heat shock proteins potentiate amyloid dissolution by protein disaggregases from yeast and humans. *PLoS Biol.* **2012**, *10*, e1001346.

[290] Murray, A. N.; Kelly, J. W. Hsp104 gives clients the individual attention they need. *Cell* **2012**, *151*, 695–697.

[291] Shorter, J. The mammalian disaggregase machinery: Hsp110 synergizes with Hsp70 and Hsp40 to catalyse protein disaggregation and reactivation in a cell-free system. *PLoS One* **2011**, *6*, e26319.

[292] Kuo, Y.; Ren, S.; Lao, U.; Edgar, B. A.; Wang, T. Suppression of polyglutamine protein toxicity by co-expression of a heat-shock protein 40 and a heat-shock protein 110. *Cell Death Dis.* **2013**, *4*, e833.

[293] Song, Y.; Nagy, M.; Ni, W.; Tyagi, N. K.; Fenton, W. A.; López-Giráldez, F., et al. Molecular chaperone Hsp110 rescues a vesicle transport defect produced by an ALS-associated mutant SOD1 protein in squid axoplasm. *Proc. Natl. Acad. Sci. U.S.A.* **2013**, *110*, 5428–5433.

[294] Kim, Y.; Park, J. -H.; Jang, J. -Y.; Rhim, H.; Kang, S. Characterization and Hsp104-induced artificial clearance of familial ALS-related SOD1 aggregates. *Biochem. Biophys. Res. Commun.* **2013**, *434*, 521–526.

[295] Vacher, C.; Garcia-Oroz, L.; Rubinsztein, D. C. Overexpression of yeast hsp104 reduces polyglutamine aggregation and prolongs survival of a transgenic mouse model of Huntington's disease. *Hum. Mol. Genet.* **2005**, *14*, 3425–3433.

[296] Cushman-Nick, M.; Bonini, N. M.; Shorter, J. Hsp104 suppresses polyglutamine-induced degeneration post onset in a *Drosophila* MJD/SCA3 model. *PLoS Genet.* **2013**, *9*, e1003781.

[297] Lo Bianco, C.; Shorter, J.; Regulier, E.; Lashuel, H.; Iwatsubo, T.; Lindquist, S.; Aebischer, P. Hsp104 antagonizes alpha-synuclein aggregation and reduces dopaminergic degeneration in a rat model of Parkinson disease. *J. Clin. Invest.* **2008**, *118*, 3087–3097.

[298] Jackrel, M. E.; DeSantis, M. E.; Martinez, B. A.; Castellano, L. M.; Stewart, R. M.; Caldwell, K. A., et al. Potentiated Hsp104 variants antagonize diverse proteotoxic misfolding events. *Cell* **2014**, *156*, 170–182.

[299] Torrente, M. P.; Shorter, J. The metazoan protein disaggregase and amyloid depolymerase system. *Prion* **2013**, *7*, 457–463.

[300] Pereira, C.; Lopes-Rodrigues, V.; Coutinho, I.; Neves, M. P.; Lima, R. T.; Pinto, M., et al. Potential small-molecule activators of caspase-7 identified using yeast-based caspase-3 and -7 screening assays. *Eur. J. Pharm. Sci.* **2014**, *54*, 8–16.

Index